Total
Improvement
Management

Total Improvement Management

The Next Generation in Performance Improvement

H. James Harrington

with James S. Harrington

McGraw-Hill, Inc.
New York   San Francisco   Washington, D.C.   Auckland   Bogotá
Caracas   Lisbon   London   Madrid   Mexico City   Milan
Montreal   New Delhi   San Juan   Singapore
Sydney   Tokyo   Toronto

1 2 3 4 5 6 7 8 9 0 DOH/DOH 9 0 9 8 7 6 5 4

ISBN 0-07-026770-7

The sponsoring editor for this book was James H. Bessent, Jr., the editing supervisor was Fred Dahl, and the production supervisor was Donald F. Schmidt. It was set in Palatino by Inkwell Publishing Services.

Printed and bound by R. R. Donnelley & Sons Company.

This book is printed on recycled, acid-free paper containing a minimum of 50 percent recycled de-inked fiber.

*I dedicate this book to three men
who have significantly contributed
to my personal life over the years.*

*I thank Walter L. Hurd, Jr., Robert Maas,
and Armand (Val) Feigenbaum
for their unwavering support and friendship.*

*Knowing them has made Marguerite's and my lives
a lot more rewarding and enjoyable.*

Contents

Acknowledgments   xvii

Introduction   1

Introduction   *1*
We Cannot Lose Our Smokestack Industries   *2*
General Motors' Downturn   *3*
Ford Motor Company's Quality Improvement Program   *4*
IBM's Problems   *6*
Wallace, Inc.   *7*
Summary   *7*

Overview   9

Introduction   *9*
Losers, Survivors, or Winners   *10*
Government Needs to Improve   *13*
Characteristics of the Winners   *13*
Does the Customer Want Improvement?   *19*
So You Want to Improve   *19*
Confusion Reigns Supreme   *19*
Management's Improvement Dilemma   *23*
Looking at the Total   *30*
Improvement Methodologies' Impact on Each Other   *31*
Blending Together the Improvement Methodologies   *32*

Improvement's Impact on Stakeholders   *38*

How TIM Affects the Organization   *41*

Do the Total Management Methodologies Pay Off?   *42*

Benchmarking—The Worst Pays Off Big   *43*

How Does the United States Measure Up?   *48*

Japan, Incorporated   *49*

Moving Production Back to the United States   *52*

Return on Investment for TIM Expenditures   *52*

Summary   *54*

References and Sources   *56*

1. Top Management Leadership: The People Who Need to Change First

59

Introduction   *59*

How Does Top Management Feel About
Improving Quality and Productivity?   *60*

What Makes Believers Out of Top Management?   *61*

Why Top Management Keeps at Arm's Length   *62*

It Doesn't Have to Be the President and/or the
Chairman of the Board   *63*

How Does Top Management Show Leadership?   *64*

Committing and Giving Freely of Top Management's Time   *65*

Personal Performance Indicators   *66*

Top Management's Personal Support of the Improvement Process   *67*

Supplying the Required Resources   *68*

Releasing Supporting Policies and Procedures   *68*

Organizational Impact   *76*

Measurements of Improvement   *79*

Summary   *83*

References   *86*

2. Business Planning Process: Aligning the Organization and the People

87

Introduction   *87*

What's the Problem? Don't We Have Enough Already?   *88*

What's in a Good Business Plan?   *89*

Setting Direction   *91*

Establishing Expectations (Measurements)   *94*

Defining Actions   *96*
What's in an Effective Planning Process?   *98*
How Do You Use a Business Plan?   *101*
Adaptation from Experience   *102*
Summary   *102*

**3. Environmental Change Plans: Best Practices
for Improvement Planning and Implementation**   ***104***

Introduction   *105*
Business Plans versus Environmental Change Plans   *105*
Why Do You Need an Environmental Change Plan?   *106*
What Creates Your Organization's Culture?   *107*
Assessment of Today's Personalities   *108*
How Do You Change an Organization's Personality?   *110*
Organized Labor Involvement   *110*
Establish Environmental Vision Statements   *111*
Setting Performance Improvement Goals   *114*
Desired Behavior and Habit Patterns   *114*
Three-Year Improvement Plans   *116*
Combined Three-Year Plan   *118*
Rolling 90-Day Improvement Action Plan   *119*
Making the Improvement Process Work   *120*
Organizational Change Management   *120*
Working Definition of OCM   *121*
OCM Best Practices   *122*
Identify and Orchestrate Key Roles   *125*
View Resistance as a Natural Reaction That
    Must Be Expected and Managed   *132*
Recognize the Levels of Commitment That Are Required   *135*
Understand the Strategic Importance of the
    Organization's Culture   *137*
Summary   *138*
References   *140*

**4. External Customer Focus: Best Practices for
Outstanding Customer Relationships**   ***141***

Introduction   *141*
Why Single Out External Customers?   *141*

Today's Customers   *142*

Customer Focus   *142*

Customer-Related Measurements   *143*

Using the Right Words   *143*

Customer Perception   *145*

Needs versus Expectations versus Desires   *146*

When the Customer Remembers Your Name   *147*

Customer Satisfaction   *149*

Marketing's Impact on the External Customer   *149*

Sales and Delivery Staff Impact on the External Customer   *151*

Other External Customer Contacts   *153*

The Customer Satisfaction Process   *154*

Customer Data   *154*

External Customer Data Systems   *156*

External Customer Satisfaction Measurements   *157*

Customer Complaint Handling   *158*

Getting and Staying Close to Customers   *160*

Designing for Customer Satisfaction   *161*

Developing Strategic Customer Partnerships   *161*

Summary   *166*

References   *167*

5. Quality Management Systems: ISO 9000 and More   **168**

Introduction   *168*

What Is a Quality Management System?   *169*

The Development of Quality Management Systems   *171*

The ISO 9000 Series Standards—An Overview   *174*

ISO 9000 or QMS Implementation Tips and Traps   *179*

Who Do Organizations Get Certified?   *187*

Summary   *187*

6. Management Participation: Management Must Set the Example   **190**

Introduction   *190*

What Do We Call Them?   *192*

Why Start with Management First?   *194*

Managers Are Ultimately Held Accountable   *196*

Why Is Management the Problem?   *198*

Why Managers Fear the Improvement Process   *199*
Management's New Role   *199*
Building Trust and Understanding   *201*
Recognizing Good and Bad Performance—
    The Feedback Process   *204*
Basic Principles   *205*
Tomorrow's Managers   *205*
Understanding the Customer   *209*
Participation/Employee Involvement   *210*
Organized Labor's Involvement in Participative
    Management   *216*
Overcontrolled and Underled   *218*
The New Middle Manager   *221*
Management's Change Process   *222*
Developing the Desire to Change   *223*
Management Education   *224*
Job Descriptions   *226*
New Performance Standards—Error-Free Output   *227*
Measurement Systems and Performance Plans   *228*
Upward Appraisals   *229*
Management Improvement Teams (MIT)   *230*
The Down Side to Improvement   *232*
Management and Employee Opinion Surveys   *234*
Management Self-Assessments   *234*
Summary   *235*
References   *237*

7. Team Building: Bringing Synergy to the Organization

238

Introduction   *238*
Elements of a Team   *240*
The Problem-Solving Process   *247*
Types of Teams   *250*
Training Teams   *253*
Using the Team Approach to Organize and Run Meetings   *254*
Evaluating Team Meetings   *255*
Basic Problem-Solving Tools   *256*
Macro Improvement Tools   *258*
The Seven New Management Tools   *259*

Reaching and Managing Decisions   *261*

How to Implement a Team Process   *262*

How to Measure Team Success   *263*

How to Deal with Problem Teams   *264*

The Future of Teams   *265*

Summary   *266*

References   *268*

8. Individual Excellence: Going Beyond Teams   **269**

Introduction   *269*

Training—Opening the Door to Individual Excellence   *272*

Improvement-Related Training   *275*

Job-Related Training   *275*

Career Growth Training   *276*

Developing Individual Performance Plans   *277*

Performance Evaluations (Appraisals)   *278*

The New Employee   *279*

Career Building   *281*

Building a Bond with Your Manager   *282*

Reinforcing Desired Individual Behavior   *283*

Cross-Discipline Training   *285*

Turning Employees' Complaints into Profits   *286*

Getting Ideas Flowing   *291*

Problems Without Known Solutions   *291*

Safety   *293*

Empowering the Individual Closest to the Customer   *294*

The Start of Individual Excellence   *295*

Creativity   *296*

Creativity for the Individual   *297*

Self-Managed Employees   *305*

Summary   *308*

Suggested Reading   *311*

References   *311*

9. Supplier Relations: Developing a Supply Management Process   **313**

Introduction   *313*

Approach   *314*

Current State Assessment *314*

Material Goals and Strategies *315*

Definition and Scope of Supply Management *316*

What Is Supply Management? *317*

Simple Classifications, Strategies, Tactics, Tools,
  and Techniques to Get Started *318*

Supply Management Process *324*

Generic Supply Management Model—Ten Steps *325*

The Commodity Team *326*

Application of the Supply Management Process (SMP)
  Leading to Certification *327*

Commodity Team—Yearly Activities/Responsibilities *332*

Guidelines and Models for Implementation *333*

Pitfalls to Avoid During Implementation *335*

Supply Management—A New Competitive Advantage—
  Yes and Yes *336*

Summary *336*

References *338*

10. Process Breakthrough: Jump-Starting Your Process *339*

Introduction *339*

How to Improve Your Business Processes *340*

Phase I—Organizing for Improvement *340*

Phase II—Understanding the Process *343*

Phase III—Streamlining the Process *345*

Phase IV—Implementation, Measurements, and Controls *350*

Phase V—Continuous Improvement *353*

Does BPI Work? *353*

Summary *354*

**11. Product Process Excellence: The Production Side
of All Organizations** *356*

Introduction *356*

Product Processes *357*

Product Development Phase *358*

Product and Process Design and Innovation *369*

Production Phase *372*

Information Technology *375*

How Motorola Woke Up   *376*

Summary   *377*

12. Service Process Excellence: How to Best Serve Your Customers   ***380***

Introduction   *380*

What Is a Service Industry?   *381*

Importance of Service Industries to the U.S. Economy   *382*

Service Is America's Number One Problem   *382*

Characteristics Involved in Service   *382*

Major Classifications for Service Organizations   *385*

Service Quality in the Banking/Financial Industry   *386*

Service Quality in the Health Care Industry   *393*

Service Quality in the Utilities Industry   *401*

Overview of Improvement in the Service Industry   *408*

Service Sector Summary   *412*

Banking Source References   *415*

13. The Measurement Process: The Balanced Score Card   ***416***

Introduction   *416*

Using Measurements   *418*

Benefits of Measurement   *419*

Understanding Measurement   *421*

Measurements Are Key to Improving   *425*

Types of Measurement Data   *427*

Clear Performance Data   *427*

Measurement Characteristics   *428*

Poor-Quality Cost   *431*

Surveys as a Measurement Tool   *434*

Using National Quality Award Criteria to Measure Improvement   *436*

Management Information System Measures   *438*

Planning the TIM Measurement System   *441*

Summary   *442*

References   *443*

14. Organizational Structure: Restructuring the Organization for the Twenty-First Century   ***444***

Introduction   *444*
The Evolution of the Organizational Structure   *445*
The Vertical Organization   *446*
The Bureaucratic Organization   *447*
The Decentralized Organization   *447*
The Network Organization   *449*
Two Models of Network Structures   *452*
Implementation of Network Structure   *455*
Organizational Structure Design   *457*
What Are the Barriers to Implementation?   *461*
Summary   *463*

15. Rewards and Recognition: Rewarding Desired Behavior
465

Introduction   *465*
Ingredients of an Organization's Reward Process   *467*
Reward Process Hierarchy   *468*
Why Reward People?   *468*
Key Reward Rules   *469*
Types of Rewards   *470*
Financial Compensation   *474*
Monetary Awards   *475*
Group/Team Rewards   *475*
Public Personal Recognition   *476*
Private Personal Recognition   *477*
Peer Rewards   *478*
Customer Rewards   *479*
Organizational Awards   *479*
Implementation of the Reward Process   *480*
Summary   *481*

Index   *483*

Acknowledgments

I want to acknowledge the many contributions to this book made by the team at Ernst & Young.

To Linda Gardner and Debi Guido, who converted and edited endless hours of dictation into the finished product.

To the following Ernst & Young professionals or alumni, who helped develop the concepts or who prepared chapters or parts of chapters:

- Craig A. Anderson
- Chuck Bayless
- Charles Cheshire
- C. Keith Cox
- Jeff Ellis
- Mark B. Hefner
- Norm Howery
- Paul C. Kitka
- Ken Lomax
- Ralph Ott
- M. Melanie Polack
- Jose R. Rodriguez-Soria
- Ben T. Smith
- Jennifer Whalen
- William D. Wilstead
- Don K. Yee

I would be remiss in not acknowledging Terry Ozan for his support on this project and his leadership in the development of the "Performance Improvement" concepts. In addition, I would like to thank Dorey J. (Jim) Talley, President of Talley-Ho Enterprises, for preparing the chapter on measurements.

But, most of all, I want to recognize and acknowledge the contribution of my wife, Marguerite. She has stayed up late many nights and weekends proofreading, correcting grammar, and standardizing the format in this book. She was always there when I needed her, with love and a kind word.

It is important to state that this book is based on my 1987 book, *The Improvement Process*. As a result many of the concepts were not repeated. This book reflects an additional eight years of applying these basic concepts to many organizations, plus hundreds of employee-years of Ernst & Young's experience. This book therefore brings to light a whole set of new dimensions that goes way beyond Total Quality Management.

H. James Harrington

Total
Improvement
Management

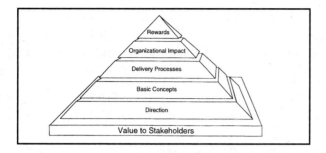

Introduction

Stop worrying about quality, productivity, cost, and cycle time. Focus your energies on organizational performance improvement and all the rest will follow.
DR. H. JAMES HARRINGTON

■ Introduction

We are playing in a World Series ... U.S. Businesses versus The Rest of the World, and the score is:

	1	2	3	4	5	6	7	8	9	Total
World	0	2	1	0	2	1	0			6
U.S.	0	0	0	0	0	0				0

In the second inning, the world scored twice by capturing the clothing and steel industries. In the third inning, they scored by taking over the shipbuilding industry. In the fifth inning, they scored twice with the auto and commercial electronic industries. In the sixth inning, they scored again with the semiconductor industries.

Where and when will they score next? Will it be commercial airlines, banking, or computers? Many think the computer industry will follow the same path as the semiconductor industry. In 1980, America owned 57 percent of the chip market. By 1989, our share had decayed to 35 percent, and it is continuing to decay. By the end of 1994, Japan and the United States were running neck and neck. At the present time, America is losing its market share in computers almost twice as fast as it did in the semiconductor industry. In a market that should be creating jobs, the U.S. is losing jobs. *Workplace Trends* Newsletter reported that U.S. computer companies have cut or announced plans to cut 196,729 employees since 1986, a 40 percent reduction in the U.S. computer workforce. How important is this to America? Well, the electronics sector is America's No. 1 business, even bigger than the auto industry. Losing command of this industry will have grave economic impact on everyone throughout the United States. In 1993, there were 27 computer manufacturers in the U.S. that employed 733,656 people. These organizations' profit per employee was minus $9220. No industry can exist for long with this magnitude of losses.

■ We Cannot Lose Our Smokestack Industries

American business has decided to focus its efforts in the service industries. But what makes us think we have something special to offer in service? The truth of the matter is that if we cannot win in the basic technologies, we won't be able to win any place. Backing away from our manufacturing industries threatens our ability to defend ourselves. Today we depend on components manufactured abroad to assemble our military radar units. Moving into a service environment only is hurting us financially. You make a great deal less money working at McDonald's than you do at McDonnell-Douglas. In the three-year period from 1990 to 1993, the average real hourly wage (in 1982 dollars) dropped from $7.92 to $7.39.

Now it is the seventh inning, and the U.S. is up to bat. We have to change our game plan if we are going to win. In the 1970s, we realized that the management processes that made us world champions were out of date. Our competition had copies of our game plan and had developed competitive strategies to offset our every move. The 1980s were a period of experimentation. Everyone was looking for the quick fix, the magic pill, the fountain of youth, the silver bullet. Management was looking for a simple tool that would improve the organization's performance without requiring them to change any of their basic habits. They tried Quality Circles, and when that did not work, they switched to statistical process control. When that did not solve the problem, they tried the next flavor of the month.

Well, in the 1980s, we did make improvements, but not fast enough. The auto industry is a good example. Quality of U.S. automobiles improved over 10 percent per year. Cadillac won the Malcolm Baldrige National Quality Award, and we all patted ourselves on the back. But, to look at it from another direction, it meant that Lincoln, Pontiac, Oldsmobile, Buick, Chevrolet, Chrysler, Dodge, Plymouth, and Ford were not good enough to win the National Quality Award, and if the award were open to organizations outside the United States, probably Cadillac would not have won it. More than three years after Cadillac won the Malcolm Baldrige National Quality Award, *Consumer Report* (April 1994) could rate any of its models' reliability as better than below average and the highest rating any model got for owner satisfaction was average. Not good for a car that costs $45,000. U.S. auto companies' market share decreased from 51 to 43 percent, a loss of 15.6 percent of their market in 5 years. In the past 10 years, G.M. alone lost almost nine points in market share, which is costing G.M. the equivalent of $13 billion in retail sales per year. The three major U.S. auto manufacturers are producing less today than they were five years earlier, while the seven major auto manufacturers in Japan and Europe are producing more. In 1978, G.M.'s U.S. operations employed

612,000 employees. By 1992 this was cut back to 368,000 and going down fast. To add to the problem, for every auto worker who is laid off, two parts workers are put out of work.

Detroit's quality has improved greatly since 1981 when they were shipping approximately 7.0 defects per auto. By 1992 this had decreased to 1.5 defects per auto based upon J. D. Power and Associates annual study. This means that the average American car has only 0.4 more defects per auto than the Japanese average. In fact, Ford is producing better quality than Nissan, Mazda, and Mitsubishi, but they are behind Toyota and Honda.

■ General Motors' Downturn

In the early 1980s, General Motors' vehicles were so bad that in 1983, the Federal Trade Commission ordered G.M. to adopt broad arbitration guidelines for an eight-year period. The owner of any G.M. vehicle that had or needed engine or transmission repair could request arbitration for reimbursement regardless of the vehicle's age or mileage. They could even ask G.M. to buy back or replace the vehicle.

How can we take pride when the largest automobile manufacturer in the United States, General Motors, takes out a full-page advertisement in *USA Today* on July 11, 1990 showing graphically that their cars are only 25 percent worse that the average car made In Japan. Is it any wonder that G.M. has lost more than 25 percent of their market share; going from almost 50 percent of the U.S. market to 36 percent.

This downward trend resulted in a $4.5 billion loss in 1991, the biggest annual loss ever reported by a U.S. corporation until February 1992, when General Motors broke its own record. G.M.'s North American operations lost $12 billion during 1991-92. Add to that a $970 million loss for the first nine months of 1992 for the total corporation, of which $753 million was reported in the third quarter. This put G.M. in a precarious position; it had $5.2 billion in cash at the end of September 1992, but it needed a minimum of $3 million to open its doors each day and at that rate, G.M. would run out of money during the next nine months. G.M. had already dipped deeply into their $15 billion pension fund to cover some of their cash shortages over the past two years, leaving $8.9 billion of their pension obligation underfunded. So that source of internal cash was exhausted.

The corrective action taken by General Motors was the biggest top-level shake-up since the group was formed 84 years earlier. This clean sweep of the executive ranks took place - in November 1992, removing Mr. Robert Stempel, G.M.'s Chairman and Chief Executive Officer, along with a host of other executives loyal to Mr. Stempel. The chairman who replaced Mr. Stempel was Mr. John Smale, the first-ever nonexecutive chairman for General Motors. Mr. Smale came to General Motors from Procter & Gamble.

What does this tell us about quality? Well, G.M.'s executives certainly understood quality. The fact that a General Motors' division won the Malcolm Baldrige Award proves that. Their top management was committed to quality. Mr. Robert Stempel was the Chairman of the American Quality Foundation from its conception.

The difference in performance between G.M. automobiles and Japan's is so small that it cannot be observed by an individual customer. The big difference is in the way the companies are managed, how they spend their R&D monies, and their productivity.

G.M.'s Saturn car, which is their benchmark for customer satisfaction and performance, racks up a net financial loss for every car sold. James Harbour, a Michigan-based analyst, reported that G.M.'s labor costs were $795 per car greater than Ford's—a comparative disadvantage of $4 billion per year. G.M. takes almost 39 employee hours to build a mid-size car, compared to 22 hours for Ford and 16 to 18 hours for Toyota. Japanese companies also move faster and spend less resources in designing a new vehicle. In Japan, a new auto takes 1.7 million employee hours to design, and In the U.S. and Europe, it takes three million employee hours.

G.M.'s big management error, as well as the other U.S. auto companies in the 1980s, was its investment strategy. In cooperation with the Japanese government, the Japanese auto manufacturers imposed quotas on themselves to give the U.S. auto industry a chance to recover. This allowed the U.S. auto industry to make big profits for the majority of the 1980s. With profit increasing. the U.S. auto industry embarked on a diversification strategy that, on the whole, didn't pay off. At the same time, Japan's auto industry continued to invest heavily in their core competencies so that when the restrictions were removed and the recession of the early 1990s hit, they were in a good position to weather the storm.

■ Ford Motor Company's Quality Improvement Program

Recently, I ran across a very interesting article in *Quality Engineering* Magazine (Vol. 5, Issue 2), written by William H. Smith and C.R. Burdick, both of Ford Motor Company. The second paragraph reads, "Every product possesses quality in some degree. The greater the degree of quality, the higher the demand for that product. This fact has been expressed very well in the slogan. 'Quality and Demand Go Hand-In-Hand.' We know that our sales. and hence the security of our company and its employees, are dependent upon quality. Upon that basis has been founded the entire Ford Motor Company Quality Improvement Program." The authors continue on to list many highlights of the Ford Quality Improvement Program.

Poster Campaign.   Realizing that they must involve the employees in the movement, Ford embarked upon a campaign promoting quality consciousness. Their reasoning was simple, i.e., "Good personnel morale and quality performance are cast in the same die. To have one you must have the other." They began their promotion with a poster campaign, designed to integrate the idea of quality into the individual employee's work effort in terms of pride of performance, job security, and cooperative effort. These posters were not only visible in Ford facilities, but they also infiltrated the plants of many of their suppliers.

Slogan Contest. Every hourly wage earner was given the chance to submit a slogan pertaining to the issue of Quality. Participation was excellent, and judging was extremely difficult. Finally, John Hislop of the Chester Plant received a new Ford automobile for his first-place entry. "Quality and Demand Go Hand-In-Hand."

Quality Control Publication. The purpose of this monthly publication was to give credit for achievements, to announce upcoming events, and to inform Quality Control personnel of new developments.

Rewards and Recognition. Ford sponsored its own Quality Control Exhibition. They implemented an award system that would give recognition in plants that achieve outstanding quality and improvements. A "Quality Queen" contest was held as a way to "stimulate quality thinking on a nationwide, family basis."

Statistical Quality Control. A major step forward was taken with the application of statistical methods to the control of product quality. "First of all we recognized in Statistical Quality Control a tool which would give us the economical benefits of controlled operations, and at the same time give our customers better products, products not simply better than the rest, but products better than the BEST."

Education. In order for the company to enter this era of statistical control, they had to make sure that their employees were on board. This required that a vast employee education process be implemented. The best student is a motivated student, and to be motivated one has to want to learn. So the first job of the extensive education program was to sell Statistical Quality Control to the employees.

"Our Training Department, in conjunction with the Quality Control organization, developed several training courses suitable for the instruction of both hourly personnel and management. It has been proven beyond doubt that the average inspector and production man can readily assimilate, understand, and apply the principles of Statistical Quality Control. It is equally essential, of course, for management to take full cognizance of the new methods, and for this reason all of our inspection and production supervisors are required to take training … We now have nearly 3000 control charts in active use throughout our manufacturing and assembly divisions and 5000 applications in receiving inspection areas, and this number is being increased daily."

Supplier Improvement. Statistical Quality Control (SQC) was not only used at the Ford plants, but also in forming customer-supplier relationships. Not only did they require their suppliers to use SQC, but Ford went out and taught classes for the suppliers on SQC.

"If we do a little summing up, we find that Quality, economy, and worker morale are so closely interrelated that they might be compared to a three-legged stool. If one leg is missing, the whole program topples over. But if we have all three of those things—quality economy, and worker morale—in high enough degree, our competitive position in business is secure."

After reading this overview, you may wonder why I have chosen to highlight this particular improvement process. It is no different than what many companies are already doing today. The interesting hidden fact in this example is that it was originally published in *Industrial Quality Control* Magazine in 1950, detailing events that had occurred at Ford in the 1940s. One of the numerous things that needs to be learned from this example, is that the management team must keep focused on the improvement process to hold the gains that it provides. It is an ongoing process, not a program that ends in one, two, or even five years. Ford got off to a good start, but did not follow through during the 1950s and 1960s.

■ IBM's Problem

IBM, under the lackluster leadership of John Akers, has indisputably led the decay of America's leadership in the computer world market (IBM's share of the world-wide computer industry dropped from 30 percent in 1985 to 18 percent in 1992. At the same time, IBM was extremely active in the quality movement for years. IBM started its quality improvement process in the late 1970s. By 1984, all corporate executives and key managers had attended Phil Crosby's Quality College in Winter Park, Florida. Most employees had received a minimum of 32 hours of quality training. IBM had established its own quality university, and developed, released and widely distributed a quality policy and goals for the 1980s. They also appointed a Corporate Vice President of Quality, Directors of Quality for their divisions, and Quality Champions at most locations. They then renewed their commitment and focus on quality in the last half of the 1980s with a program called "Market-Driven Quality." Their Rochester, Minnesota group won the Malcolm Baldrige Award in 1990. While Chairman of IBM, John Akers served as the chairman for the National Quality Month activities in 1991, proving his personal commitment to quality. IBM is recognized as the quality leader around the world. In 1992, IBM Sumare in Brazil, IBM Guadalajara in Mexico, and IBM Jarfalla in Sweden were recognized by winning national awards for their major accomplishments in quality achievement and management. IBM Malaysia was recognized with the Industry Excellence Award for its achievements in customer focus and quality management.

The results have been disastrous. IBM's share of the world market has continued to decline, with no end in sight. In the last three years, IBM's revenues have dropped over 8 percent, even though the industry had surged by 19 percent. About 150,000 employees will have been released through early retirement or layoff programs by the end of 1993. Their unwritten lifetime employment policy has been dropped. Stock prices plunged about 68 percent, from approximately $160 to below $46 per share. IBM's financial performance has continued to decline until it finally went negative in 1991. In January 1993, it announced its biggest ever one-year loss of $4.97 billion for 1992, outdoing even G.M.'s 1991 loss. Based on this poor performance, a group of its 772,000 shareholders banded together to pressure IBM's board of directors to take immediate corrective action. As a result of this group's actions and IBM's poor financial performance, a special meeting of the board of directors was held on January 26, 1993. At this meeting, the board of directors cut

IBM's once sacrosanct annual dividends of $4.84 a share by 50 percent. Immediately following the board of directors' meeting, IBM's CEO, CFO, and President took early retirement—a clean sweep of all key top management personnel. For the first time, IBM looked outside the corporation to find a new chairman, Louis Gerstner, formerly CEO of R.J.R. Nabisco. This indicates that their succession planning is also not working. When John Akers took over as president of IBM it was at a crossroads much the same as it was when T. J. Watson Jr. took over the reins from T. J. Watson Sr.—Watson Jr. wanted to bet IBM's future on computers and Watson Sr. thought calculators were the future. T. J. Watson Jr. won out and the 360 series was born. John Aker's alternatives were main frame or P.C. Akers took the main frame route. It is this lack of vision and future direction compounded with many bad decisions and reorganization to cover up and delay their impact was the real IBM problem, not quality improvement. The quality problem IBM faces is quality management decisions.

Mr. Louis Gerstner is making history. This was the first time that one of America's premier companies has hired a CEO from outside of its industry. If Mr. Gerstner is going to turn around IBM it is going to be by making the hard business call and developing the correct corporate vision not by trying to change the culture. Gerstner needs to bring IBM culture back in line with where it was in the early 1980s. IBM culture is very achievement oriented. Give the employees a goal and they will get it done. IBM has some of the best people in the world. They are doers when their goals are clear. Let's hope Mr. Gerstner does a better job for IBM than he did for R.J.R. Nabisco at improving stock prices (R.J.R. Nabisco was at $8 per share when he left in March 1993, down from over $12 per share in the last half of 1992). In his first year with IBM, he set a new world record with more than an $8.1 billion loss for 1993. IBM must do much better to survive.

■ Wallace, Inc.

A Malcolm Baldrige Award winner in Houston, Texas, Wallace, Inc., a distributor of steel and plastics, went Chapter 11 within the first year after they received the award.

■ Summary

It becomes obvious that the key to profitability and survival is more than just quality. It depends on how we use all of our resources to improve the quality of our outputs and the productivity of our operations, and how we integrate the correct technologies while making the optimum use of the capital and facilities available to the organization. Yes, to bring about optimum improvement to an organization,

- Quality,
- Productivity,
- Technology, and
- Cost

must be balanced to ensure that all the stakeholders' needs and expectations are met. As C. Jackson Grayson, Chairman of the American Productivity and Quality Center, put it, "We need to be lean, hungry, quick, dedicated to excellence, oriented toward inventive action— meaner than a junkyard dog, but smarter than a barnyard cat."

The effective use of a Total Improvement Management (TIM) process provides a "Win-Win" scenario for all the organization's stakeholders. Organizations that are serious about improving march around the 12 Steps in the Improvement Win-Win Square (see Fig. I.1).

Step 1. Increased dedication to improvement

Step 2. Increased owner investment

Step 3. Increased management attention

Step 4. Improved process

Step 5. Increased employee satisfaction

Step 6. Increased trust

Step 7. Increased cooperation

Step 8. Better products and services

Step 9. Fewer customer complaints

Step 10. Increased customer loyalty

Step 11. Increased growth and profit

Step 12. Increased owner returns

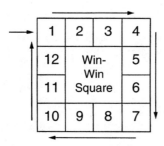

Figure I.1. The win-win square.

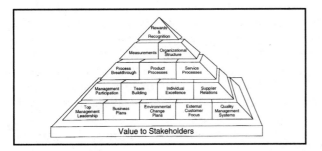

Overview:
The Essence
of Improvement

*Don't start an improvement process to improve
customer satisfaction or employee morale. It will do
that, but the real reason you need an improvement
process is to increase the organization's performance
PROFITS).*   DR. H. JAMES HARRINGTON

■ Introduction

Wow! What an exciting time the last 15 years have been for the quality professional.
It is almost as though management suddenly realized that there was a direct
correlation between quality and profit, when for years, they had thought that they
were opposing forces. Every place you look, you see quality being promoted. You
open up prestigious magazines like *Fortune,* and you see sections devoted to quality.
You open your newspaper and you find Tom Peters' column on quality. You turn
on the TV and the commercials promote the quality of their products, using catch
phrases like, "Quality Is Job One." You go to conferences and CEOs expound upon
their personal, as well as their organizations' strong commitment to quality, followed by an explanation of what they are doing to improve their quality. They may
not understand exactly what they are saying, but they are using the right words.
They are talking about: "doing it right every time," "we need to prevent errors,
rather than react to them," "we need to continuously improve everything that we
do," etc. This surely is a step forward.

You ride down the street, and you see stores using quality in their names.
Names like Quality Dry Cleaning, Quality Delicatessen, Quality Repair Shop,
Quality Insurance Company, etc. In fact, I was going down a street in London,

and ran across a store called, "Quality Seconds." Yes, quality scrap. Maybe that is the problem. Everyone is talking about quality and organizations are throwing big bucks at the problem, but the results have been less than acceptable for many organizations. Some organizations, like Florida Power & Light, are disappointed with the results of their improvement efforts. A 1992 Gallup survey indicates that this could also be true of as high as 10 percent of the organizations which have initiated an improvement process.

The actual number of organizations that have initiated an improvement process and failed is not important. Whether it is 20 percent or 1 percent, America is not improving fast enough. In one decade, we went from the world's richest nation to the world's largest debtor nation, and our outlook for the future is bleak. Will the United States have to follow the Soviet Union's example and break up into small countries to get out from under its huge national debt? The United States' share of total world exports dropped from 13.4 percent in 1973, to 12.1 percent in 1983, to an estimated 11.6 percent in 1992. Today's college graduates face the poorest job market in recent history and they will receive lower wages. The average real hourly wage of a college graduate is going downhill. After adjusting for inflation, in 1973, it was $16.45 an hour; in 1987, it was $15.24 an hour; in 1991, it was $14.77 an hour; and In 1993, it was at $14.21 an hour. This represents a 13.6 percent decrease in their buying power.

You cannot pick up a newspaper without reading about new layoffs. Organizations like Hewlett-Packard, AT&T, Sears, IBM, and General Motors are downsizing. Since 1989, 440,000 defense industry workers have been laid off, and 100,000 civilian defense department employees have been let go. By 1995, the downsizing of the United States Armed Forces will put more than 500,000 veterans into a job market that is already in dire straits. Cities are declaring bankruptcy. This is the first time since the great depression that our next generation will live in a less affluent environment than their parents did. Has the United States reached its zenith and is now slipping downward, or is it on the back side of a wave, soon to rise again? All indications are that our economy has crested, and we are now going downhill.

Sixty percent of Americans' buying power decreased an average of over 33 percent between 1976 and 1992. A two-wage earner situation has become essential just for people to keep from falling further behind. This additional income has allowed the median family income (adjusted for inflation) to increase 0.67 percent per year for the last ten years. In the United States, 30 million people are going hungry every day2. We need to stem the tide and start climbing back up the hill to undo the damages we have imposed upon our nation during the last 25 years. This means that we need to turn our losing organizations into surviving organizations, and our surviving organizations into winners. We must improve at a much faster rate than we did in the 1980s if our way of life is to survive.

■ Losers, Survivors, or Winners

The Pentagon's budget should shrink down to about $234 billion in today's dollars by 1997. That would have a rippling effect on defense, military, and nonmilitary jobs. The Federal Reserve estimates that would result in the loss of 2.6 million jobs.[3]

Your department, your organization, and your country fall into one of three categories. There are winners, survivors, or losers. Those are the choices. Check one of the following that best describes the organization you work for.

- Loser   _____
- Survivor   _____
- Winner   _____

How do you know if your answer is right? That's easy. Look at your organization's relative performance in the following areas:

- Return on Assets (ROA)
- Value-Added per Employee (VAE)
- Market Share
- Customer Satisfaction

A relative performance analysis needs to be conducted to evaluate the organization from a short- and long-term performance standpoint. In addition, you need to evaluate the organization's performance against itself and its best competition (see Fig. O.1).

Short-Term Analysis

A short-term analysis compares the present performance to the performance of the organization 12 months prior. A plus (+) indicates that it has improved, a zero (0) indicates that there is no change, and a minus (–) indicates that there has been a negative change. An "NU" rating means that the data is "not used."

	Short term			Long term		
	Actual value	Self	Competition	Average value	Self	Competition
ROA	0	+	+		+	+
VAE	+	+	+	+	+	+
Market share	NU	–	–	NU	+	-
Customer satisfaction	0	NU	NU	–	NU	NU

Figure O.1. Relative performance analysis table.

Actual Value—For actual value, give the organization a plus, minus, or zero rating using the following table.

ROA			VAE (in United States dollars)		
0 to 2 percent	=	Minus	Below $47,000	=	Minus
2 to 6 percent	=	Zero	$47,000 to $74,000	=	Zero
Over 6 percent	=	Plus	Above $74,000	=	Plus

The customer satisfaction rating is based upon the percentage of external customers that rate your service and/or product in the upper percentile (on a scale of one to ten, one equals very poor, ten equals the best). Upper percentile is defined as the percentage of external customers that rate your organization eight to ten.

Below 50 percent   =   Minus

51 percent to 75 percent   =   Zero

76 percent to 100 percent   =   Plus

Short-Term Self Analysis—The "self" column compares the organization to where it was 12 months before.

Short-Term Competition Analysis—The "competition" column compares the organization's performance to the average of the top 10 percent of its competition.

(Note: Customer satisfaction data Is not used for either the *self* or the *competition* categories because market share change provides a better indication of customer acceptance of the organization's service and products.)

Long-Term Analysis

The long-term analysis looks at the organization's average performance over the last five years. Scoring rules that were used in the short-term evaluation apply here also.

Rating the Organization

Figure O.2 provides you with a guide to rating your organization's performance. Your organization is a winner if it has one or less minuses or two or less zeros, or a total of no more than two zeros and minus ratings combined. Your organization is a survivor if it has 14 or less zeros, or six or less minuses, or a combined total of zeros and minuses equal to less than 14. Organizations that have 15 or more zeros, or 7 or more minuses, or a total of 15 or more minuses and zeros are classified as losers (see Fig. O.2).

Classifications	Zeros	Minuses	Combined Total
Winners	2 or less	1 or less	3 or less
Survivors	14 or less	6 or less	14 or less
Losers	15 or more	7 or more	15 or more

Figure O.2. Performance Ratings

Using this criteria, there have been very few winners over the last 10 years in the United States. There are a lot of survivors and a lot of losers, but very few winners. This document is written primarily for the survivors and the losers, although even the winners can benefit from it. For if the winners do not continuously practice the principles and methodologies presented, they will soon slip back to the survivor classification, and eventually end up in the loser classification. When you get right down to it, there are really only two types of organizations—those that are making significant improvement, and those that are pushed out of the way by the ones that are improving.

Monopolies and the Government

Do the same classifications apply to monopolies and the government? Yes, very much so. This evaluation applies equally well to organizations like gas companies, water companies, telephone companies, and all privately owned companies. Even monopolies have indirect competition. Florida Power & Light's competition consists of all the power companies in every other state. Each monopoly needs to compare itself to the best organization outside its monopoly. Do not just use organizations in your marketing area. You should compare your organization to similar organizations around the world. Gas and electricity organizations in the United States should compare themselves to like organizations within the United States and in countries like Japan and Germany. In these cases, they will probably need to replace market share with price of unit delivery, and evaluate customer satisfaction in all six areas (see Fig. O.1).

■ Governments Need to Improve

Our top government officials have been talking about quality improvement for the last 20 years. For example, Ronald Reagan, past president of the United States, stated, "A commitment to excellence in manufacturing and service is essential to our nation's long-term economic welfare." On February 25, 1986, he issued Executive Order #12552 that stated, "There is hereby established a government-wide program to improve the quality, timeliness, and efficiency of services provided by the federal government. The goal of the program shall be to improve the quality and timeliness of service to the public, and to achieve a 20 percent productivity increase in appropriate functions by 1992. Each executive department and agency will be responsible for contributing to the achievement of this goal."

In October 1992, George Bush, then President of the United States, stated, "My budget indicates that my administration is launching quality demonstrations in the IRS, Social Security Administration, and the Department of Veterans Affairs. The purpose of this effort is to demonstrate and evaluate what works in order to improve federal quality in programs that touch millions of Americans."

Bill Clinton, President of the United States, stated, "Continued emphasis on quality by American companies is critical. It is what makes 'Made in the USA' something to be proud of." Stressing the importance of quality improvement in government performance, President Clinton stated, "Innovative management techniques such as TQM, should be considered as one of the many approaches to make government more effective and efficient. We can no longer afford to pay more and get less for our government. The answer to every problem cannot always be another program and more money."

These certainly are good words, but let's look at the results. Probably the best indicator of the improvements undertaken by the federal government is our growing annual national debt. For the first 200 years, our accumulated federal debt was $0.7 trillion. In the last 15 years, it grew to over $3.0 trillion. Another excellent indicator is the number of days an average person works per year to pay his or her tax bill. In 1950, that figure was 93 days. Today, it is over 128 days. Now that is really negative productivity!

I recently spent a week in South America, where the only English-speaking channel on television was one that was broadcasting the activities within the United States Congress. They were in the process of debating the annual budget. After watching Congress in action for a short period of time, I am firmly convinced that there are two things you never want to see made: One is sausage, and the other is law in the United States. They would call for a 10-minute recess just before a vote, and the room would empty out. At the end of 10 minutes, no one would return to start business. Fifteen, 20, 30 minutes later, people would start to trickle in. At other times, Congressmen or women would give brilliant presentations on a specific issue and when the camera would pan the audience, only one or two Congressmen or women were sitting at their tables. The rest of the room was empty.

One thing is for sure: There is a great deal of opportunity for our government to improve, to eliminate the bureaucracy, to cut waste, to streamline procedures, to make effective use of resources, and to provide better service. In all probability, a concentrated improvement effort could reduce administrative costs as much as 50 percent and improve quality of service by over 100 percent. This is true for most city, state, and national governments. In 1992, then President George Bush reported, "As a result of cost-effective quality improvement incentives, the level of satisfactory IRS responses to correspondence rose from 60 percent in 1988 to 85 percent in 1991." Clearly this is a marked improvement, but to look at it another way, 15 percent of the IRS customers were dissatisfied with their responses. Should anyone brag about a 15 percent error rate? In manufacturing, we expect our employees to perform at a parts-per-million level, while the IRS brags about not satisifyng its customers 15 percent of the time. In 1994, between February and April 15, the line was busy for

75 percent of the callers. When the IRS does answer the phone, it is estimated between 20 and 35 percent of the answers are wrong. There certainly is opportunity for great improvement.

President Clinton pushed a major tax hike through Congress in 1993 to help slow down the growing national debt, but even with this tax increase the national debt would not be reduced, and in six years we would be worse off than we are today. On the other hand, if we cut our national government's poor-quality cost in half, we would generate more savings than we would generate with a tax increase 100 times as great, and it would continue for years to come. We must improve the way our government is managed, and we must do it now.

President Clinton was a strong supporter of Total Quality Management when he was governor. Now he has focused his attention on "Reinventing Government."

This approach to improvement makes use of the Business Process Improvement methodology that should reduce costs on the government's critical business processes from 40 to 90 percent. If a process redesign approach is used, it should take about 90 days to redefine the new process. If the reengineering approach is used, it takes approximately one year. Vice President Gore has been put in charge of this effort with a commitment to spend upwards of 50 percent of his time on this project. With Vice President Gore's focus on the "Information Super Highway," it looks like this effort is going the way of President Reagan's improvement efforts. It is surprising that none of the results of their business process improvement efforts over the last year have been reflected in the national budget.

Executive order 1286Z (presidential document 48257) dated September 1993 entitled "Setting Customer Service Standards" was a clear indicator that the national government is redirecting its focus back to TQM methodologies that should provide a 10 to 20 percent per year improvement in the federal government's efficiency and effectiveness. The measurement of improvement in our federal government is its ability to sustain or improve the services they provide while reducing the percent of gross national product that the federal government consumes. A reasonable short-range target is getting back to the percent of gross domestic product consumed by the federal government in 1980 before President Clinton's term in office is complete. This can only be accomplished with a major focus on TIM. It will require that we stop expending our resources to slow down spending like we are in health care, and truly understand why health care and other governmental costs have not stayed in tune with inflation, and then bring them back in line with historical cost levels of the 1970s.

The government presents a unique problem. Most of us view the government as a monopoly that we must live with, but that is far from true. I know of no other organization that puts all subcontracts out for open bid every four years, and that is exactly what we do when we have an election. Our elected officers are really subcontractors with whom we have signed contracts to manage our government, and they do have competition. I am sure that past President George Bush will agree, after Bill Clinton succeeded in "beating the Bush and flushing out the Quayle." I will admit that most of us do a very poor job of evaluating these suppliers (elected government officials) before we sign contracts with them. In fact, if we ran our industries like we run our government, with suppliers that promise many things they never intend to deliver, our industries would all be rated as losers and our country would be in much worse condition than it is today. Our products would not even be able to compete with the products manufactured in Cuba.

We need to determine if our government organizations are winners, survivors, or losers in much the same way we evaluate our businesses. We have turned our backs on this critical portion of our lives for far too long, causing mismanagement at all levels of our government. We are faced with disgraceful, runaway national, private, and governmental debt. Our accumulated national debt has jumped from $4 trillion to $10 trillion in just 10 years. United States national debt is double the GNP. During the 1980s, the debt-to-GNP ratio increased by 30 percent. Something drastic needs to be done about the way our suppliers are running our government for us. This mismanagement has resulted in individual cities like San Juan Bautista, California and New Haven, Connecticut declaring bankruptcy.

Local city departments should also compare themselves to other departments doing similar activities in other city organizations, states to other state organizations, and our national government to other national governments. These evaluations should be based upon comparative change in performance of the organization itself and a comparison to the change of the top 10 percent of similar type organizations. Top government officials need to run these areas like a business. They should release quarterly reports providing key measurement data like the following:

- Percent of Gross Area Products Consumed
- Customer Satisfaction
- Percent of Campaign Implemented
- Net Favorable or Unfavorable Balance
- Ratio of Income to Expense

This is the type of data that the voter needs when contracts come up for renewal (election of candidates). Candidates running for office or reelection should campaign on how they will improve these types of key measurements. We, the United States people, are not measuring our elected officials as we should.

Past President Reagan put it well when he wrote, "The need for and importance of improving the efficiency with which the federal government delivers goods and services to the American public cannot be overstated. The federal government now accounts for 24.6 percent of the GNP."[4]

■ Characteristics of the Winners

Behind the winners is a story of common and uncommon sense that cuts a wide swath across the organization's operation and business practices. To gain real insight into the winning organizations, you need to compare the winners to the losers. Based on survey results from Ernst & Young's report, "The American Competitiveness Study," four marketplace position factors were identified to separate the losers (−) from the winners (+) (see Fig. O.3).

Marketplace position factors	Relative profitability	
	Losers	Winners
Relative Quality	Worse	Better
Relative Cost	Higher	Lower
Relative Price	Lower	Higher
Relative market share	Smaller	Larger

Figure O.3. Marketplace position factors.

The winning organizations are those organizations that have:

- Better relative quality

- Lower relative cost

- Higher relative price that, combined with reduced cost, provides a very significant profit advantage

- Larger relative market share

Figure O.4 was based on this Ernst & Young study and relates key business characteristics to better relative performance.

A Broad Agenda of Consideration Aimed at Market Leadership

The ultimate selection of improvement initiatives and the allocation of funds to accomplish renewal objectives are products of business planning, and two aspects of planning were found to be directly related to better performance (see Fig. O.5).

Figure O.4. Impact of key business characteristics on relative performance.

| | Relative profitability | |
Planning factors	Losers	Winners
Breadth of focus	Narrower	Broader
Leadership intent	Follower	Leader

Figure O.5. Planning factors.

First, organizations with planning processes that were broadly focused across a range of external and internal considerations were more successful than those with more narrowly focused planning agendas.

Winning organizations were more likely to address matters of internal organization and external competition than were their counterparts. Second, these broader planning agendas were aimed at achieving the more demanding ambition of market leadership rather than just self-improvement. Here, winning organizations were more likely to be pursuing product and service quality superiority, which helped form the necessary foundation for pricing leadership.[5]

Product and Market—Scope of Business— as Broad as Manageable

For the majority of organizations, the planning agenda most often included product and market considerations, and a number of choices related to product and market strategy were found to influence overall performance (see Fig. O.6).

More successful organizations offered broader product lines than their competitors and were more active in upgrading these product lines through innovation. But these organizations also believed that their customers criteria for quality extended well beyond today's physical product quality to include their general reputation for better products and services. More successful organizations were also more vertically integrated, more likely to have some involvement in international markets, and were less likely to find themselves in market situations where customer bargaining power was the principal catalyst for increased competitive pressure. In summary, organizations were rewarded for their ability and willingness to manage complexity and drive innovation.

| | Relative profitability | |
Improvement factors	Losers	Winners
Product line scope	Narrower	Broader
Product innovation	Infrequent	Frequent
Quality criteria	Product	Reputation
Vertical integration	Less	More
International scope	Less	More
Competitive pressure	Customers	Competitors

Figure O.6. Product and market strategy improvement factors.

Business results are the product of an organization's good and bad habits. Habits are easy to form and hard to break. A total improvement management process focuses on underlying bad habits and replaces them with winning ones. Too often we focus our improvement efforts on special improvement methodologies that are separate from the day-to-day activities. As a result, they are never incorporated into the basic fiber of the organization so that they become an automatic habit that occurs without thought. Success is only obtained when these tools and methodologies become habits and are no longer recognized as improvement activities.

■ Does the Customer Want Improvement?

In the 1980s, there was a great deal of improvement in the manufacturing industries, while some improvement was seen in the service industries. As a result, many people feel that this improvement has eliminated the need for further improvement because they believe their organization is meeting its customers' needs. Research conducted by the Opinion Research Corporation proves this to be wrong. There is a major gap between the average product performance for all industries and their customers' expectations. There is even a significant gap between the best performing organizations and customer expectations. The organization that reduces these gaps will obtain a very significant competitive advantage.

■ So You Want to Improve

Improvement is not part of the game—it *is* the game today. Everyone wants things to change for the better. Top management wants employees to stop making so many errors. Engineering wants marketing to give them better forecasts. Marketing wants sales to improve their sales record. Sales wants manufacturing to produce better products so they will be easier to sell. Manufacturing wants engineering to give them designs that are more manufacturable. Everyone wants everyone else to change, but too often they are unwilling to change themselves. You can no longer wait for someone else to change. The improvement process must start with you. The question is: How does an organization make the process work for them? There are many approaches. Suddenly there are hundreds of consultants knocking on management's door with the *single right answer* for you, and they are all different and in some ways the same.

■ Confusion Reigns Supreme

Is it any wonder that management is confused? Even the individuals who were recognized as the gurus in the continuous improvement process cannot agree on how an organization should implement the improvement process.

Philip B. Crosby's "14 Steps" focused on motivating the individual, documenting their commitment to quality by having them sign pledge cards and measuring progress through the use of quality cost (a concept developed by A. V. Feigenbaum in the 1950s). His 14 Steps of quality improvement are:

1. Management Commitment
2. Quality improvement Teams
3. Measurement
4. Cost of Quality
5. Quality Awareness
6. Corrective Action
7. Zero Defect Planning
8. Employee Education
9. Zero Defect Day
10. Goal Setting
11. Error-Cause Removal
12. Recognition
13. Quality Councils
14. Do It Over Again

Dr. W. Edwards Deming introduced Japan's top management to the statistical process control methods developed by Walter Shewhart in the 1920s. Japanese management were quick to realize that this was the "secret weapon" that allowed the United States to mass-produce the vast quantities of high-quality weapons that defeated Japan In WWII. Dr. Deming developed a different "14 Point" program just for the United States.

Just before he passed on, Dr. Deming began to advocate a system he calls, "Profound Knowledge," that is made up of another 14 points. They are:

1. Nature of variation.
2. Losses due to tampering (making changes without knowledge of special and common causes of variation).
3. Minimizing the risk from the above two (through the use of control charts).
4. Interaction of forces, dependence, and interdependence.
5. Losses from management decisions made in the absence of knowledge of variation.
6. Losses from the successive application of random forces that may be individually unimportant (such as workers training other workers).
7. Losses from competition for market share and trade barriers.

8. Theory of extreme values.

9. Statistical theory of failure.

10. Theory of knowledge in general.

11. Psychology, including intrinsic and extrinsic motivation.

12. Learning theory.

13. Need for the transformation to leadership from grading and ranking.

14. Psychology of change.

Dr. Armand V. Felgenbaum focuses his effort on 10 benchmarks that direct the improvement effort. His "10 Benchmarks for Quality Success" are:

1. Quality is a company-wide process.

2. Quality is what the customer says it is.

3. Quality and cost are a sum, not a difference.

4. Quality requires both Individual and teamwork zealotry.

5. Quality is a way of management.

6. Quality and innovation are mutually dependent.

7. Quality is an ethic.

8. Quality requires continuous improvement.

9. Quality is the most effective, least capital Intensive route to productivity.

10. Quality is implemented with a total system connected with customers and suppliers.

Dr. Feigenbaum is the father of Total Quality Control and published the first book on the subject in 1951. He also originated the concept of Quality Costs. He looks at the total product value cycle and applies systems engineering approaches to bring about improvement.

Dr. Joseph M. Juran, on the other hand, fosters the belief that an improvement effort is driven by many small, step-by-step improvements. Each saves the company approximately $100,000. He uses pareto analysis to define the critical few problems and assigns teams to solve these problems. Dr. Juran defines quality as "fitness-for-use." He looks at what he calls, "The Spiral of Progress in Quality." The quality function is the entire collection of activities through which we achieve fitness-for-use, no matter where these activities are performed. It includes:

1. Market Research

2. Product Development

3. Product Design/Specification

4. Purchasing/Suppliers

5. Manufacturing Planning

6. Production and Process Control

7. Inspection and Test

8. Marketing

9. Customer Service

Dr. Kaoru Ishikawa was the leading quality expert from Japan and the originator of the quality circle concept. He espoused that the best way to improve performance is through the empowerment and enlightenment of the employees. Dr. Ishikawa's concepts fueled the unparalleled explosion in employee team skills and problem-solving training. Although Dr. Deming and Dr. Juran are given credit for the miraculous transformation of Japan, Inc., I believe that Dr. Ishikawa was the real genius because he took many concepts, put them together, and implemented them all effectively. Without Dr. Ishikawa's activities, I believe Deming's, Felgenbaum's, and Juran's work would have had little effect on the Japanese. Dr. Ishikawa looked at quality as a way to manage the total organization. He saw the management transformation as six categories:

1. Quality first—not short-term profit.

2. Consumer orientation—not producer orientation. Think from the standpoint of the other party.

3. The next process is your customer—breaking down the barrier of sectionalism.

4. Using facts and data to make presentations—utilization of statistical methods.

5. Respect for humanity as a management philosophy—full participatory management.

6. Cross-function management.

Dr. Deming's popularity expanded because he was given credit for the success of Japanese quality programs. When comparing his and Dr. Juran's approaches, he stated, "I'm not interested in stamping out fires. That's what Juran does. I'm creating a system of profound knowledge that will still be good a century from now." But many think that Juran contributed more to the success of Japan, Inc. than Deming. "Juran was more important to Japan than Deming," says Junji Noguchi, executive director of JUSE. "SQC (statistical quality control) applies only to technicians." In 1969, JUSE developed a super prize for organizations that had won the Deming Prize and had demonstrated continuous quality improvement over a five-year period. To recognize Joe Juran for his major contribution to Japan's quality movement, JUSE asked Dr. Juran if they could name this prize the Juran Medal. Juran responded with a noncommittal answer that the Japanese felt was a polite turndown. They named the award, "The National Quality Prize."[6]

In 1987, McGraw-Hill's landmark book, *The Improvement Process*, explored the importance of:

1. Relating the organization's improvement efforts to the business plan.

2. Engineering the total improvement effort.

3. The need for both continuous and breakthrough improvement through the use of teams and business process improvement methodologies (later to be called Process Reengineering or Redesign).

4. The importance of empowerment and creativity to allow the individual to excel.

Along with the approaches sold by these gurus, other consultants and professional organizations develop even more improvement approaches. The engineering community stresses the need to invest in R&D to improve the technologies to be more competitive. The financial community talks about using Total Cost Management to improve profits. Productivity Centers around the world promote improving productivity to become more competitive. The United States Department of Defense is pushing a program called Total Quality Management as a way of improving the level of customer satisfaction.

In a very short span of time, all of these methods and tools, as well as many others, were brought to management's attention. Many of these concepts were tried to one degree or another in most of the progressive organizations during the 1980s. Each of the approaches were presented to management as the best way of obtaining a competitive advantage. Today, there are more than 180 different improvement tools and/or methods available.

■ Management's Improvement Dilemma

Management's dilemma is the fact that they have a limited amount of resources to dedicate to the improvement effort (see Fig. O.7), and they have at least five different methodologies all competing for these limited resources:

- Total Cost Management

- Total Productivity Management

- Total Quality Management

- Total Resource Management

- Total Technology Management

But as profitable as each approach seems, it is obvious that the organization still had to use most of its resources to provide the products and/or services to their external customers that fund the organization's operation. Top management's job is to divide

Figure O.7. The competition for resources.

the limited improvement resources among the five improvement approaches to get the maximum results. The winning organizations have done an excellent job of distributing these improvement resources among the five approaches, shifting emphasis at the correct time. Most of the survivor organizations have adopted one approach and held dogmatically to it, ignoring the others. The losers have shifted randomly among each approach, without explaining to their employees why they were changing direction. Consequently, employees were left with a feeling that they can wait it out. Why change when the next time top management attends another conference, they will come back with another new approach that will change the organization's direction? Management needs to understand all five methodologies to be able to make correct decisions and to stop changing direction so often.

Total Cost Management (TCM)

In the mid-1980s, a collaboration of financial-type employees from major U.S. organizations developed a technology called "Activity-Based Costing." From this basic start evolved a new methodology called "Total Cost Management (TCM)."[7] It was designed to obtain a step-function improvement in key processes by analyzing every activity within the process, classifying its cost as valued-added or no-value-added, and then taking positive action to eliminate the no-value-added cost.

The TCM methodology can be divided into five phases:

1. *Assessment.* Define which business process the methodology should be applied to.

2. *Organization.* Involve and train management and process improvement teams (PIT).

3. *Analysis.* Flowchart the process, conduct a process walk-through, and do a value and root-cause analysis.

4. *Design.* Lay out a new process with as many of the no-value-added activities removed as possible, and perform a cost/benefit analysis.

5. *Implementation.* Implement the proposed process changes and measure the results.

Typical tools that are part of Total Cost Management are:

- Activity-Based Costing
- JIT Cost Accounting
- Process Value Analysis
- Performance Management
- Responsibility Accounting
- Integrated Financial Reporting
- Poor-Quality Cost

Total Productivity Management (TPM)

One of the priority projects which General Douglas MacArthur undertook when he was put in charge of the occupation of Japan was to improve Japan's productivity. The Japanese Productivity Center was formed to spearhead this movement. Their success soon got the attention of American management. As a result, the American Productivity Center, Inc. was formed, led by C. Jackson Grayson in Houston, Texas. In the 1970s and 1980s, more productivity centers sprang up around the world to collect data and promote productivity improvement within the country.

In the late seventies and early eighties a flood of organizations and consultants got productivity improvement programs started based upon the assumption that Japan and West Germany were taking away our market because their productivity growth rate exceeded ours.

All of these approaches had tools and methods that would improve the quantity of output per unit of resources consumed, whether the resource was people, dollars, or equipment. The movement, which could have been called "Total Productivity Management (TPM)," focused on improving productivity by automating time-consuming, boring, repetitive activities, and eliminating waste.

A typical Productivity Improvement Program consisted of five phases:

1. Awareness
2. Information (Education)
3. Planning
4. Action
5. Follow-Up

In IBM's Technical Report TR 02.911 dated January 15, 1981 were listed the following steps to improve productivity.

1. Lessening of government regulations.
2. Invest in capital equipment.
3. Invest in research and development.
4. Make all management aware of the problem.
5. Make effective use of creative problem solving.
6. Increase use of automation and robotics.
7. Increase teamwork and employee involvement.
8. Expand international markets.
9. Do the job right the first time.

These programs recommended that they start by getting the Chief Executive's active support first. From there they would get top management involved, then appoint a Productivity Steering Committee. The next step was to train middle management and supervisors on productivity concepts and provide them with the tools and techniques used to improve productivity. These programs then focused on establishing measurements and involving people at all levels through communication and training. I can remember when all IBM managers and key employees went through a course taught by the American Productivity Center, Inc. on how to improve productivity and the follow-on program that it inspired.

In 1980, the American Productivity Center, Inc. (APC) predicted that if American organizations did not implement a productivity improvement activity, by 1991 France, Japan, and Germany would be outproducing the United States by over 8 percent per year per person. Well, I guess we did do some things right in the 1980s because the United States is still the most productive nation in the world. And according to new APC predictions, no country is projected to be ahead of the United States as we enter the twenty-first century. By the year 2010, however, Japan will be the most productive nation in the world (see Fig. O.8).

For productivity to keep increasing in the United States, we need to increase the output per capita. Big layoffs like the ones at IBM and G.M. are having should improve their productivity, but it will not help the nation since there is no increase in output and the number of available employees remains the same. For productivity to grow, we need not only to eliminate waste, but also to invest in better equipment and create outputs that were not there before. Professor Frank Lichtenberg of Columbia University claims that a dollar invested in R&D yields productivity gains approximately eight times greater than a dollar invested in plants and equipment. Jack Welch, former CEO of General Electric, made his feelings about the need for productivity improvement very clear when he said, "For a company and for a nation, productivity is a matter of survival."

Total Quality Management (TQM)

In the early 1980s, quality became the magic word, driven by Japan, Inc.'s success in capturing world markets as a result of better design and production quality. For

Country	1989 level (U.S. = 100)	Hypothetical relative levels*		
		Average % growth 1979 - 89	2000	2010
United States	100.0	1.1	100	100
Canada	94.0	1.3	96	98
France	85.9	2.0	95	103
Italy	87.3	1.9	95	103
West Germany	82.0	1.7	87	93
United Kingdom	71.5	1.8	77	83
Japan	72.7	3.0	89	107
Korea	39.8	5.2	62	92

Using 1979-89 Growth Rates
Source: Bureau of Labor Statistics

Figure O.8. International labor productivity with the U.S. level equal to 100 percent (GDP/employee, purchasing power parity).

example, they earned over 30 percent of the United States automotive market because Japanese manufacturers were able to produce cars that, when compared to cars built in Detroit, had less than one quarter of the defects after they were delivered to the customer. This drove the quality discipline to new heights. In the mid-1980s, the Department of Defense popularized the term "Total Quality Management (TQM)," extending the quality discipline into all areas of the business. Fueled with the belief that W. Edwards Deming and Joseph M. Juran were responsible for Japan's post-World War II manufacturing marvel, TQM became the "in thing" to do. Management embraced TQM in a blind leap of faith, guided by an onslaught of self-acclaimed quality consultants. Everyone who was out of work and had taken a class in SPC or problem solving, hung out their shingle as a quality consultant.

In a survey conducted by the American Society for Quality Control of organizations that are using TQM, 31 percent said they had made some mistakes. The most frequent mistakes were:

1. Not beginning sooner.

2. Failing to make quality a priority.

3. Making quality a project, not a continuous process.

4. Expecting immediate financial results.

5. Not having everyone involved.

6. Not focusing on measurements.

TQM is much more difficult to define than TPM because it was never clearly defined to start with. Many books have been written on the subject, but each one is a little different.[8,9] It seemed that any program that anyone wanted to get approved in the 1980s and early 1990s was called TQM.

The general, basic elements of a TQM process are:

- Start with top management involvement.

- Educate all levels of management.

- Understand your external customers' requirements.

- Prevent errors from occurring.

- Use statistical methods to solve problems and control processes.

- Train all employees in team and problem-solving methods.

- Focus on the process as the problem, not the people.

- Have a few good suppliers.

- Establish quality and customer-related measurements.

- Focus on the internal as well as external customers.

- Use teams at all levels to solve problems and make decisions.

In a survey conducted by MAPI (Manufacturers' Alliance for Productivity and Innovation) of major organizations using TQM, the following results were reported:

- 40 percent—significant improvement
- 45 percent—some improvement
- 15 percent—marginal improvement
- 0 percent—no improvement

Total Resource Management (TRM)

Driven by the major gains Japan made in the way it used resources (like inventory, floor space, and its employees), a methodology began to gain favor with management that could be called "Total Resource Management (TRM)." Human Resource groups, in an effort to empower the workers, began the most aggressive educational program ever undertaken by business. These educational programs included teaching team skills, problem-solving capability, and job-related training. Technical vitality became a priority in many organizations as management realized that a B.S. degree could be obsolete in as little as five years. The objectives of these training programs were to increase employee loyalty, productivity, and skills. Management realized that their employees were their most valuable asset and, if the organization was going to keep pace with the technologies, they needed to keep upgrading the skill level of their employees. Training and empowering the employees at all levels also helps decrease the costly turnover rates that American organizations were realizing. The worker turnover rate in the United States is more than 10 times that of Japan.

At the same time, the industrial engineering organization was redesigning the manufacturing layout and the stocking areas to balance work flow and minimize parts movement and stocking levels. This activity was largely based upon the results Toyota had with their "Just-in-Time" production process. This movement challenged the basic belief that the most effective and efficient manufacturing strategy was to process large lots of the same units through the manufacturing process together. This effort focused on making better use of other key resources; among them, floor space and inventory cost. Equipment utilization was often traded off for reduced inventory.

We realized that storage was using up to 75 percent of the manufacturing area. If the lot size could be reduced, storage could be reduced. But large lots were required because of the long set-up time. This need drove a new approach to the way tools were designed and how the production areas were laid out. Set-up cycles which had taken four hours, were reduced to less than 10 minutes at companies like Toyota and Ford. This approach minimized the amount of set-up time required and drove the production operations toward a single-unit build concept. In many cases, in-process inventories were reduced from four months to four hours. This was accomplished through the use of tools like Just-in-Time, zero stock, one-minute die change, single-unit build, and so on.

Total Technology Mangement (TTM)

By the early 1980s we also began to realize that our technologies were falling behind our international competitors. A lower percentage of our students were attaining the more difficult B. S. degrees and instead were taking the easier way out of college by obtaining a B.A. degree. This approach soon saturated the job market, driving many of our young people back into school to obtain a Masters in Business Administration. This is driving an effort to upgrade our educational system to do a better job of teaching the "three Rs" and reestablishing science as a desirable career.

At the same time, the product life cycle was being drastically reduced, requiring more new products to be developed and brought to the marketplace in half the previous time. It also meant that the cost of developing a new product had to be reduced, because it was now being spread over fewer units. In the 1960s, a product life cycle was 14 years, and in the 1970s, it was seven years. In the 1980s, it dropped to four years, and by the year 2000, it will be measured in months. These shorter life cycles and high customer performance expectations eliminate the luxury of shipping initial products with problems that could be corrected by later engineering changes.

This required a new approach to the way we managed technology that could be called "Total Technology Management (TTM)." The TTM approach was spurred on when American management realized that Japan, Inc. was able to develop a new product and bring it to market in almost half the time and cost. Many companies got so caught up with TTM so that the total organization was reorganized around the available technology, creating situations where the technology drove the business, rather than the business driving the technology.

TTM focused its activities on staying ahead of the competition by having the most advanced technology in its products. It also applied new technologies to the development process to reduce cycle time and cost. TTM advocates discovered that Japan was issuing a higher number of patents per capita than the United States. Investigation revealed that Japan, Inc. was investing its Research & Development dollars with an emphasis on applications development; the United States was spending most of its R&D dollars on basic research. This has resulted in the United States winning more Nobel Prizes, and Japan winning more customers. In light of this, President Clinton is directing the government's support away from basic research to applications development.

Since the late 1970s, there has been a major trend in the United States to implement Concurrent Engineering using a team made up of development, manufacturing, customers, suppliers, and process personnel who work together during the development cycle. As a result, the product design and the process that will produce the product are created and developed simultaneously. Starting in the 1970s, TTM also capitalizes on using Information Technology to bring about improvement and reduce cycle time. (Examples: Computer-Aided Manufacturing, Computer-Aided Design, etc.) TTM is based upon the fact that you can have the best price and the best quality, but if you don't have a product that the customer wants, you will go out of business. You cannot prosper in today's market selling buggy whips or tube TVs.

■ Looking at the Total

Fortunately, there is a lot of overlap between the five individual methodologies (see Fig. O.9).

The dark area in the center of Fig. O.9 indicates activities that are part of four total management methodologies (TCM, TPM, TQM, and TTM). These methodologies are shown as being on top of the TRM circle because all the other four methodologies have a direct impact on the organization's resources. Typical activities that are used in all five areas of total improvement are:

- Top Management Involvement
- Team Problem-Solving
- Process Improvement Methods
- Strategic Planning
- Education

In Fig. O.9, the areas with a 4 in the center of them indicate activities that have a positive impact on four of the five methodologies. For example, by eliminating design review you may cut cost, reduce cycle time to product release, and increase productivity, but you cause quality problems when the product reaches manufacturing.

The areas in Fig. O.9 with a 3 in the center of them represent activities that improve only three of the five improvement methodologies. For example, a new material may be developed that reduces costs but has no impact on quality or the hours required to produce the product. Or, an employee may suggest a way to do an activity that improves productivity, thereby reducing costs, but has no impact on technology or quality. (Ernst & Young's technical report 93.001 lists the major tools used by each methodology.)

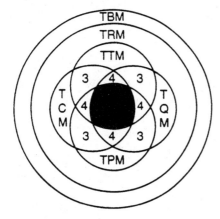

Figure O.9. Improvement methodology relationships.

Total Business Management (TBM)

The MBA schools teach that the rapidly improving communication systems and the constantly changing environment have added a whole new dimension to the way upper management looks at and plans for the organization. To compete, they need to continuously question whether they are providing the right products and services, if they need to move manufacturing locations, whether to diversify or consolidate based on their core strength, and what alliances they need to develop. More and more it is becoming imperative to develop partnerships with key customers and suppliers. Even organizations which have been competitors for years are forming alliances to compete in the international and local markets (example: IBM and Apple). This new way of directing the organization's future could be called "Total Business Management (TBM)."

Enclosing Fig. O.9 is an outer circle called TBM. It is a key consideration that must be handled well if the organization is going to continually grow, increase profits, and survive in today's competitive environment. TBM activities focus on the overall business to identify areas of opportunities or constraints. It addresses business issues such as:

- Should the organization diversify?
- Should it consolidate?
- What products should be dropped?
- What key technologies will direct the future business opportunities?
- Where shall we expand and what locations should be closed?
- How shall we invest our net favorable balance for the short- and long-term good of the organization?

These are all key business issues and must be handled correctly if the organization is going to survive. To ignore them in your improvement effort will condemn it to failure. Because of its importance, TBM is shown as the ring which holds the whole improvement effort together.

■ Improvement Methodologies' Impact of Each Other

The key is to realize that each improvement activity can have any one of four results:

1. A positive impact upon all methodologies.
2. A positive impact on one or more of the methodologies, and a negative impact on others.
3. A positive Impact on one or more of the methodologies, and no impact on others.
4. A positive Impact on one or more of the methodologies, no impact on one or more of the methodologies, and a negative impact on one or more of the methodologies.

Remember that all improvement is change, but not all change is improvement. The total interaction of all changes must be evaluated before the change is implemented. To make the process even more complex, there are many different definitions of what tools and methods make up each of the improvement methodologies. For example, some proponents of TQM claim that it is doing everything perfectly, always making the very best decision, not just a good decision. Others claim that TQM is the elimination of errors. These are two very different concepts. In the first case, it's a degree of performance, and in the second case, it's a level of performance. For example: You decide to go out to dinner In a specific town. If you have a meal which completely meets your expectations and the service was excellent as well, then you could say you had a quality meal. In this case you are defining "quality" as a level of performance. On the other hand, if quality is a degree of performance, you would not know if you had a quality meal unless you were sure that you could not buy an equivalent or better meal within that city for the same or lower price.

■ Blending Together the Improvement Methodologies

To blend together the many improvement facets, we have developed a combined methodology called "Total Improvement Management (TIM)." The five-tier pyramid in Fig. O.10 represents this new methodology.

Tier 1—Direction.   The building blocks (BBs) in this tier develop the strategy that will set the future direction of the improvement process and focus the energy of the organization on key business relationships.

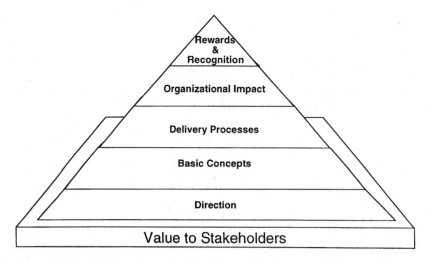

Figure O.10. Total improvement management pyramid.

Tier 2—Basic Concepts.   The BBs in this tier introduce the organization to the basic improvement methodologies and integrate them into the normal business activities.

Tier 3—Delivery Process.   The BBs in this tier focus on the processes that drive the product and service industries, make the organization more effective, efficient and adaptable while reducing cost, cycle time, and variation.

Tier 4—Organizational Impact.   The BBs in this tier develop new organizational measurement and structure.

Tier 5—Reward and Recognition.   The BB in this tier focuses on developing a reward and recognition system that provides both financial and nonfinancial rewards. This rewards and recognition system is designed to reinforce the importance of the other tasks within the pyramid.

The pyramid was selected to represent the TIM methodology because it is synonymous with strength and longevity. The pyramids were also built so that they set the absolute direction (north, south, east and west). What could better symbolize the strength, consistent direction, and long-lasting endurance of an organization that correctly implements TIM? If your organization uses the TIM pyramid's concepts, as time passes by, it will see competitors come and go and economic conditions ebb and flow, but your organization will grow and prosper. Figure O.11 shows the building blocks for each tier.

The Foundation of the Improvement Pyramid

The purpose of any progressive, long-lasting organization is to provide products and services to its customers that have more value, better quality, and are less costly

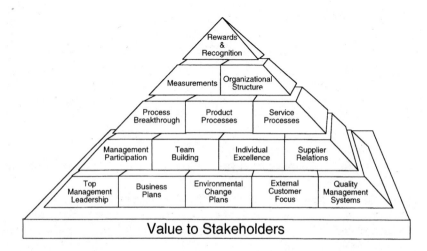

Figure O.11.   The tasks that make up the total improvement pyramid.

than other organizations offer. But it also has an obligation to all its stakeholders, which include investors, management, employees, suppliers, and the community. Truly great organizations provide ongoing security and value to all their stakeholders, not just their customers. TIM is built upon establishing strong stakeholder partnerships with the organizations undertaking the improvement activities. The word "partnership" infers that all parties involved will mutually benefit from their relationship. Without building a strong stakeholder foundation, your improvement process cannot sustain itself. It is like building your home on sand close to the ocean. No matter how well you put the building blocks together on top of a bad foundation, sooner or later the sand will shift and your house will come tumbling down. One of the most difficult jobs all organizations face is to balance the needs of all its stake holders so that the organization is perceived as value-added by these stakeholders.

Tier 1—Direction

The first tier in the pyramid is used to set the direction of the Improvement process. It consists of five building blocks (BBs), which are:

BB1: Top Management Leadership.   Top management must do more than just support TIM. They must be part of the process, participate in designing the process, assign resources, and give freely of their personal time. The start of any improvement process is top management leadership.

BB2: Business Plans.   All employees need to understand why the organization is in existence, what the behavioral rules are, and where the organization is going. This direction must be well communicated to the stakeholders, and there needs to be an agreed-to plan on how to get there. That is what a Business Plan does for an organization. It sets the direction of the business, what products are going to be provided, what markets are going to be serviced, and what goals need to be reached in the future. Without an agreed-to, well-understood business plan that is implemented effectively, the organization has no direction. It is like an automobile screaming down the road at 100 miles per hour without a steering wheel. If the organization does have a Business Plan, but it is not communicated throughout the entire organization, it is not much better off. Management is behind the steering wheel of that car screaming down the road at 100 miles per hour, but now the steering wheel is not connected to the front wheels.

BB3: Environmental Change Plan.   The only thing that management has control over is the environment within the organization. If we are going to improve the organization, it means that we must change the environment within the organization to produce the desired results. Environmental Change Plans first develop a set of vision statements that define the desired future environment. Individual vision statements and desired behavior patterns are developed for every influencing factor (example: Management Leadership, Business Processes, Customer Partnerships, etc.). Then a three-year plan is developed to bring about the desired transformation. The long-term effect of changing the environment is a change in the organization's culture.

A Change Management Plan is also developed and implemented. This plan paves the road for effective implementation of the environmental changes which are required to bring about the desired environment and behaviors within the organization. It is very important to prepare the stakeholders for these changes before, during, and after their implementation. Even the very best improvement effort can be shot down if the stakeholders have not been prepared to embrace the required changes. As a result, the Change Management Plan is a crucial part of the direction-setting activities.

Whenever you do anything, you have four options. You can do the wrong thing effectively (Option I), or do it ineffectively (Option II). You can do the right thing effectively (Option III), or do it ineffectively (Option IV). In the 1980s and early 1990s, many organizations were doing a number of good things, but doing them ineffectively because they did not prepare their stakeholders to embrace the changes. Often the losing organizations' stakeholders spent their efforts trying to define why the change would not work and/or sabotaging the change, instead of trying to make it work. As a result, many of the changes failed to meet expectations or accomplish the improvement that they should have. The winning organizations tended to prepare their stakeholders for the changes. Because the stakeholders were prepared for the changes, they embraced them and spent their efforts making the changes work. As a result, these change programs often exceeded expectations.

BB4: External Customer Focus.   Organizations are formed to service customers. As John Young, past president of Hewlett-Packard, put it "Satisfying customers is the only reason we're in business." The primary ingredient for the success of any organization is an excellent understanding of, and a close working relationship with their external customer/consumer. All planning must be based upon improving this relationship, for it is this relationship which generates the means to meet the needs and expectations of the other stakeholders.

BB5: Quality Management Systems.   This building block is used to establish Quality Management Systems that are in keeping with good business practices. This basic level of minimum operating systems is necessary before more sophisticated improvement methods can be effectively implemented. The Quality Management Systems should be in compliance with the International Standards Organization IS0-9000 series, or the appropriate military or commercial specification (example: MIL-S-9858A). These systems are the "blocking and tackling" of the improvement process. They are an essential building block for the rest of the structure. Usualiy as TIM is implemented, some of the controls that are required initially in these systems are replaced because they are no longer needed.

Included in the Quality Management System are all the quality of life impactors. This enables safety, security and environmental issues to be addressed as part of the Quality Management Systems. Requirements, procedures and audits of the quality, security, safety and environmental impacts should be combined. Management's number one priority is not satisf~ng their customers, but ensuring the safety of their employees and their customers.

The Direction Tier of the pyramid is extremely important and is the one most loser organizations have paid too little attention to. Ignoring or quickly passing through this phase is the reason why most organizations did not progress at the rate they should have in the 1980s and early 1990s. Not paying the proper attention to each building block in this tier results in a haphazard approach to improvement that often confuses rather than helps the employee, and in the long run slows down the progress made by the entire organization.

Tier 2—Basic Concepts

The second tier in the pyramid is directed at integrating the basic concepts into the organization. It consists of four building blocks, which are:

BB6: Management Participation.   This building block is designed to get all levels of management actively participating in the improvement effort. Having management feel comfortable in a leadership role is essential to the success of the total process. It is important that you bring about the proper change in top, middle, and first-line managers and supervisors before the concepts are introduced to the employees. Most organizations have done a poor job of preparing management for their new leadership role.

BB7: Team Building.   The use of management and employee teams to solve the organization's problems and to be involved in the organization's change process is a key ingredient in today's competitive business environment. This building block develops team concepts as part of the management process, and prepares all employees for participating in a team environment.

BB8: Individual Excellence.   Management must provide the environment as well as the tools, that will allow and encourage employees to excel and take pride in their work, and then reward them based on their accomplishments. This is another key ingredient in every winning organization's strategy. You can have a *good* organization using teams, but you can have a *great* organization only when each employee excels in all jobs they are performing. Care must be taken to have a good balance between team cooperation and individuals who strive for excellence in all their endeavors. The two concepts need to work in tandem, not compete with each other.

BB9: Supplier Relations.   Winning organizations have winning suppliers. The destiny of both organizations is inevitably linked. Once the improvement process has started to take hold within the organization, it is time to start to work with your suppliers. The objective of this partnership is to help them improve the performance of their output and increase their profits, while reducing the cost of their product and/or service to you.

The Basic Concepts Tier provides the infrastructure for improvement. It is designed to help management change from their role as "bosses" to "leaders." This results in an environment where all of the skills of the organization's employees are better utilized and challenged. From the employees' standpoint, it demonstrates to them the advantage of being part of the team. It also shows them how to balance their personal needs for success with the needs of the organization, while at the same time increasing the personal satisfaction they gain from being more creative. These building blocks develop a new set of relationships between the employees and their internal and external customers and suppliers is developed. The building blocks that make up Tier 2 are the fundamental ingredients in a continuous improvement process.

Tier 3—The Delivery Processes

The third tier is the Delivery Processes level. This tier of the TIM pyramid focuses on the organization's processes and the output that its customers receive. It is made up of three building blocks, which are:

BBl0: Process Breakthrough.   This building block uses cross-functional Process Improvement Teams (PITs) to make a quantum leap forward in the critical business processes (overhead-type activities). It focuses on making these important parts of the organization more efficient, effective, and adaptable. This building block makes use of many different streamlining techniques, including bureaucracy elimination, value-added analysis, benchmarking, and information technology, carefully woven together. This approach brings about drastic improvements in the processes to which it is applied. Improvements between 400 to 1000 percent are being realized in a period as short as six months.

BB11: Product Process Excellence.   This building block focuses on how to design and maintain product delivery processes so that they consistently satisfy external and/or internal customers. It is directed at the product design activities and the production process. All organizations, where they are classified as service or product industries, have production processes.

BBl2: Service Process Excellence.   The delivery processes for products and services are very different. These differences make it necessary to apply different improvement methods and common methods in different ways in the delivery of service. This building block focuses on how to design, implement, and improve the service delivery process in the service and product industries.

Tier 4—Organizational Impact

The fourth tier of the pyramid is the impact level. By now the improvement process is well underway within the organization, and it will soon start to impact the organizational structure as well as its measurements. This tier consists of two building blocks, which are:

BB13: Measurement Process.   This building block highlights the importance of a comprehensive measurement plan in all improvement processes. It helps the organization develop a balanced measurement system that demonstrates how interactive measurements like quality, productivity, and profit can either detract from or complement each other. Only when the improvement process documents positive measurable results can we expect management to embrace the methodology as a way of life. A good measurement plan converts the skeptic into a disciple. As the process develops, the measurement system should change. When you start the improvement process, you measure activities. About six months into the process, you start to measure improvement results, and about 18 months into the process, the normal business measurement should start to be impacted.

BB14: Organizational Structure.   As the smokestack functional thinking and measurement systems begin to change to a process view of the organization, bureaucracy is removed from the processes and decisions are made at much lower levels. In this new environment, employees are empowered to do their jobs and are held accountable for their actions. With these changes, large organizations need to give way to small business units that can react quickly and effectively to changing customer requirements and the changing business environment. Functions like Quality Assurance and Finance take on new roles. The organization as a whole becomes more process-driven rather than functional organization-driven. In this environment, the organization needs to become flatter and decentralized, requiring major changes to the organizational structure. This building block helps an organization develop an organizational structure that meets today's needs and tomorrow's challenges.

Tier 5—Rewards and Recognition

The fifth and top tier of the pyramid is the Rewards and Recognition level. The top of the pyramid has only one building block, which is:

BB15: Rewards and Recognition.   The Rewards and Recognition process should be designed to pull together the total pyramid. It needs to reinforce everyone's desired behavior. It also needs to be very comprehensive, for everyone hears "Thank You" in a different way. If you want everyone to take an active role in your improvement process, you must be able to thank each individual in a way that is meaningful to him or her. There is a time for a "pat on the back" and a time for a "pat on the wallet." Your rewards and recognition process should include both.

■ Improvement's Impact on Stakeholders

The stakeholder is any individual or group of individuals impacted by an organization or a process. It is becoming more and more accepted that all organizations need to consider all their stakeholders in every decision that is made. If you accept this premise, it is easy to see that your improvement process must consider more than just the end customer. Certainly it is easier and less complex if you can direct your efforts at maximizing the positive impact on just one or two stakeholders. But that is not possible for most organizations today, since most organizations have six different stake holders with very different priorities. A typical organization has the following stakeholders:

Stakeholders	Priority points
• Its management	+48
• Its investors	+41
• Its external customers	+27
• Its suppliers	+1
• Its employees	–3
• Its community/mankind	–24

These six stakeholders have very different needs and expectations. Trying to satisfy six different stakeholders with such different needs is a very significant challenge to any management team. For what is good for one, may be bad for another. For example: It would be good from the investors' and management's standpoint to move a manufacturing operation to Mexico, since it would reduce cost, as well as having the benefit of less strict pollution requirements; however, for obvious reasons, this change is not advantageous from the employees' and community's standpoint. It also could increase the pollution in Mexico.

To live with this dilemma, many management teams have prioritized the importance of their stakeholders. Typically, the way management consciously or unconsciously prioritizes the six stakeholders is shown in the previous stakeholder list. Management is the top priority, and the community is the lowest. This unwritten prioritization has resulted in our government passing laws to protect the general public, the environment, and employees. In order to understand the complexities of satisfying all the stakeholders, we need to understand each stakeholder's priorities. The following table lists the six stakeholders, their top five priorities, and how the six methodologies impact those priorities. The priority points are used to indicate the favorable or negative impact the methodologies have on each stockholder.

Table O.1.

Legend	Priority points	
● = Direct/impact	+2	TBM = Total Business Management
◗ = Indirect impact	+1	TCM = Total Cost Management
○ = Little or no impact	0	TPM = Total Productirity Management
N = Negative impact	–2	TQM = Total Quality Management
		TRM = Total Resource Management
		TTM = Total Technology Management

Management's measurements of improvement	TBM	TCM	TPM	TQM	TRM	TTM
• Return on assets	●	●	●	●	○	●
• Value-added per employee	●	●	●	●	◗	●
• Stock prices	●	◗	●	●	◗	●
• Market share	●	◗	○	●	◗	●
• Reduced operating expenses	●	●	●	◗	◗	1
❑ Total priority points = +48						

Investors' measurements of improvement	TBM	TCM	TPM	TQM	TRM	TTM
• Return on investment	●	●	●	●	◗	●
• Stock prices	●	◗	●	●	◗	●
• Return on assets	●	●	●	●	○	●
• Market share	●	◗	○	●	○	●
• Successful new products	●	N	○	◗	○	●
❑ Total priority points = +41						

Customers' measurements of improvement	TBM	TCM	TPM	TQM	TRM	TTM
• Reduced cost	●	●	●	●	◗	●
• New or expanded capabilities	●	○	○	○	○	●
• Improved performance	○	○	○	●	◗	●
• Ease of use	○	○	○	◗	○	●
• Improved responsiveness	○	○	○	◗	◗	●
❑ Total priority points = +27						

Suppliers' measurements of improvement	TBM	TCM	TPM	TQM	TRM	TTM
• Increased return on investment (supplier)	○	○	○	◗	○	○
• Improved communications/fewer interfaces	○	○	○	●	○	○
• Simplified requirements/fewer changes	○	○	○	◗	○	●
• Longer contracts	◗	○	○	●	○	○
• Longer cycle times	N	N	○	N	○	N
❑ Total priority points = +1						

Employees' measurements of improvement	TBM	TCM	TPM	TQM	TRM	TTM
• Increased job security	○	N	N	◗	●	N
• Increased compensation	N	N	○	◗	●	◗
• Improved growth potential	◗	N	N	◗	●	◗
• Improved job satisfaction	N	N	N	●	●	◗
• Improved morale	○	N	N	●	●	○
❑ Total priority points = –3						

Community/mankind's measurements of improvement	TBM	TCM	TPM	TQM	TRM	TTM
• Employment of people	N	N	N	N	○	N
• Increased tax base	○	N	N	◗	○	○
• Reduced pollution	N	N	N	◗	●	N
• Support of community activities	N	N	N	N	◗	N
• Safety of employees	○	○	○	◗	●	○
❑ Total priority points = –24						

The ideal improvement process would improve the organization's performance in all the stakeholders' priority issues. In these tables the most frequent impact was noted, but sometimes one methodology can have more than one impact, depending on the circumstances. For example, total quality management can have a positive or a negative impact on job security. If improving the product increases the organization's market share, resulting in increased workload, job security is improved. But if TQM reduces waste, thereby improving productivity, but does not increase market share to the point that it offsets the productivity gains employees could be laid off. This results in a negative impact on job security.

By analyzing the previous tables, it is easy to see why all six improvement methodologies need to be combined to at least indirectly impact all of the stakeholders' top five priorities. The stakeholders' priorities are listed based upon an average organization. We realize that these priorities can change based upon products and circumstances within the organization. For example, if the organization is a nuclear power plant, safety would be the number one priority for the community and the employees.

■ How TIM Affects the Organization

The TIM process, when implemented correctly, has many positive impacts on the organization, some of which follow:

Legend:
◗ = Sometimes
○ = No
● = Yes

	TBM	TCM	TPM	TQM	TRM	TTM
• Increases market share	●	●	●	●	●	●
• Increases return on investment	●	●	●	●	●	●
• Increases value-added per employee	●	●	●	○	●	○
• Increases stock prices	●	●	●	●	●	●
• Improves morale	○	○	○	●	○	○
• Improves customer satisfaction	○	○	○	●	○	●
• Improves competitive position	●	●	●	●	○	●
• Improves reliability	○	○	○	●	○	●
• Improves maintainability	○	○	○	○	○	●
• Improves safety	○	○	○	○	●	○
• Decreases waste	○	●	●	●	●	○
• Decreases overhead	●	●	●	●	●	○
• Decreases inventory	●	●	○	○	●	○
• Causes layoffs	◗	●	●	◗	◗	◗
• Increases the number of employees	◗	○	○	◗	◗	◗
• Increases profit	●	●	●	●	●	●

■ Do the Total Management Methodologies Pay Off?

- Caption: *Electronic Business* Magazine, October 1992 issue, page 48—"Probably 95 percent of (TQM) programs fail, but no one has the guts to say so," charges industrial veteran Luigi D'Angola, now a Professional Fellow at Dartmouth College.

- Rath & Strong polled 95 corporate senior managers and reported that 38 percent gave their quality improvement falling grade.

- An Arthur D. Little survey of 500 United States executives revealed that only 64 percent believe that their quality programs have improved their competitiveness. (Source: *Fortune* Magazine, May 18, 1992.)

If you use these examples as your yardstick, one of the Total Management Methodologies Total Quality—management—is in trouble.

On the other side of the coin, the same issue (October 1992) of *Electronic Business* Magazine reported on page 47 that of the 70 companies they surveyed, none of them have scrapped their TQM efforts, and 91 percent reported that their quality had improved compared to their competition. In a Gallup survey conducted by the American Society for Quality Control in 1992, they determined that 63 percent of U.S. consumers think that U.S. products are better than they were five years ago. In yet another 1992 survey conducted by Gallup of executives from 604 organizations, only 8 percent said they had a quality improvement program in place but that they were frustrated with the results. Only 1 percent stated that there was no need to use quality improvement as a business strategy. These are only three reports which confirm the business success of the improvement efforts going on within the United States today; there are many more.

The conflicting data reported is confusing management. However, it is sufficient to say that when you look at hard data, you can state with a better than 95 percent confidence level (plus or minus 4 percent sampling error) that the TQM efforts are positively making a contribution to most U.S. organizations, and are making them more competitive within the country as well as internationally.

But even if only 1 percent of the organizations that undertake a TQM effort are unhappy, this is too many. I agree that somewhere between 90 to 98 percent of the organizations that undertook an improvement effort in the 1980s did the right thing. The problem that some organizations have had with TQM is not what they did, but how they did it. Their implementation in many cases was poor, and they did not receive the full return on their investment.

Billions of dollars were spent on training that was not put to use. I estimate that 5 to 10 percent of the organizations did a very poor job of implementing the improvement process, and received little or no return on these huge investments. About 10 to 20 percent of the organizations implemented the improvement process very effectively and documented a return on their investment of as much as 40 to 1. Between 70 to 85 percent of the organizations that implemented TQM fell some-where in the middle. They improved at least 5 percent per year, making it worthwhile, but these organizations did not obtain the results they should have. Many of these same organizations did not see their market share grow even though they reduced waste, cut defect levels between 15 to 20 percent, cut cycle time, and increased customer satisfaction.

The reason for the lack of bottom-line results is that these organizations are improving exponentially, but their competition is also implementing the improvement process at the same time, causing these organizations to stay at a par with their competitors. As a result, their market share did not increase. In some cases, the organizations that are utilizing the improvement process are even losing market share. This usually occurs when organizations observe that their competition is implementing an improvement process and they decide to copy it. These organizations usually start one to two years behind their competition.

Figure O.12 shows the exponential improvement curves for two different organizations. Curve "A" is the improvement curve for the competition, and Curve "B" is the organization that started the improvement effort one year later. Both curves are essentially the same, but the second curve is offset by one year. You will note that as both organizations progress with their improvement process, the difference between the two curves becomes greater, due to the exponential nature of the improvement curve.

You have just learned Improvement Rule No. 1: You cannot copy your competition. For when you get to where you want to be, they will be far ahead of you. You must improve at a steeper rate than your competition in order to be competitive.

Rule No. 2: Do not go to your competition and give them all your improvement secrets—they may listen to you.

■ Benchmarking—The Worst Pays Off Big

In all businesses, there are winners and losers. The smart competitor studies both to maximize its total performance. Many organizations have documented return on investments in one year of 40 to 1. Globe Metallurgical, Inc., for example, has documented a 40 to 1 return on its improvement efforts. At the other end of the spectrum are organizations like Florida Power & Light. In 1989, they were the winner of the first Deming Award ever presented outside of Japan. In 1990, Florida

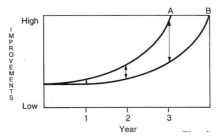

Figure O.12. Two identical improvement curves offset by one year.

Power & Light's net loss was $391 million, yet their rate per 1000 users remained higher than for the three other major private utilities in Florida.

When James Broadhead became CEO of Florida Power & Light, he met with employees and discovered widespread resentment for the quality effort. In the July 1, 1991 issue of *Fortune* Magazine, he was quoted as saying, "I was most troubled by the frequently stated opinion that preoccupation with process had resulted in our losing sight of one of the major tenets of quality improvement, namely respect for the employees." He found "less recognition for maktng good business decisions than for following the Quality Improvement Process." He also stated that "It (FPL) pushed itself to the brink of nervous collapse in the months before being chosen in 1989."

Even though less than 5 percent of the organizations lose money as a result of implementing an improvement process, you cannot consider an organization a winner unless it receives a minimum of 3 to 1 return on investment. Using this as a guideline, about 20 percent of the organizations that undertake an improvement process can be classified as losers. These losers are hard to find. They do not write articles about their failures, and their CEOs do not go to conferences and lecture on how badly they are doing (if they know they are failing). Today, there are very few virgin organizations that have not undertaken some type of improvement effort. Of the organizations that contact me, I see the following cross-section:

Status of the organization	Percent of organizations
Organizations that are so excited about what they have accomplished that they are committed to continuously expanding their efforts and are looking for more ways to improve.	25%
Organizations that are happy with what they, have accomplished, but it is not becoming part of the way they do business, and they feel there is more potential than they have received from their activities.	35%
Organizations that are unhappy with their progress.	30%
Organizations that want to start an improvement process and have not already been involved in the use of some of the improvement tools. (Example: Teams or Statistical Process Control)	10%

Why Organizations' Improvement Efforts Fail

The organizations that are unhappy with their progress have many things in common. The following are the primary reasons why organizations are displeased with the progress of their improvement process:

1. Change in top management (new top management).
2. Change in top management's priorities and/or direction.
3. The theory has been taught in class but not put into practice.
4. Downturn of the economy caused them to discontinue their effort.
5. Middle management did not buy into the process.
6. Other, higher priorities within the organization kept it from being effective.
7. The consultants they hired did not understand their business.
8. They are not improving fast enough to keep pace with the competition.
9. Lack of hard, measurable results. There is a need to show management the return on investment.
10. The teams are not solving meaningful problems.
11. The improvement process is interfering with getting the job done.
12. Lack of focused strategy to integrate all efforts.
13. The quality efforts are not reflected in the bottom line.
14. Lack of organized labor support.
15. The methodologies used did not work.
16. A big layoff killed the activities.

These are all symptoms, not root causes. The root causes for these failures are:

Root Cause No. 1.   Upper management did not believe that they needed to change. They were giving lip service to the change process, instead of leading it. They wanted everyone else to change, but they did not want the improvement process to impact them. This usually occurs because top management accepted the improvement activities in a blind leap of faith. A business case was not developed and a viable improvement plan was not developed and embraced by the entire top management team.

Root Cause No. 2.   Lack of trust between management and the employees is the biggest single cause of improvement process failures. In about 65 percent of the organizations we work with, we find that lack of employee trust in management and lack of management trust in the employees is one of the top priority problems that needs to be addressed first. Years of secrecy, suspicion, and seeming lack of interest in the employees has caused the employees to distrust upper management. Most employees believe that top management know much more than they do about what is going to happen to the organization and purposely keep the employees in the dark. They see the organization's president getting paid millions of dollars per year, while they get about $20,000 and work just as hard as the president does. When things get tough, the lower-level people get laid off, while top management get increases. Employees reason that top management must have a master plan that they are following; otherwise, they would be fired for their poor performance.

Management, on the other hand, reasons that the employees don't care about the organization and that they cannot be trusted with the organization's property or information. Using an organizational loyalty index, with 100 being the best, I rate the United States 56; West Germany 64; and Japan 85.

Root Cause No. 3.   The organization's improvement champion is the third major cause of failure. The champion or czar is the person within the organization selected to lead the improvement process. Too often, management selects the wrong person for the wrong reasons. Frequently, top management selects a senior executive who is about to retire, reasoning that he or she has the needed prestige and has earned the right to be assigned to what is perceived as a low-stress assignment. Often management believes that there is no need for the improvement champion to have any knowledge of the improvement methodologies. After all, anyone knows what good quality is. Even illiterate consumers know what quality is. So why does the improvement champion need to have a background in the improvement methodologies?

This type of reasoning is very much like saying that everyone knows the difference between a good and a bad design. Even illiterate consumers would not buy a television that has a fuzzy picture. So why not make anyone the vice president of Research & Development? An engineering degree should not be necessary. Sure, you can take any competent executive and with enough time, training, and a lot of errors. eventually he or she will be capable of providing minimum performance in either assignment. But what organization today has that type of luxury? It takes 4 to 6 years for a good manager to become an exceptional improvement expert.

Too many organizations that failed took a senior manager and made that person responsible for the improvement effort as a way for him or her to slow down before retirement. IBM is a good example of this type of approach. Their first two vice presidents of quality retired from their jobs within three years of being assigned as Director or Vice President of Quality. They were both excellent people, but they retired at the point where they were just really beginning to become effective.

Other organizations select bright, upcoming executives to become the improvement champions. This is a better approach if you can afford the years of delay in the implementation process or the cost associated with a good, full-time consultant to partner with the improvement champion. Other organizations select someone from the quality assurance group for this job. These individuals have usually had a number of years' experience with some of the improvement tools, having already implemented them within parts of the organization. They also have already established direct contact with customers and suppliers.

Five years ago, I felt that assigning someone from Quality Assurance to be the champion was the wrong approach because the quality organization was viewed as only being responsible or interested in the manufacturing process. But after reviewing the results obtained by many organizations that used all three approaches, I have changed my mind. I now believe that it is easier, faster, and more effective to get the quality assurance professional to expand his or her role to total improvement than it is to start over with someone who has little or no experience in the elimination of errors. The exception to this would be an individual who does not have the respect of the management team.

Root Cause No. 4. Both successful and unsuccessful organizations base their improvement process on a consultant's methodology. Often their improvement process was based upon a book written by a consultant. Others used consultants who had limited knowledge of how to convert theory into practice in different types of organizations. Let's be practical about it. A consultant's knowledge is his or her product. How can you expect them to give away their product at a conference for free or in a book for which their income is $2.65. Most authors of improvement books lose money on the books they sell. On an average book, the author invests just under 200 days to prepare and they receive about $12,500 in royalties.

No one can afford to lose money on their products and stay in business. On the other hand, it does make good sense to use consultants to help you implement your improvement process. The best consultants have spent 30 to 40 years understanding and implementing the improvement process methodology. They have spent hundreds of thousands of dollars developing individual tools. They have made some errors which they have learned from, and have had a lot of successes. Why shouldn't you learn from the mistakes the consultants have already made? Why spend the hundreds of thousands of dollars to develop a class and the supporting materials when you can buy them for much less than it costs to do the job yourself? As far as training and implementation assistance, there is a time to use consultants and a time to stand on your own two feet. To start the process or to implement a new tool, a consultant can prevent you from making the mistakes that can shoot down the total process.

The problem many organizations have had with consultants is that they have selected the wrong ones or terminated the relationship with the consultant too soon. Many teachers consider themselves consultants. This is far from true. It is much easier to teach a class and discuss theory than it is to implement the theory. A consultant who teaches a class on subjects like QFD, SPC, etc., has only done 15 percent of the job. Helping implement it is 85 percent of the consultant's job. Other consultants have only limited understanding of the total improvement process, which prevents them from looking at the total picture and making the best suggestions.

There are over 400 improvement tools available today. Select a consulting organization that understands all the tools and can help you make the best choice. Few consultants can be really good at more than 10 improvement tools, but a really experienced consultant can understand the strengths, weaknesses, and interactions of all of them. These experienced consultants can direct you to a consultant who has the skills to help implement the tools that best meet your needs.

Yes, using consultants is good business. Most of the winners do it. High-tech companies like H-P, IBM, and Martin-Marietta use consultants; low-tech companies like United States Steel, Reebok, and Campbell Soup do it. The following are key traits to look for when you select an improvement consulting firm.

1. Are the key people certified by the appropriate professional society? (Example: quality and reliability engineers by the American Society for Quality Control, professional managers by the Institute of Certified Professional Managers, etc.)

2. Are the key people state-registered as professional engineers?

3. Do they have experience with implementation for at least two complete product cycles before they started in the consulting field?

4. Are they Fellows or honorary members in the appropriate national associations?

5. Are they capable of training and implementing (can they do the job without your people, or are they just theoretical)?

6. Do they cover the full range of improvement tools?

7. How many improvement books have they published?

8. What type of research database do they have?

9. Do the leaders of the consulting organization have a worldwide reputation and a worldwide communication system, ensuring they know the latest developments?

Do not select a consulting firm based upon the color of the consultant's eyes, the way they dress, or the glitter in their promotional materials. Select a consultant based upon his or her cumulative knowledge over the past 20 to 30 years. The consultant needs to know not only what the latest trends are, but why they have developed. Too many consultants sell the latest fad, not the balance between today's fad and the solid foundation of past business experience.

Root Cause No. 5.   Forgotten Middle Management—Of all the people who have been impacted the most by United States improvement processes, it is middle management. Middle management are the ones who have felt the pinch of cutbacks, layoffs, and flatter organizations. The percentage of middle managers during the 1980s who were pushed, pressured, fired, or retired (to save face) from their jobs was twice as high as employees, supervisors, or top management. Is it any wonder that it is this group that is the most suspicious of the improvement process? In most organizations that have failed, top management did not take the time to prepare middle management for their new role in a participative environment. Top management wanted the employees to be empowered, but did not want to empower middle management. The organization had failed to keep these key people technically competent in the skills they brought to the organization (example: engineering marketing, programming, etc.). So when middle management jobs were eliminated, they had no place to go but out. Not because they had failed, but because the organization had failed to provide a technical vitality program. Many organizations turned their back on this problem and took their improvement process directly to the employees. This turned needed allies into saboteurs.

■ How Does the United States Measure Up?

- *Percent of the world's total export.* Since 1950, the United States has lost over 43 percent of its share of the world's market mostly to Japan and West Germany.

- *Return on assets.* Our largest organizations are performing poorly. In 1992, the United States had five out of the ten world's largest industrial corporations. Three of them lost $9.3 billion; $1.6 billion more than the combined profits of the five non-United States organizations. In 1993, the top five organizations' total ROA was 0.8 percent. Anything below 2 percent is poor (losers).

- *Nondefense R&D expenditures.* Since 1980, the United States R&D expenditures have been remaining constant at about 1.8 percent of our gross national product. Japan's R&D efforts have continued to increase from about 2.3 percent in 1980 to just over 3 percent today.

- *Profits as a percent of GNP.* This measure has continuously dropped off in the United States from a high of 12 percent in 1965 to below 6 percent today (a 50 percent decrease). In 1993, the top five organizations' profits, as a percentage of sales, were 1.4 percent.

- *Hourly compensation.* In 1975, the U.S. worker was the best paid in the world ($6.36 per hour). Today, the German worker is the best paid at $24.20 per hour. Workers in France, Italy, Japan,and Sweden are just some of the ones that are paid more than the U.S. workers today.

- *Annual earnings.* The average earning in true dollars peaked in the early 1970s within the United States and have been going down since.

- *Overhead.* Overhead is much greater in the United States (26 percent of manufacturing costs) compared to 21 percent in Germany and 17.5 percent in Japan.

- *Federal debt.* The U.S. federal debt has grown to 50 percent of our gross national product. This is an increase of more than 26 percent in the last 12 years. Of the federal income for 1992, 21 percent was borrowed to cover deficit. This is 3 times the total income from corporate taxes. Today, the United States has the world's largest national debt.

- *Public and private debt.* Not only the public, but also the private sector is going in to debt as never before. Savings are way down and personal debts have skyrocketed. Since 1980, the debt growth rate has increased 300 percent over the previous 40 years.

- *Human rights.* One of the most basic human rights is the right of protection of one's being and property. Here, the United States is one of the worst of the developed nations. Robbery rate in the United States is 130 times that of Japan's; violent crimes are 360 times that of Japan's; and motor vehicle thefts almost 22 times that of Japan's. We have a lot to learn from countries like Japan and Singapore in this area.

■ Japan, Incorporated

The early 1990s took its toll on Japan, Inc. The NIKKEI stock average market price fell from a high of 38,985.87 on December 29, 1989, to below 18,000 in April 1992 (it continued even lower), a 54 percent drop. The drop in the NIKKEI wiped out $2.6 trillion in stock value. Along with the NIKKEI decline, Tokyo's real estate prices have dropped drastically. In some areas, they have dropped 30 percent and are still going down. The annual percentage change in pretax profits has gone progressively negative for the past three years, and in 1992 the negative change in pretax profit exceeded 20 percent. In 1992, Sony lost $158 million, its first-ever operating loss. Toshiba's profits dropped by 60 percent. JVC announced it would cut 3000 workers.[10] Nissan plans to elimInate about 4 000 jobs over the next three years.

As Nobuhiko Kawamoto, president of Honda Motor Co., put it "Suddenly we have to change our industrial structure for slow growth. Many Japanese don't understand this. This is not the way they have done things for 40 years. We have to change or we will not survive." Even mIghty Toyota has slipped its goal to reach an annual sales rate of six million vehicles from 1995 to 2000.

Japanese engineers have been under a great deal of pressure and are getting discouraged. When asked how they viewed their company, only 40 percent said they were satisfied, 42 percent said that they were not content, 44 percent said they were dissatisfied, and 10 percent said that they were thinking about quitting.[11]

Bankruptcy jumped from 6468 companies in 1990 to 11,385 in 1992, over a 75 percent increase. Michio Nakajima, president of Citizen Watch Company, stated "In the bubble economy (1980s), Japanese companies lost their way."

Japan, Inc. is beginning to see a new breed of worker which it calls *Shinjinrui* ("new human beings"). These young Japanese are very different from their hardworking parents. They are more interested in playing hard than working hard. They like to travel and want more free time. These habits will cramp Japan's productivity and savings.

Some Japanese think that Japan could lose its lead in quality. The quality improvement rate in Japan was much lower than it was in the Unite4d States during the 1980s. Today, Ford is producing autos at a higher quality level than half of the Japanese auto manufacturing firms. Quality problems are slipping out to the consumer. Matsushita, Pioneer, Sony, and Toshiba have recalled dozens of models that smoked and caused fires. Lexus was recalled for cruise control and brake light problems. Seiko Epson laptop computers were recalled for poor circuit soldering. A Japanese-manufactured medication for treating obesity, insomnia, and other problems became contaminated during the manufacturing process and caused hundreds of illnesses and several deaths in the United States.

In a Brouillard Communications poll a higher percentage of United States consumers rated United States products as having higher quality (53 percent) than Japanese products (48 percent) or German products (39 percent). The fourth runner-up was Britain (9 percent). Junji Noguchi of JUSE stated, "(J.M.) Juran has said the United States can catch Japan in the 1990s, and I think that's possible." Tadashi Kagawa senior managing director at Daini Denden, Inc. said "Japanese quality is going downhill. In my estimation, Japan's quality is still better than U.S. quality in some areas, but the United States is closing the gap fast. In other areas, Japanese products have never been as good as the United States."

Japanese companies are fierce competitors and will become even more competitive as they try to offset these trends and a weaker domestic market by boosting their exports. These pressures have driven Japan, Incs'. trade surplus to over 125 billion (United States dollars) for 1992[12] and in 1993 it went still higher with no end in sight.

Some of the things that the world has believed about Japan, Inc. were shaftered during the first half of this decade.

- There is an end of the cheap money supply for Japan, Inc.
- Market share is not all that counts.
- People are being laid off in Japanese companies.
- The world did survive the NIKKEI dropping below 20,000.
- There are things that can stop Japan, Inc. in expanding in all countries.

One of the key advantages that Japan, Inc. still has over the United States is its approach to developing and delivering new products. They have consistently been able to bring a new product from the concept stage to the customer delivery stage in half the time and at half the cost as the average American organization. Japanese organizations spend more time planning (Japan, 40 percent; U.S., 25 percent) and as a result, suffer development setbacks on a smaller percentage of their products (Japan, 28 percent; United States, 49 percent). Japanese organizations also waste less of their time debugging the finished product (Japan, 5 percent; United States, 15 percent).

Japanese organizations develop and communicate clear visions of where they want to be 10, 20, 30 or more years in the future. Sony's chairman, Akio Morita, said, "American companies struggle to create a vision for the next quarter; Japanese companies have a vision for the next decade." With a 30-year vision that the organization is working toward, management is more motivated to invest in R&D. Japan started the 1990s investing 50 percent more of its GNP in nondefense R&D than the United States (Japan, more than 3 percent; United States, less than 2 percent).

In a move to stay ahead in technology, Japan is assigning new product development teams made up of people from research, marketing, product engineering, and manufacturing to develop three different levels of the same product. The low-level product design provides an incremental upgrade of the present product. The second level is directed at a major improvement. The third level is real innovation. Peter F. Drucker, management consultant and professor at Claremont University, writes "The idea is to produce three new products to replace each present product, with the same investment of time and money, with one of the three then becoming the new market leader."

Where Japan is clearly ahead of the United States is in:

- *Precollege education*. Japan rates No. 1; the U.S., No. 17.
- *Child development*. Most Japanese families accept their responsibility for the young children and one of the parents stays home to help the child develop and keep focused on education.
- *Process improvement*. Japanese organizations have more than three times the focus on improving their processes through tools like Business Process Improvement, Process Simplification, and Cycle-Time Analysis, than Canada, Germany, and the United States.

■ Moving Production Back to the United States

Things are in line for a major shift of production back to the United States. Some organizations have already done it, and many more will be following. As Robert Collins, CEO of G.E. Fanue North America, said, "We've exploded the myth that you can't produce competitively in the United States, and with better quality." They moved their programmable logic controller from Koyo, a Japanese company, to their plant in Charlottesville, Virginia. ATT's Applied Digital Data Systems moved their computer terminal (typical unit sells for $600) manufacturing from Asia to Long Island. This allowed them to reduce finished goods and raw material inventories to one-eighth their previous level. IBM produces products in their Rochester, Minnesota plant cheaper than they can procure them from Asia. The list of organizations that are moving production back to the United States is growing every day.

Today, with the labor content running between 5 percent to 12 percent of the total cost, management needs to concentrate on the total impact of producing overseas. For example, direct labor was 25 percent of the cost of producing an electronic module at Modecon's plant in Andover, Massachusetts in 1988, so it was probably a good business decision to move it to Hong Kong at that time. Since that time, automation, simplified designs, and streamlined manufacturing have caused them to rethink this decision. In 1990, they moved production back to Andover. Today, direct labor costs are 5 percent of the product cost. Motorola CEO George Fisher put it this way, "The days of chasing low-cost labor are over."

■ Return on Investment for Improvement Expenditures

The return on investment for improvement activities is hard to predict since it is so dependent upon what and how effectively the improvements are implemented. It is also very dependent upon the status of your organization. The organizations that were classified as losers have much more potential for a higher return on investment than the winning organizations. Another problem is: How do you measure the indirect savings of changing a market share trend from going negative to positive? In spite of these measurement problems, return on investments between 8 to 1 and 50 to 1 in a three-year period are being recorded when the improvement process is implemented properly.

Bob Praegitzer, CEO of Praegitzer Industries Inc. in Dallas, Texas, asked Matt Bergeron, their chief financial officer, to evaluate the returns they were receiving for the effort and money invested in the improvement process. He found out there was a definite payback and a favorable relationship between gross profits and the money spent on their improvement process. Bergeron uses as an example that there was a six-times return from training.

Bethlehem Steel's improvement results also have been impressive: Steel production cost per ton has been reduced 24 percent, employment cost as a percentage of sales has gone from 50 to 38 percent, inventory required per dollar of sales has decreased 54 percent and the dollar amount of sales per employee has increased 70 percent.

Now let's take a look at the results of the mid-sized, young organization. Iomega Corporation, formed in 1980, evolved and pioneered the Bernoulli Technology for removable disk drives. With just over 1100 employees, their 1991 revenue was $136 million and they had a pretax profit of nearly $18 million. Their improvement efforts resulted in:

- 136 percent productivity improvement
- 81 percent reduction in annual scrap
- 54 percent increase in revenue
- 86 percent improvement in cycle-time
- 75 percent reduction in inventories
- 35 percent reduction in product costs
- 28 percent reduction in quality costs
- 41 percent reduction in manufacturing space
- 99 percent customer satisfaction index

Globe Metallurgical, Inc., a producer of iron-based metals with just over 200 employees, headquartered in Cincinnati, Ohio, was faced with serious foreign competition and declining profits in the early 1980s. In 1985, they started their improvement process. The following improvements were noted over the next three years.

- Productivity jumped 367 percent
- Waste reduction of over $10 million per year
- $ 1 million saved in reduced transportation costs
- 77 percent reduction in inventory
- Absenteeism reduced to .029 days per employee per year
- They won the Malcolm Baidrige National Quality Award in 1988
- Customer complaints dropped 91 percent (Source: 1993 Shingo Prize Guidelines, page 22)

Arden C. Sims, chief executive of Globe Metallurgical, Inc. estimates that Globe's investments in quality have produced a 40 to 1 return. From 1986 to 1988, Globe cut operating expenses by a hefty $11.3 million, and its quality efforts continue to pare operating costs by about $4 million a year. Recently, Sims told the *Harvard Business Review* that annual savings from quality efforts for his $115 million company should increase to about $13 million by 1995.

Another example is Motorola's improvement crusade, which has saved the company almost $2.4 billion.

The health care industry is currently actively seeking new and different ways to improve quality and reduce costs. A hospital in the southeast, for example, decided to form and train five cross-functional teams. These groups included both clerical and non-clerical representatives from departments such as radiology, nursing, laboratory, and pharmacy. In less than one year, these teams were instrumental in achieving nearly $1 million in cost savings while improving the quality of health care. Moreover, an additional $1.5 million of benefits had been identified for future implementation.

Typical results achieved in the banking industry are:

- Reduced cost to prepare statements by 40 percent
- Reduced handling of statement exceptions by 60 percent
- Reduced costs of Treasury handling by 20 percent

Typical results achieved in the insurance industry are:

- Reduced backlog claims inventory from 370,000 to 70,000
- Reduced claim processing time from 28 to 5 days

The aforementioned results are impressive, but they could have been even better. For even in these organizations, there is still room for improvement if they use Total Improvement Management effectively.

■ Summary

There is no doubt about it. The United States is the blue-ribbon country of the world—the best place to live, work, and raise a family. We are more productive and have the best standard of living of anyone in the world. People are more satisfied with their jobs in the United States than in Canada, Europe, or Japan.[13]

- U.S. Index   40
- Canadian Index   39
- European Index   29
- Japan Index   16

Money Magazine evaluated the standard of living in the 16 wealthiest nations. It compared them in five areas: health, solid job prospects, comfortable income, upward mobility, and adequate leisure time. The United States ranked No. 1; Japan, No. 7; Germany, No. 8; and the United Kingdom, No. 15. We are the envy of the rest of the world, and when you are No. 1 everyone is using you as a benchmark to beat. As a result, the gap between the United States and other countries around the world has decreased during the last quarter century.

How does it look for the future of the United States? We are positioned well in the products that will lead the next decade. The is recognized as being among the very best in microelectronics biotechnology, new materials, civilian aviation, tele-communications, computers, and software.

After World War II, our production capabilities were the only ones that were not out-of-date or bombed out, ensuring us immediate success. As a result, we gained a false sense of confidence. We began to believe it was our management style that set us apart, not because the war was not fought on our soil. In Europe, an MBA Degree began to stand for a *Manager who has Been to America*. The rest of the world was quick to learn from us. People around the world set a personal objective to exchange the rice or potatoes on thefr plates for the steak that was on ours.

As a result, we slept through the 1960s. The alarm clock rang in the 1970s, but we rolled over and turned it off. In the 1980s, we woke up, showered, dressed, and drove to work. Now in the 1990s, we have our shirt-sleeves rolled up and we are committed to not losing more ground. This new, leaner, informed America is transforming itself from a sleeping giant into a customer-oriented team that will do anything to satisfy its customers.

International customers are attracted to your organization for four reasons, in the following order:

Win customers	Lose customers
1. Capabilities	1. Trust
2. Trust	2. Quality
3. Price	3. Capabilities
4. Quality	4. Price

Product and service capability is driven by using the latest technology and/or using present technology in more creative ways. Trust is based upon experience and reputation. It reflects the faith that the customer has in your ability to meet your cost, schedule, and performance commitments. Price today ties in directly with value. Customers are looking at getting the best performance at the least cost. Quality reflects more than just the initial view of the products or services purchased. It reflects the quality of the total organization, the reliability of its products, and the capability of its sales and service personnel. You lose customers for the same four reasons that you attract them, but in a different order.

For an organization to survive in today's competitive international environment, there must be improvement efforts in both the continuous and break-through improvement methodologies. Management need to make the correct business decisions so that the correct products are available at the time they are needed, while making the most of everyones efforts. There needs to be a high level of cooperation between government, business, labor, and academia. Each must improve the value of its products and/or services as viewed by its customers. This means that all functions in all organizations must use the most appropriate technology to improve their effectiveness, efficiency, and adaptability. In addition, all organizations need to have a well-communicated, agreed-to plan that merges together the many improvement methodologies to provide the greatest value to all of their stakeholders.

As we prepare for the future, we need to be able to compete fairly with all organizations in all countries. Our preoccupation with Germany and Japan needs to give way to an international concern. I believe that by the year 2010, Japan's predominance as a competitor will be replaced by China. China's productivity growth rate is twice that of Japan and the United States. Their real GDP growth rate between 1986 to 1991 was about 8 percent, and has gone up since then. Since 1986, they have more than doubled their total export value, and for the first time since World War II, they are seeing growing trade surpluses with the United States ($20 billion in 1993 alone). China is now in the top five countries that the United States has the biggest sale unbalance with. If the present trends continue, the United States will have a bigger negative balance of trade with China in 10 years than we have with Japan. If you think Japan is a fierce competitor, it's a pussycat in comparison to China.

The other alternative to China is the European Common Market. They could come together, combining their specialties, to make a manufacturing super power. A combination of the Soviet Union's scientific capabffitles, Germany's craftsmanship, Italy's design flair, and Great Britain's financial management, would be hard to compete with.

The last, and probably the most important, competitor is all the emerging industrial nations. They are modernized, upscale, and hungry. They pay lower wages, have higher work ethics, are improving at a faster rate, and their standard of living is improving faster than Japan, Germany, and the United States. I believe there is a direct correlation between a person's work ethics and the last time he or she went hungry.

Although things in the United States today, on an average, are good and will continue to be during the 1990s, it could be a very different story in the twenty-first century. As Lester Thurow, economist at MIT, wrote, "No one at the end of the twentieth century is less prepared for the competition that lies ahead in the twenty-first century." The United States has five years to prepare, and today, it is not ready. It has not prepared itself as it should have. The United States needs to really get serious about making a major improvement in the way its government, business, and schools perform.

Today, I would rate our production organizations a B– and our service organizations C–, because they have started to improve; our government a D–. because it is not really trying; and our educational system, an F, because it is a dismal failure at the grade school and high school levels. We spend a higher percentage of our GDP (6.8 percent) on education than Japan (6.5 percent) and Germany (4.6 percnet), yet only 7 percent of our 17-year-olds are prepared for college-level science courses.

As Former President Bush put it "A dedication to quality and excellence is more than just good business. It is a way of life, giving something back to society, offering your best to others."

■ References and Sources

1. *Fortune,* February 14, 1992, p. 40.

2. *USA Today,* November 20, 1992.

3. *Fortune,* February 8, 1993, p. 84.
4. Letter to Congress dated July 31, 1985.
5. Ernst & Young's report entitled "The American Competitiveness Study."
6. *BusinessWeek,* 1991 special quality issue.
7. Ernst & Young, *Guide to Total Cost Management*, John Wiley & Sons, 1992.
8. *Total Quality Management* by Dorsey J. Talley.
9. *TQM: A Step-by-Step Guide to Implementation* by Charles N. Weaver.
10. *Electronic Business,* October 1992, p. 158.
11. Survey by Recruit Research, Inc.
12. Economic Planning Agency.
13. Steel Cases Calc 1992 Worldwide Office Environment Index— The Higher the Number, the Greater the Satisfaction.

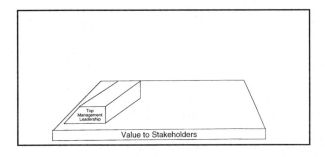

1

Top Management Leadership: The People Who Need to Change First

If you're going to sweep the stairs
always start at the top.   MY GREAT GRANDMOTHER

■ Introduction

We often hear that Japan, Inc.'s improvement process is a "bottom up" movement. Nothing could be further from the truth. The first major improvement breakthrough in Japan occurred when Dr. Kaoru Ishikawa decided that Japanese top management should get a better understanding of how statistics could help them improve their operations' performance. As a result, Dr. Ishikawa personally contacted key corporate executives, inviting them to a meeting to listen to Dr. W. Edwards Deming. Because of the executives' respect for Dr. Ishikawa, there was a good turnout for the meeting. Even through Dr. Deming had been lecturing for some time in Japan, this meeting with top management was the key meeting in getting the Japanese improvement process started. This was followed up with a series of top management meetings featuring people like Dr. Joseph M. Juran and Dr. Armand V. Feigenbaum that helped to develop the total quality control process now used in Japan.

It makes no difference where an organization is located. Top management sets the direction, middle management develops the implementation strategy, and line managers and employees implement the plans. In today's environment, there is a lot of overlap. Employees may even be involved in setting directions, but it is top management who has the final responsibility. When we look at the TIM process and its objectives, it is designed to provide a new way of managing and operating the total organization. Keeping this in mind, the decision to implement TIM must be made by top management. There is no way that it could start at the bottom. Confusion would reign supreme if we left it up to each individual or small work group to decide what products they would produce, how they would operate, and who their suppliers and customers would or would not be.

The one thing that all the gurus agree on is that the improvement process needs to be led by top management. This does not mean, however, that you cannot try to improve the way you are doing your job while you are waiting for top management to get its act together.

■ How Does Top Management Feel about Improving Quality and Productivity?

Hewlett-Packard's CEO, Dean Morton, at the Deans of Business Schools Conference in Las Vegas, Nevada, stated "Unless I am misreading the signs, I believe that we have reached the end of the golden age of shabby products."

Edwin L. Artzt, chairman of the board and chief executive officer of Procter & Gamble, stated "1992 tests our leadership and ingenuity as never before. A commitment to continuous quality improvement must be a top priority in our companies, our educational institutions, and our country to excel in an increasingly competitive global marketplace."

> Looking at it from a national level, the improvement of quality in products and the improvement of quality in service—these are national priorities as never before.
>
> Past U.S. President George Bush

> Our task is to react quickly to consumer demand...the quantity, range, and quality of goods, that is, just what people need, will be the main thing, and not gross output.
>
> USSR Past President Mikhail S. Gorbachev

> Quality first is our long-term strategic policy in our economic construction.
> Former Premier of the People's Republic of China, Zhao Ziyang

Yes, management is all for improvement. I don't know of any organization head that doesn't want things to get better. I don't know of any president that doesn't care about their customers or if their employees are doing their jobs correctly. I have never heard a CEO who came on the loudspeaker and said "Let's see how much scrap we can make today." Or one who has attended a sales meeting and said "Let's see how many sales we can lose this month. I will give a big, fat bonus to the man or woman who loses the most sales." No, quite the contrary. Managers since the beginning of time have been telling their teams to work harder, faster, better, and smarter. So what's new or different?

The real problem is that not much is different. Management today find them-selves caught in a vice. One jaw of the vice is low quality, productivity, and profits; the other jaw is increased cost, competition, and regulation.

The management styles that made us so successful in the 1960s will not work today. In a survey conducted by the American Management Association/Ogilvy, Adams, and Renehart of 128 corporate executives, 90 percent of the manufacturing and service organizations have started an improvement process. Nearly half of these were started in the last two years. Quality improvement was top or high priority for 69 percent of the organizations surveyed. For the last 10 years, there has been a lot of talk about improvement by CEOs, but when all is said and done, there has been more said than done. As Bill Clinton said when he was Governor of Arkansas, "You can't build a reputation on what you are going to do."

■ What Makes Believers out of Top Management?

Unfortunately, the thing that makes most top management believers is when their organizations fail or are near failure. Just look at the organizations that are leading the improvement movement within the United States They are organizations like Xerox, IBM, and General Motors. All of them lost over a quarter of their market and as their market disappeared, they finally decided that they needed to do something different to bring about the improvement needed to stop their organizations' decline. Even the federal government started its improvement process when the nation's national debt became overwhelming.

Today, more and more organizations have been forced into starting their im-provement process by their customers. For example, when Motorola won the Malcolm Baldrige Award, they sent out letters to their suppliers informing them that if they wanted to continue to do business with Motorola, they needed to apply for the Malcolm Baldrige Award.

Sure, there are a few organizations that started their improvement process because their top management were far-sighted enough to see the benefits the organization would receive from the process, but these are few and far between.

Over 80 percent of the large and mid-size organizations within the United States have started some type of improvement activity, but the great majority of them do not have their top management truly leading the transformation. Sure, they want you to change, them to change, the suppliers to change, but top management change? No way! After all, they got to be in top management by performing in a specific manner, so why change now? Why should they spend more of their precious time to help in the implementation of the improvement process? Management truly gets involved in the improvement process for many different reasons. Some of the more common are:

1. Loss of market share

2. Customer-imposed

3. A new senior manager

4. Tough competition

5. Customer complaints

6. Poor performance

7. It's the "in" thing to do

8. High costs

9. Poor return on assets

10. Missed schedules

■ Why Top Management Keeps TIM At Arm's Length

Too often we approach management and ask for their support of the improvement process based on intangibles. We explain that it will improve customer satisfaction, reduce waste, or improve morale. This approach gets their support, gets them saying "Hallelujah!", but does not get them up in the pulpit preaching or out working with the congregation. What we need to do is provide top management with tangible data that proves the business case for implementing the improvement process. Some of the best ways to convince management that they should support and actively lead the improvement process efforts are:

1. Competitive Benchmarking

2. Market Studies

3. Customer Surveys/Focus Groups

4. Poor-Quality Cost Analysis

5. Improvement-Needs Analysis

6. Customer-Loss Analysis

To get top management to provide real leadership in the improvement process, you need to provide them with hard data that convinces them that it is in their best interest to expend the most valuable resource they have—their personal time. For when the improvement process is implemented correctly, it does take considerable amounts of top management's time to get it started and to keep it properly focused. For example, when Edward F. Staiano, president of Motorola, discussed his involvement in the improvement process, he stated "You cannot manage the business by looking at the score card." Ed does not go to the P&L meetings and he does not look at the financial reports. He does go to all the quality meetings. Ed stated "I work on quality 100 percent of my time. Financial reports are after the fact." More typically you will find that the improvement process will require about 20 percent of top management's time during the first year.

It is important at the start to clearly define why the organization needs to improve. It should not be because another organization is doing it or to improve employee morale. It should be one or more of the following:

- Get a "bigger piece of the pie" (increase market share)
- Create a "bigger pie" for everyone
- Improve the bottom line
- Ensure long-term survival and growth
- Increase job security
- Combat competitive pressure

■ It Doesn't Have to Be the President and/or Chairman of the Board

Everyone seems to be waiting for someone "up there" to get serious about the improvement process. I talk to the employees and supervisors. They tell me that middle management only gives lip service to the improvement effort and that the organization should get serious about it right now. I talk to middle management and they tell me that they are all for it, but top management does not really mean what it says. What top management says is "Do the best job you can." What they really mean is "Do the best job you can at minimum cost and in half the time required." I talk to the president and he or she tells me he or she realizes how important it is to improve, but the chairman of the board is more interested in quarterly profits than in improving the organization's long-range performance. I talk to the chairman of the board and he or she tells me that the organization must improve or it will lose its stockholders, but the board of directors doesn't understand so it won't support the long-term investment required for the improvement effort. I talk to the board members and they tell me that they know how important it is to improve for the growth of the organization, but they just represent the shareholders and the shareholders are focused on short-term profits. I then ask them who the shareholders are and they tell me it is the employees.

Today is the day to stop waiting for someone else "up there" to take the lead. Step up to the problem and start improving your own performance. Do something to improve within the guidelines set by your management and do it a little better today than you did yesterday and better tomorrow than you did today.

Individuals cannot make TIM work, but a business unit can. Any group of people who are measured on their performance can start the process. It can be the total organization, a division within the organization, a plant, or a sales office. As long as the top management within that organization will step up to the leadership responsibilities for the TIM process, it will work. It is always best to start with the president of the organization as the leader of the TIM process, because the process will be implemented much faster and with greater consistency, but if you cannot get it started at the very top, get it going some place. TIM is a lot like the measles. If it catches hold in one part of the organization, it will spread throughout the rest of the organization very quickly.

■ How Does Top Management Show Leadership?

The magical triangle for success is quality, cost, and schedule. It is easy to get one leg of this three-legged stool at the sacrifice of the other two. The real challenge is to improve one or more of the three legs of the stool without harming the other ones. This means that upper management must have a balanced approach to the way they make decisions and operate their organizations. But no matter how hard they try, there will be an occasion where one or more of the three elements must be sacrificed for the good of the others. In these cases, top management can never forget that these decisions have a long-term impact on their organization, its reputation, and on their external customer. Because of this, management must look beyond the simple dollars and cents analysis of the individual situation. Often it is far better to lose money on a particular situation in order to gain long-term profitability. This brings us to the 3 key directives for top management leadership. They are:

- Never sacrifice quality for cost and/or schedule.
- Never sacrifice schedule for cost.
- Lose money on an individual transaction if you have to in order to build long-term relationships.

Not only the president and chairman of the board need to become actively involved in the improvement process, but all of the top managers must be contributing also. Top management must have a unified position on the improvement process or weaknesses will develop and spread like a cancer. For example, if all the top managers except the vice president of engineering are leading the improvement process, the word soon gets around that engineering is not trying to improve, so why should manufacturing and sales strive for excellence? The same is true of any other function in the organization. A chain is as strong as its weakest link, and the same is true of the improvement process. Careless, negative comments by any top manager can cause vibrations throughout the organization.

Steve Moksnes, president of AccuRate, pointed out "The way I provided leadership in the quality process is that at the very start when we were just looking at quality improvement, I wanted to clearly understand just what it was all about. Certainly I'm not in a position to lead unless I really understand the concept very well. That's why I read a lot of books and went to seminars."

It may be very hard for some of the old-timers to accept and adjust to a new way of managing, but this adjustment must be made, or they must be removed. Because of this, your organization needs to take adequate time to obtain total buy-in of all your top managers before you start to disseminate the activities to your middle managers, front-line managers, and employees. As the process moves forward, demonstrated upper management commitment and support of the improvement process will be critical to its long-term success. Top management can demonstrate its commitment and show its leadership of the improvement process in the following ways:

1. Committing and giving freely of their personal time.
2. Providing the required resources.

3. Releasing and enforcing pertinent directives, policies, and procedures.

4. Setting the example for organizational change.

5. Developing and following up on improvement measurements.

6. Building a team environment.

7. Tying their compensation into the improvement measurements.

■ Committing and Giving Freely of Top Management's Time

The one resource that we cannot get more of is time. All people have 24 hours a day that must be stingily dealt out. The very best way that top management can demonstrate their belief in the importance of the improvement process is to dedicate their time to the process. Remember, your employees listen to your lips and say yes, but they react based on your actions. Top management must champion the improvement process within their areas.

In an internal communication, Bob Talbot, IBM Director of Quality, points out "John Akers has spent more time on quality than any other CEO in the history of IBM. The Management Committee and Corporate Management Board have spent more time on quality and process-related topics than any other topics during the last 18 months."

Corning, Inc. Chairman James R. Houghton is out talking to his team about continuous improvement 50 times a year. His theme has been very much the same for the last eight years: Quality, Customer-Focus, and World-Class Performance. Houghton stated "After eight years, if I stop talking about quality now, it would be a disaster." He continued to point out that their focus on continuous improvement had increased operating profits by 111 percent in five years.[1]

Sam Walton, founder of the Walton Retail Store Chain, was never afraid of making an example of himself if it proved a worthwhile point. For instance, back in 1983, he was convinced that the company was going to have a down year. As a result, he promised his employees that he would dance the hula down Wall Street if they managed to beat his down-scale projections. When his team came through and performed better than expected, he lived up his promise, grass skirt and all. Now this type of leadership and commitment may be hard to obtain, but upper management must believe in the process for it to work. As Pope John Paul II said "Some things have to be believed to be seen."

A study sent to *Fortune* 1000 companies pointed out that 50 percent of the respondents felt top management had a great or very great facilitation impact on their current employee involvement process, compared to 26 percent for middle managers and 19 percent for first-line supervisors.[2]

Two factors seem to have a major impact on the success of an improvement activity. They are:

- Perceived management support
- Perceived individual benefits

■ Personal Performance Indicators

Top management are all for change, as long as it is someone else who changes. The number one top management improvement rule is: Top management must change first.

Why should top management change? Aren't they already successful? Look at all the money they are making. In truth, more than 99.99 percent of top managers have a lot of opportunity to improve. In fact, none of us are perfect and most of us have many opportunities to improve our own personal behavioral patterns.

To prepare their personal performance indicators, top managers need to define what they do. I don't mean things like "motivate employees" or "manage R&D." These are their assignments. Examples of what management do are:

- Attend meetings
- Read and answer mail
- Answer telephone calls
- Make decisions
- Delegate work
- Chair meetings
- Etc.

Once top management has completed a personal list of what they do, they need to define behavioral patterns related to these activities that could be improved. For example:

- Start meetings on time.
- Do not attend meetings that can be delegated to employees.
- Return all telephone calls within eight hours.
- Don't set items aside that can be done quickly.
- Always show up on time for meetings.
- Read all new mail each day.
- Don't use overnight mail if the project can be finished earlier and sent by regular mail.
- Talk to three customers each day.
- Make a minimum of three one-hour tours of the employee work areas each week.
- Read five technical articles each week.
- Have the office organized so well that there is no need to search for lost or misfiled items.
- Have a clean desk at the end of the day.
- Stop doing things that can be delegated.

- Stop using bad language in the workplace.
- Arrive at work on time.
- Have an agenda in each attendee's hands before any meeting is held.
- Etc.

They then should select a maximum of eight of these behavioral patterns that they want to improve. Each selected behavioral pattern should be recorded on a card similar in size to an airplane ticket, which can be carried easily in a purse or a coat pocket. Each time the executive does not behave as defined, he or she has made an error, and a check mark should be placed behind the appropriate behavioral pattern. Once a week the total number of check marks should be counted and plotted on a run chart. We recommend that each executive set an improvement target of 10 percent of the first month's average error rate. When this target is reached, it is time to add eight more behavioral patterns to the list and start over again. In organizations where top managers have a high degree of confidence and credibility, each top manager is encouraged to post his or her personal performance run chart in his or her office, demonstrating to their fellow workers that top management accepts the fact that they personally need to change and improve. Eventually this same approach will be used by all managers and employees to measure their personal improvement.

Each executive is also encouraged to define his or her personal set of values that govern their behavior at work and in their personal life. These values are seldom mirror images of the organization's values or principles because they reflect the total person. We encourage every top manager to post these value statements in a very visible place in their office and/or conference room. Sharing these personal values with their fellow managers and employees and posting them where they can influence the executive's behavior has a major positive influence on the individual and the organization. Don't be afraid to let your employees know you are human.

■ Top Management's Personal Support of the Improvement Process

Now let's look at some of the ways top management should expend their efforts to make the improvement process a success.

1. *Take classes.* Top management should attend all the improvement classes first to ensure they understand what they are asking their employees to do.

2. *Participate on the Executive Improvement Team (EIT).* The improvement process should be directed by a group we will call the Executive Improvement Team (EIT). This team is chaired by the top officer in the organization and is made up of all department heads and key staff personnel reporting to him or her.

3. *Give talks to employees.* The entire top management team needs to go out of its way to talk about the improvement process with all the organization's stakeholders.

4. *Join professional improvement organizations and get involved.* Top management needs to understand what is going on outside the organization in the improvement methodologies. To do this, they should join one or more of the organizations that support these types of efforts.

5. *Attend and give presentations at conferences outside your organization.* There are hundreds of organizations around the world that are putting on conferences that discuss different improvement methods and the results that have been achieved.

6. *Audit the improvement process.* As the process develops, start an active top management audit of the functions that report to each of the top managers to determine how effectively the improvement process is being implemented. Don't just rely on verbal communication.

7. *Conduct focus group meetings.* Hold frequent focus group meetings with your employees, external customers, and internal customers. Top management should conduct these focus group meetings.

■ Supplying the Required Resources

One of top management's major responsibilities is to ensure that the *seven M's* are available to properly support the process.

- Men/women (properly trained, skilled employees)
- Machines (equipment well maintained and up to date)
- Methods (processes that use appropriate technologies)
- Materials (component parts, ingredients, documentation)
- Media (environment, time)
- Motivation (creativity, helpfulness, attitude)
- Money (financial support)

As Tom Malone, president of Milliken, said "It takes two or three years for it to really happen. You've got to get the scoreboards in place and then the environment of applauding people for improving the scoreboards. You can document all that up-front cost. The payback is later." You must remember that an investment in the improvement process is truly an investment, not a cost. The benefits of this investment will be increasingly evident as time goes on.

■ Releasing Supporting Policies and Procedures

One of the many ways top management provides direction is through the release of organizational policies and procedures. These documents provide employees and management with direction and the boundaries that govern their work effort. Top management should prepare and distribute to employees appropriate written direction so the employees understand what is expected of them. Training should also be provided as required. Typical documents used to help provide direction for the improvement process are:

1. Improvement Beliefs and Concepts
2. Improvement Policy
3. New Performance Standards
4. Corporate Instructions and Directives
5. A "No-Layoff" Policy

Improvement Beliefs and Concepts

One of the activities that the top management team or the EIT needs to accomplish early in the improvement process is to agree on a set of basic improvement beliefs and/or concepts that will be followed throughout the process. The following definitions will ensure that there is a common understanding of terms.

- *Principle.* The ultimate source; a fundamental truth; the motivating force upon which others are based.
- *Concept.* A generalized idea or general notion.
- *Belief.* An acceptance of something as true.

As you can see, all principles are concepts or beliefs, but not all beliefs and concepts are principles. For example, the statement "It is always better to do it right the first time" is a concept that many people believe in, but it is not a principle because sometimes it is not true. For this reason, we have used the terms "beliefs" and "concepts" rather than "principles", for although in the vast majority of instances they are correct, some of them in a small percentage of occasions (less than 3 percent) may be incorrect in a specific circumstance. The following is a list of basic improvement beliefs and/or concepts.

- Quality is defined by the customer.
- The customer is king.
- Improved processes are the key to error-free products and services.
- Everyone has suppliers and customers.
- Prevention is better than reaction.
- Anything short of error-free performance needs to be improved.
- Decisions should be based on fact and the risk understood.
- A participative management style provides the best results.
- Everyone should be provided with a way to contribute to the organization's success.

- Everyone and everything must continuously improve.
- Symptoms need to be transformed into root causes before a problem can be permanently solved.
- All employees should be involved in solving problems.
- Management must set the example.
- The person who knows the job best is the person who is performing it.
- Management and employees must have trust and confidence in each other.
- A rewards and recognition process should reinforce desired behavior using the *four P's*: (*P*ats on the back, *P*laques for the wall, *P*resentations to management, and *P*ay for the wallet).
- Everyone in every job contributes to the organization.
- We all need to respect the rights, dignity, and efforts of all individuals.
- Our people are our most valuable resource. In order to maintain them, we must train them.
- People need to be empowered to control their own assignments.
- Improvement requires a well-thought-out plan that has broad ownership.
- Small business units perform better than large business units.
- All stakeholders should be considered in every decision.
- The flatter the organization, the better its performance.
- The organization is obligated to provide a fair return on the stockholders' investment.
- The customer is looking for value in the products and services they purchase.
- Strategic alliances with other organizations will strengthen the organization.
- Cycle-time reduction provides a key competitive advantage.
- Strengthening our core capabilities and competencies is key to our future growth.
- Our suppliers are our second most valuable resource.
- Nothing is too sacred to be questioned.
- The customer, although not always right, is never wrong.
- Appropriate use of the best technologies, combined with our creativity, is a key to profitability.

Normally, these basic improvement beliefs and concepts are developed following the completion of the organization's environmental visions (see Chap. 4). It is suggested that the EIT do a preliminary draft of the basic improvement beliefs and concepts and review them with focus groups of employees to gain their views of the statements and to ensure that they are complete and comprehensive. The list provided here is too long for most organizations. We suggest you limit your basic improvement beliefs and concepts to a maximum of 15. Following the focus group reviews, the EIT should regroup and develop a finalized basic improvement beliefs and concepts list, then communicate this to the employees verbally and in writing.

Lockheed Corporation uses the term "principles" in place of beliefs and concepts. They selected the following six principles:

1. Focus on the customer

2. Be led by management

3. Make a long-term commitment to quality

4. Involve all employees and suppliers

5. Work as a team

6. Establish high standards

Xerox's principles are:

1. Continuous improvement

2. What if this was my company?

3. Empowerment

4. Simpler organization

5. Managers as coaches

6. Improved communication

7. Focus on training

8. Teamwork

Improvement Policy

An improvement policy is designed to explain how the improvement process will impact the business. For many years, organizations have been preparing quality policy statements. For example:

Xerox Quality Policy

Xerox is a quality company. Quality is the basic business principle for Xerox. Quality means providing our internal and external customers with innovative products and services that fully satisfy their requirements. Quality improvement is the job of every Xerox employee.

You will note that the aforementioned quality policy statements are directed at the customer and focus on meeting customer expectations by providing them with high-quality products and service. Direction related to improving cost, schedule, and profit was omitted from many of these quality policies.

The improvement policy statement takes on a little different dimension, since it relates not only to quality, but also includes productivity, cost, and technology. A typical improvement policy would read:

> We will constantly improve our quality, productivity, and creativity, allowing us to provide on-schedule products and services that represent the best overall value to our present and potential customers. Our improvement process' objectives are to increase our volume of business, thereby providing greater job security to our employees and better return on investment for our stockholders.

In this case, the improvement policy focuses on the need to continuously improve so that the organization provides greater value to all its stakeholders. Most quality policy statements made a major error when they did not document a concern for the organization's employees and stockholders. This error has been corrected in the improvement policy statement.

Care must be taken in organizations that have a well-communicated quality policy statement already in place. When top management prepares the improvement policy statement, they must take time to be sure that both policy statements support each other and then communicate to the employees the additional direction that the improvement policy provides.

The improvement policy should be carefully explained to all employees and become part of the indoctrination sessions for all new employees. The improvement policy should be kept visible to remind everyone of the direction the organization is taking. The policy should always be signed by all the top managers in the organization. We like to see it prominently displayed in each manager's and supervisor's office. In these cases, the managers or supervisors and all management in that management chain should also sign the policy statement, indicating their approval and support of the document.

New Performance Standards

One of management's major jobs is to set the performance standard for the organization. In the 1980s, a new "error-free" performance standard gained favor throughout the world. AOQs (Average Outgoing Quality) and AOQLs (Average Outgoing Quality Limit) lost favor and began to be viewed as a way of allowing shabby products to reach the customer. One supplier, when delivering an order for 1000 parts at an AOQL of 2 percent, sent two boxes, one containing 980 parts marked "to specification," and another containing 20 parts marked "not manufactured to specification." The two boxes were accompanied by a note saying that they had to readjust their equipment specifically to make the 20 bad parts, and would the company mind if the next time they sent all good parts.

The concept of 95, 99, or even 99.9 percent being good enough is being discarded around the world. Just take a look at what it would mean if we were right only 99.9 percent of the time.

- At least 10,000 incorrect drug prescriptions each year
- 50 newborn babies dropped at birth by doctors or nurses each day

- One hour of unsafe drinking water each month
- No electricity, water, or heat for 8.6 hours each year
- No telephone service or television transmission for nearly 10 minutes each week
- Two short or long landings at O'Hare Airport each day (also New York, Los Angeles, Atlanta, etc.)
- Nearly 500 incorrect surgical operations per week
- 2000 lost articles of mail per hour
- 22,000 checks deducted from the wrong account each hour
- Your heart fails to beat 32,000 times each year

There is no problem so small that you cannot lose a customer over it, so how can any manager tell their employees how many errors are acceptable? Could you possibly tell your son or daughter that it is all right to drive through the back wall of the garage once in every 1000 times he or she borrows the car? Motorola uses a six sigma program. They are trying to reduce errors to .00034 percent, but is that good enough? The answer is *no.* Consider the truck driver who has 10,000 opportunities for a head-on collision every day. Certainly, six sigma is not an acceptable performance standard in that case. On the other hand, consider the office worker that has something wrong in three out of 10 documents they generate a day. Telling this employee that they should work for two years and make only one error is putting them in a no win position, that is very demoralizing. What we need is a performance standard that communicates to our employees that they should continuously strive to improve until they have eliminated errors. The only acceptable performance standard is *Error-Free Optimum Performance.*

To dissect this statement, the term "error" was used in place of "defect" because everyone makes errors, while only the production floor makes defects. TIM must apply to everyone. The word "optimum" was chosen because in most decision-making situations, there are a number of options available, all of which will provide error-free performance, but only one will provide the best results. Optimum indicates that the activity is only correct when the very best alternative is selected.

Everyone is an error-free performer today. What differs between individuals is the length of time that the person can perform between errors. In one individual circumstance, a person/process combination could average one error every two hours. Is that good enough? Using the error-free optimum performance standard, the answer is obviously no. So, if he or she sets an improvement goal of a duration between errors of two days and succeeds at meeting this goal, is that good enough? No. Once he or she reaches that goal, the employee needs to set another goal. For example, it could be to extend the duration between errors to seven days, then to two weeks, then to two months, etc. The road to error-free performance is a series of successes followed with new, more aggressive goals.

Even that is not good enough today. Doing things right every time is high-quality performance, but to be world-class, you need to go beyond that. You need to do the right thing right every time. Error-free performance is selecting the best option every time and then executing it as effectively as it can be done. You are not performing error-free if you do a perfect job in ten hours and someone else can do the same job perfectly in three hours.

Everyone needs to strive to be better, and we will never reach a good enough performance level until every decision is perfect all the time. To excel is to fulfill a basic human need.

Corporate Instructions and Directives

Top management also provides direction through the release of corporate instructions and directives that set performance requirements for specific activities. For example, two of the performance improvement directives issued by IBM in the early 1980s were:

The first instruction simply told management that each new product must perform better than the product it will replace (current product) and the competition's product.

The second new corporate instruction was directed at establishing process improvement over critical business processes. The instruction required that each business process be assigned an owner, be certified, and be classified in one of five categories. The process owner is responsible for organizing process improvement teams that will bring each process up to the first level. Typical business processes are cost estimating, physical inventory management, supplier relations, engineering change implementation, product release, planning and scheduling, accounts receivable, inventory control, payroll, financial planning, fixed assets, and appropriations controls.

Employment Security

Employment security is one of the most critical and complex political and economic issues facing top management as a result of implementing TIM. The management must determine:

- If the employees are an investment or a cost?
- How much improved performance and flexibility would be gained if the organization provided employment security?
- Is it better to retrain or relocate employees or hire new employees?
- Should the organization look at other ways to handle surplus employees besides laying them off?

The employee must determine:

- What impact will TIM have on my employment security?
- Will TIM jeopardize or improve my standard of living?
- Would I be willing to change jobs or move to another location?
- Will I be more valuable as a result of what I learn during the TIM process?

How can you expect your employees to give freely of their ideas to increase your productivity and minimize waste if it means that their job or a friend's job will be eliminated? If you start a continuous improvement process and then have layoffs, what you are going to end up with is a continuous sabotage process.

Corporate America has been on a downsizing kick since the late 1980s. Their answer to business pressure is to slow down and lay off, with the hope of raising stock prices, but that does not work. In an analysis published by *USN&WR* (Basic Data: Mitchell & Co.), the stock prices of organizations that went through a major downsizing in the last half of the 1980s were compared to those within their industry groups. On an average, the organizations who downsized saw a 10 percent improvement in stock prices over the first six months after the downsizing occurred. Six months later, stock values went negative by about 1 percent. Three years after the downsizing, they were negative an average of 25 percent.

Large layoffs produce sudden, substantial stock gains. These gains occur because the impact of the removed employees has not reached the customer and the compensation has been removed from the bottom line, making the organization appear to be more profitable than it really is. But in the long run, the downsizing has a negative impact. CEO Frank Poppoff of Dow Chemical put it this way, "Layoffs are horribly expensive and destructive of shareholder value." [3]

This is because most organizations downsize rather than "rightsize." What organizations have been doing is cutting "X" percent per area—not eliminating any work, just distributing the work among the people who are left. What management is really telling its employees is that management does not believe the employees are doing a fair day's work for a fair day's pay. There are at least 16 other ways to handle the problem of surplus employees that should be looked at before the organization lays off anyone. For more information, see Ernst & Young Technical Report TR 92.0692 HJH, "The Down Side to Quality Improvement."

The cost to lay off and replace is growing all the time. Dow Chemical estimates it costs between $30,000 to $100,000 for technical and managerial type personnel. Layoffs not only cost the organization money and some of its best people, but when it comes time to hire, the best people do not trust the organization and will not come to work for them.

An alternate approach of a golden parachute or early retirement is equally bad. The people who leave are all the best performers who will not have a problem finding new jobs. The deadwood, who barely meet minimum performance, stay because they know it will be hard to find an equally good job in today's job market.

Employees can understand that the organization needs to cut back when demand for the product falls off, and they can accept that. The problem we face is what happens to the employees whose jobs have been eliminated due to TIM. We know that TIM is designed to improve productivity. But if our share of the market does not keep pace with our productivity improvement, what will management do with the surplus employees? To cover this scenario and to alleviate employees' fears, top management should release a "no-layoff policy." A typical no-layoff policy would state:

No employee will be laid off because of improvements made as a result of the TIM process. People whose jobs are eliminated will be retrained for an equivalent or more responsible job. This does not mean that it may not be necessary to lay off employees because of a business downturn.

You will note that the policy does not guarantee that employees will not be laid off as a result of a downturn in your business. It only protects employees from being laid off as a result of the improvement process. These are people who would still be working if the improvement process had not been implemented.

Federal Express Corporation has a no-layoff philosophy. Its "guaranteed fair treatment procedure" for handling employee grievances is a model for firms around the world.

I know of one organization that was able to eliminate 200 jobs as a result of its improvement process. As they started the improvement process, they put a freeze on new hires and used temporary employees to cover workload peaks. This was reviewed with the labor union leaders and they concurred with the use of temporary employees to protect regular employees' jobs. As a result, attrition took care of about 150 surplus jobs. The organization then held a contest to select 50 employees who were sent to a local university to work toward an engineering degree. While at school, they received full pay and their additional expenses were paid for by the organization. Results were phenomenal. Everyone within the organization started to look for ways to eliminate their jobs so they could go to school.

■ Organizational Impact

It is not recommended that a separate organization be established to implement TIM. Quite the contrary. The object is to integrate TIM into the everyday way the organization functions. To minimize the new jobs that are created and the formation of a new function, we recommend that the head of the organization become the chairman of an Executive Improvement Team (EIT) and that, at a minimum, all the people who report directly to him or her are also members of this team. If the organization is a union shop, a member of the local union should also be invited to be a member of the EIT.

EIT Responsibilities

The EIT is responsible for managing the implementation of TIM. Typical tasks performed by the EIT are:

- Develop a series of five-year vision statements for the organization's environment and develop an improvement plan to change the organization's environment so that it lines up with the vision statements, then assist with its implementation.

- Prioritize the use of resources so that maximum return is gained from the improvement process.

- Communicate the status of the improvement process throughout the organization.

- Lead the improvement efforts in the function that each individual is responsible for.

- Study improvement processes underway outside the organization to gain additional knowledge to determine their applicability to the organization.

- Develop measurement systems for the improvement process.

- Review the status of the improvement process to ensure it stays on schedule.

- Update the improvement plans and strategies as necessary to keep up with the changing environment.

- Resolve problems that cannot be resolved at lower levels.

Improvement Champion (Improvement Czar)

Often it is necessary to select an Improvement Champion (sometimes called an Improvement Czar) to help the EIT with the additional workload generated during the development and implementation stages of the improvement process. The improvement champion should be a temporary assignment, lasting between two to three years. The improvement champion should always report directly to the head of the organization. The improvement champion should be a person with stature, respected by the entire management team and the employees. Preferably, the improvement champion should be at the functional or vice presidential level. He or she should be an individual who embraces change and who has high personal standards. The champion should believe that the organization can and needs to do better. He or she should be anxious to take a personal role in the TIM process to help provide the organization with a competitive advantage.

The improvement champion frequently serves as an ombudsman for the employees, allowing them to come directly to him or her to discuss a specific opportunity. Many organizations will establish an improvement hot line where ideas, suggestions, and comments can be given to the improvement champion directly. In addition to working with the EIT, helping it with monitoring the implementation of the improvement process, the improvement champion will work with the teams to help them break down organizational barriers and acquire additional resources.

Improvement Steering Council

In large organizations (over 10,000 employees) where the top management team already has a lot on its plate, we find that establishing an Improvement Steering Council made up of representatives who are a cross-section of the job classifications and functions, provides a way to develop and sell new ideas. This council also provides an excellent way of evaluating the pulse of the organization. The council reports to the EIT, and the improvement champion is normally its chairperson.

Improvement Leaders

TIM is based on the belief that quality, cost, schedule, and technology are all important to the success of the organization. Based on this, the organizational structure should be balanced so that the areas responsible for each of these elements are at the same responsibility level. This means that the quality and financial officers of the corporation should be at the same level as the vice presidents of engineering and operations. Many organizations do not have a vice president of quality and this is a mistake. It tells the other officers and employees that cost, schedule and technology are more important. Today, every COO should have a quality "conscience" (a Vice President of Quality and Productivity) and a Chief Financial Officer reporting to him or her. This is important so that quality, cost and schedule are all represented at the COO's staff meetings and when critical business decisions are made.

The person who is selected as the Vice President of Quality and Productivity must have the right credentials. This person should have an outstanding background in production and service quality. The ideal individual would be a person who has a Bachelors Degree in the technology that is most applied to the products or services provided by the organization, with a Masters Degree in quality. Of course, they should also have had considerable management experience. Be careful not to take an available body who has no improvement experience to put into this critical assignment, as this can have a very negative impact on the total organization. I heard one Vice President of Quality for a *Fortune* 100 company talking at a conference. During the question and answer period he was asked "What type of AQLs are you using in Receiving Inspection?" He replied "What are AQLs?"

Board of Directors

The board of directors are the check and balance on the executive teams. To do this job correctly and not just be a rubber stamp for the executives requires a very special type of person. At least 60 percent of the board of directors should not be officers in the organization. (Example: 14 of the 17 members of G.M.'s Board of Directors are not G.M. employees.) The board should be made up of some technical people who understand the technologies the organization is involved in, consumers of products and services that the organization produces, people who have an excellent understanding of the improvement process, and financially-oriented people. Unfortunately, few boards have members who truly understand the improvement process. This is a situation that should be quickly corrected at the start of the improvement process.

Organized Labor Involvement

Although the improvement process starts by getting management involved first, eventually all the employees need to become active participants in the process. It is a fatal error to ignore the unions during the planning and execution phases of the process. Appropriate union representation should be considered for membership on the EIT, the Improvement Steering Council, and even on the Board of Directors.

Even if involving organized labor in the early parts of the improvement process slows down the process, the time and effort expended will more than pay for itself during the implementation phase.

The labor movement is undergoing a revolutionary transformation. You can see it among the communication, rubber, and textile workers. Organized labor is working with companies like AT&T, Ford, Goodyear, and Xerox to help save jobs by making the organization more competitive. G.M.'s Saturn plant's interface with the United Auto Workers is a good example of flexible collaboration. Kenneth Coss, leader of the United Rubber Workers, stated "Our goal, really, is to preserve the industrial base. We told the companies that this industry is self-destructing. We'd better work jointly using all our intelligence, our initiative, to make world-class facilities." Lynn Williams, leader of the steelworkers, stated "When it comes to dividing up the pie, we'll be adversaries. But now we have to grow the pie, and that means working together."

Xerox' relationship with the Amalgamated Clothing and Textile Workers Union (ACTWU) is an outstanding example of how unions and management can work together in implementing employee involvement. As early as 1980, contracts between Xerox and ACTWU included clauses supporting joint improvement efforts.

Working together, the clothing workers' union and Xerox have changed their interface (Source: *Fortune Magazine*, February 8, 1993).

- Xerox can use temporary workers because they granted union members job security for the duration of their contract.

- No-fault termination dropped absenteeism from 8.5 percent to 2.5 percent.

- Cooperation teams—One team working on improving the wire harness process saved $3.5 million. This allowed the process to stay in the U.S., saving 240 jobs.

Buddy Davis, St. Louis district director for USW said, "Now 100 percent of our members are on a committee, and because these committees work out day to day a lot of smaller problems that used to be saved up for the bargaining table, the bargaining goes fast and smooth.

It isn't all peaches and cream. Many hard-core, rank-and-file people still reject the whole team concept, but things are changing. If you want organized labor helping your organization improve, involve them early in the planning process and keep them informed. If you want them to agree to new work rules, then the organization needs to open up the books to them, providing cost and profit data. We need to convert what in many cases is an adversarial relationship into a partnership.

■ Measurements of Improvement

Top management knows that what gets measured gets done. I once had a department clerk tell me "If management doesn't measure it, they don't care about it." It is for this reason that top management needs to focus on how they are going to measure the effectiveness of TIM early in the process. A typical improvement measurement system should go through three phases.

Phase I: Activity Measurements. Measures the activities that are going on related to the TIM process.

Phase II: Improvement Results Measurements. This type of measurement normally starts about six months into the process.

Phase III: Business Results Measurements. This phase starts about 18 months into the process.

It is extremely important that the measurement system be defined early in the TIM process so that the impact of the new process can be compared to the previous process. Too often, measurements are put off during the first year, so when someone asks how much improvement has been realized, the starting point data is not available. One of the things the EIT should do during the first three months of the process is develop the measurement system and start collecting the data needed to drive an initial stake in the ground for each key measurement. This early focus on measurement also has the advantage of providing a focal point for the organization that will keep everyone aimed at the same target. Measurements are covered more fully in Chap. 14.

Assessments

One of the best ways to measure the less tangible impacts of the improvement process is through assessment methods. These assessments are also helpful in identifying improvement opportunities. Some of the most useful assessment methods are:

1. *Improvement-Needs Assessment.* This assessment method has already been discussed, since it is very useful in gaining top management commitment. This method can be repeated every 12 to 18 months to provide an insight into the changing environment within the organization.

2. *Quality Award Assessment.* There are a number of quality awards that have a very systematic way of measuring the performance of the organization. For example, the United States has a Malcolm Baldrige National Quality Award and a NASA Quality Award. In 1992, Europe started a European Quality Award, and Japan uses the Deming Prize to accomplish similar objectives. (The Malcolm Baldrige Award evaluates the organization's performance in seven categories and divides a maximum of 1000 points among the categories shown in Fig. 1.1.)

To conduct this type of assessment, the organization should fill out an application form and have a certified evaluator review the application. Then the evaluator should audit the organization to verify the validity of the application's data. The evaluator will then compare the organization to the criteria in each of the categories, assigning a point score to each category. The evaluator will also provide a list of the organization's strengths, weaknesses, and suggestions for improvement.

Categories	% of Total Points
• Leadership	10
• Information and Analysis	7
• Strategic Quality Planning	6
• Human Resource Utilization	15
• Quality Assurance of Products and Services	14
• Quality Results	18
• Customer Satisfaction	30

Figure 1.1. Malcolm Baldrige National Quality Award categories.

Another approach to conducting a Malcolm Baldrige-type assessment is to complete the application form and send it in to the Malcolm Baldrige National Quality Award Committee where the application will be evaluated by a number of trained professionals. These professionals will assign a point score and point out strengths and weaknesses of the organization. If the organization scores high enough, it will also receive an audit.

3. *International Quality Study Assessment.* The American Quality Foundation and Ernst & Young maintain perhaps the best benchmark databases on organizational improvement trends. This database was the result of an internationally cooperative two-year process of the major banking, health care, auto, and computer organizations in Germany, Japan, the United States, and Canada. This database looks at all elements of the management system, both past, present and future, that impact the improvement process. This benchmark database is continuously updated as additional organizations are added and compared to the existing database. For a small fee, any organization can work with the International Quality Study Team professionals to acquire the pertinent information relevant to their organization. Ernst & Young will generate a comprehensive report that compares the organization to the total database or segments within the database.

4. *Employee Opinion Survey.* All employees at all levels should be given the opportunity to participate in an employee opinion survey. It will normally consist of between 60 to 100 questions that will probe into all aspects of the business. The initial baseline survey should be conducted during the first six months of the improvement process and it should only be subdivided down to the middle manager's level. The object of an employee opinion survey is to measure changes in the employees' attitudes and identify employee-perceived problems. Typical questions asked are:

- How good a job is being done by your manager?
- How well are your skills being utilized?
- What is your overall satisfaction rating with the organization?

This type of survey should provide a number of key indices that are the average value of groups of questions. Examples of key indices are: morale index, management performance index, satisfaction index, etc.

Normally, employee opinion surveys are repeated about every 18 months. It is very important that the survey be analyzed and reported back to the employees expeditiously. Two months' delay between the survey and the report back to the employees should be the maximum.

Although the first survey is only broken down to the middle manager's level, the second survey, typically conducted 18 months later, will be divided down to the small team level (eight employees' minimum break-out). For each survey question, these small teams should be compared to their project's average, their function's average, and the organization's average for the specific question. An analysis should be made to determine if there are statistically sound differences between the small work group and the other groups it is compared to. If the small work group is trained to use the team and problem-solving tools, it should prepare corrective action plans for questions that are significantly negative compared to the rest of the organization. These corrective action plans should be presented to middle management. Middle management and upper management will then prepare corrective action plans that will improve the overall performance of the organization. These plans should be presented to the employees. The status of all plans should be tracked to be sure the corrective action is implemented.

Poor-Quality Cost (PQC)

It is often very difficult to summarize the impact an improvement process has on an organization because there are so many different types of measurements involved. For example: customer satisfaction index, mean-time-to-failure, process yields, inventory turns, etc. To overcome this problem, poor-quality cost systems have been developed that dollarize all the added costs that occur because everything is not done right every time (see Chap. 14).

In a survey of Illinois manufacturers conducted by the Illinois Management Association and Peat Marwick, the organizations surveyed indicated that they believed their direct poor-quality cost to be 6 percent of sales. The study revealed, however, that their actual poor-quality cost was approximately 25 percent of sales. Over 75 percent of the true loss to the organization was not realized. Typically, in a manufacturing organization, poor-quality costs run between 15 to 30 percent of sales. In a service organization, this runs much higher, up to 60 percent of sales.

The vice president of research and development for Saab reported at one of their conferences that R&D's poor-quality cost was 78 percent. John F. Akers, past president and past chairman of the board of IBM, said "Our studies show that more than 50 percent of the total cost (PQC) of billing relates to preventing, catching, or fixing errors."

In the July 1992 issue of *Consumer Reports*, they discussed the unnecessary costs incurred in the health care system. They stated "This year's U.S. medical bill is about $817 billion. Of that, we will throw away at least $200 billion on over-priced, useless, even harmful, treatment; and on bloated bureaucracy."

Of course, we have all heard about the waste that goes on in our government (i.e., the $150 toilet seat; the $75 hammer), but let's look at some specifics. The May 1992 *Dollars & Sense* report gave the following examples: "Delaware County, Pennsylvania used $6 million of antipoverty funds on bar tabs, banquets, gifts, and Broadway theater tickets."

"Of the 53 offices of the Department of Agriculture that pay out farm subsidies, they spent one dollar for every one dollar they disbursed."

They also reported another questionable use of our money when they pointed out that "A $5.4 million loan of government money was given to a wealthy Saudi at a 1 percent interest rate."

Select any number you want, but it is obvious that there is a great deal of opportunity to increase profits by reducing poor-quality cost. A poor-quality cost system will help management identify major improvement areas and can be very instrumental in obtaining the employees' active participation in the improvement process. In addition, it provides an outstanding way of measuring results.

■ Summary

To obtain the maximum results from TIM, it has to start with top management believing that the results achieved from the process justify them expending their personal efforts. They must be ready to make a long-term commitment to the process of personal change. The following illustrates the different characteristics of top management in different types of organizations.

What Is the Problem?

Losers: Employees do not care.

Survivors: Managers are the cause of most problems.

Winners: The process is the cause of the problems.

What Do They Think of Their Employees?

Losers: Employees need to be motivated.

Survivors: Employees need to be trained.

Winners: Employees are not the problem but can be part of the solution.

How Do They Define Customer Requirements?

Losers: We know what they need.

Survivors: Surveys are used to obtain customers' thoughts.

Winners: Customers become part of the design process.

What Is Their Management Style?

Losers: Control through hierarchical organization.

Survivors: First-line and middle managers use participative management.

Winners: Management through a vertical organization; networks rather than hierarchical approach.

What Is Their Primary Business Focus?

Losers: Quarterly profits.

Survivors: Meeting customer expectations.

Winners: Building a strong, lasting organization.

What Is Their Technical Focus?

Losers: Stay with what has been successful.

Survivors: Adapt technologies developed by others.

Winners: Create new technologies and build upon those developed by the organization and by others.

How Is Business Planning Done?

Losers: Done by top management and kept confidential so that competition is unaware of plans.

Survivors: Done by top management and partially communicated to employees.

Winners: Everyone is involved in the planning process and their personal goals are directly related to the business plan.

What Type of Communication Is Used?

Losers: Very good downward, but poor upward communication.

Survivors: Good downward and upward communication.

Winners: Very good downward, upward, horizontal, and customer communication, with a number of ways to bypass potential organizational roadblocks.

How Are Resources Controlled?

Losers: Top management control of resources.

Survivors: Management controls through annual budget.

Winners: Limits set by budget, but line managers and employees evaluate and improve options.

How Do They React to Business Downturns?

Losers: Cut discretionary spending, lay off people, and cut back in R&D and sales.

Survivors: Early retirement programs used to minimize lay-offs.

Winners: Plan for business down-turns so that most of the impact is on temporary employees. Increase training of regular employees and increase sales force during slow periods to provide increased attention on the customer, maximizing the possibility of closing sales.

What Is the Work Environment?

Losers: Cut-throat. Individuals competing with each other to keep their jobs. Decisions pushed up to top management.

Survivors: Decisions made by consensus. Teams used throughout the organization.

Winners: Individual creativity is nurtured and emphasized in an environment of cooperation and sharing. Decisions are made when appropriate by consensus. Other times, input from the affected people is considered and at times, decisions are very dictatorial. The individual circumstance governs the decision-making style used.

How Does Top Management Provide Leadership?

Losers: Support the process by assigning resources.

Survivors: Involved in the process and show their interest by attending meetings and talking about how important improvement is to the organization's success.

Winners: Excited about the process and totally dedicated to it. Aggressively search for ways to further improve the process. Recognized internally and externally as the leader of the improvement activities within the organization.

It's easy to see that there are a lot of choices for top management. Everyone is clamoring for their time. In this environment, you cannot expect top management to embrace the improvement process based on platitudes. What needs to be done is to provide them with hard data that convinces them that it is in their best interest to take a leadership role in the improvement process. When they become believers, the process will begin to hum, since they make up the heart of the critical mass required for your organization to be transformed. As a rule of thumb, the critical mass required to make the desired change occur is the square root of the population.

Why should the top management of your organization agree to lead the improvement process? Well, let's look at Motorola. In just two years, Motorola reduced its annual production costs by $250 million, mostly by eliminating rework and repair. Richard Buetow, a Motorola vice president, explains that this money all goes directly to the bottom line.

If the organization does not feel it is ready for the improvement process, it has three options:

1. Charge ahead anyway. (This is the wrong thing to do.)

2. Do not implement TIM until things get worse, and they will!

3. Take the required time to have the managers determine how TIM can have a positive impact on their part of the organization.

Don't rush through this part of the process. Take time to build understanding and commitment. Identify and work with the skeptics to be sure they have had a chance to have their fears addressed. Minimize your chance of failure and false starts by doing your homework at this time.

Today there are three kinds of organizations: losers, survivors, and winners. When it comes to top management leadership, the following are the characteristics of each.

- *Losers*—Never start the improvement process, or if they do, there are many changes in direction. Direction and focus change as top management change.

- *Survivors*—Top managers talk change, but change very little themselves. They often focus their improvement efforts on just one of the total improvement methodologies. Top management look at the process as a "bottom-up" process.

- *Winners*—Top management not only talk it, but they live it. They set the example. They show the importance of change by changing themselves, and by providing recognition to those who change and stimulate change.

Which best describes your organization's performance?

■ References

1. *Business Week*, Special Quality Issue, 1991, p. 34.
2. Employee Involvement and Total Quality Management, published by Jossey-Bass.
3. *Fortune*, June 1992.

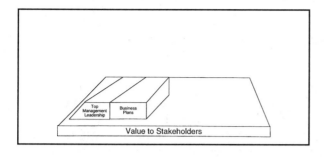

Business Planning Process: Aligning the Organization and Its People

2

Don Yee
Managing Director,
Pacific Management Partners

and

Dr. H. James Harrington
Principal, Ernst & Young

*Management who do not share their action plans with
their employees are like a ship's captain who won't tell
the helmsman what the ship's destination is, but holds
him accountable for setting the course.*
DR. H. JAMES HARRINGTON

■ Introduction

Business plans! Long-range plans! Operating plans! Marketing plans! Financial
plans! Strategic plans! Are we confused yet? No wonder many organizations don't

plan well, don't like to plan, or don't use the plans they do have. However, we do know that with proper planning we can put ourselves in the clearest and best possible position to compete in the ever-increasingly complex world of business (and to take advantage of a few moments of good fortune). We also know that success is not created by using excessively cumbersome planning language and arguing over the semantics of planning, but by the competitive focus it provides to everyone in the organization. This critical alignment enables a business to direct its resources toward common business goals that make a competitive difference. The clarity in a sound business plan will also provide the foundation for the organization's improvement initiatives, whether they be the implementation of a continuous improvement process through total quality management or through more radical business reengineering efforts.

■ What's the Problem; Don't We Plan Enough Already?

All organizations have a plan. It may not be documented and it may not even be formally communicated throughout the organization, but every organization has one. It may be haphazard and it may be arbitrary, but every organization has one. It may be aggressive or it may be passive, but every organization has one. How do we know? We can tell by how we see the most precious resources of the organization (time, people, and money) being allocated, the pattern of decisions being made and the actions being taken by the management. For example, in one organization's situation, their stated mission was to become a global information provider in order to meet the demands of their customers. Yet after three years of implementation, less than five percent of their resources were devoted to this effort, resulting in very little progress in this area. They fell further and further behind in the global market. In addition, many inside the organization realized that their stated strategy was not serious and certainly was not well thought out. In contrast, a high-technology manufacturing organization was faced with an industry slowdown and a temporary slackening in demand for their products. Quality was a cornerstone of their long-term strategy and as such, they continued to invest in this effort while other less critical initiatives were cut. The improvements they made as a result of this commitment enabled them to move significantly ahead in their competitive arena while other organizations fell back. It also served to visibly reinforce the leadership's commitment to this vital strategy.

The real purpose of business planning is to determine the external focus of your business in terms of customers served and value provided, along with identifying the areas in which the organization must excel in order to be successful. A well-crafted plan can provide the road map to success and let every individual know how they can contribute to the results. The planning process itself can be a great contributor to the development of new ideas and competitive insights, develop the process of buy-in and management commitment, and also be an integral part of the communications and mobilization effort.

In this chapter we will summarize the essence of good planning based on research into the planning practices of hundreds of organizations, and hands-on involvement with clients over 15 years of consulting in this area. In the spirit of not worrying about the semantics of planning, we will use "business planning" and "strategic planning" interchangeably. This discussion will be organized by looking at business planning along the following dimensions:

- What's in a good business plan?
- What does an effective planning process include?
- How do you use a business plan?

Along the way, we will also try to share the planning lessons learned from the winners, survivors, and losers as determined by their business results.

■ What's in a Good Business Plan?

There have been hundreds of books written on strategic planning or business planning. When we really look at what's common in all of them in terms of what's in a good plan, we can readily see the following common elements of a good, useful plan:

- A clear focus on specific markets and customers to be served.
- A vision of what the future will be like in terms of the market and the organization's leadership role in it.
- A mission for the organization.
- A clear articulation of the most important strategic initiatives or focus that the organization must undertake.
- Simple measurements or objectives related to the strategic initiatives.
- Well-defined strategies for accomplishing the above, along with the responsible leaders identified.
- Organizational values and culture which will support and enhance the implementation of the chosen strategy.

In fact, most plans can probably be a lot shorter and more precise than they are today. Far too many of us have had the experience of seeing large, voluminous plans that are prepared regularly but are never used. These "dust collectors" often do a reasonable job of describing everything, but do a terrible job of providing focus for the organization. A clear, simple set of strategic decisions for competing in a complex world will result in a core focus which can be easily summarized and communicated throughout the organization. Winners follow the planning rule, "decisions, not descriptions," while the losers tend to follow, "descriptions, not decisions."

There are three main purposes for preparing a business plan. They are:

1. To set direction

2. To establish expectations

3. To define actions

Each of these purposes interrelate and react with each other. Each of these purposes drive a number of outputs that communicates its intent to the stakeholders (see Fig. 2.1).

Market Focus

Clear market focus provides an organization and its people with an opportunity to accomplish great things through understanding the customer's needs better than its competitors. This understanding then provides the foundation for the development of superior products and services. Knowing who you're serving is the first step in developing a good business plan and achieving the ultimate business results that the winners in business experience. The losers more often are so unfocused that they cannot serve anyone well with the demands on their limited resources. The really successful organizations will understand markets in terms of what actual customers want and will pay for, thus reaching a level of understanding few others achieve. As one leading computer manufacturer said, "We feel that one of our significant competitive advantages over other companies is our ability to better understand and anticipate customer needs, whether it's through our sales force or through superior market/customer research." You certainly couldn't argue with their results!

Purposes	11 Outputs (time frame)	
Direction	• Visions	10-20 years
	• Mission	Open-ended
	• Values	Open-ended
	• Strategic focus	5 years
	• Critical success factors	3 years
Expectations (measurements)	• Business objectives	5-10 years
	• Performance goals	1-5 years
Actions	• Strategies	1-5 years
	• Tactics	1-3 years
	• Budgets	1-3 years
	• Performance plans	3-12 months

Figure 2.1. Business planning elements and timing.

■ Setting Direction

The principle role of top management is to set the direction for the organization. This can best be accomplished and communicated to the stakeholders through the business plan. The outputs that are used to provide this direction are:

- *Visions of the Organization.* These are usually prepared by top management and are directed at what the organization's output will be like and/or how it will be used 10 to 20 years in the future.

- *Mission.* The stated reason for the existence of the organization. It is usually prepared by the chief executive officer and seldom changes, normally only when the organization decides to pursue a completely new market.

- *Values.* The basic beliefs that the organization is founded upon, the principles that make up the organization's culture, are often called values. These are prepared by top management. They are rarely changed, because they must be statements that the stakeholders can depend on as being sacred to the organization.

- *Strategic Focus.* These are the key things that will set the organization apart from its competitors over the next five years. This list is defined by top and middle-level managers.

- *Critical Success Factors.* These are the key things that the organization must do exceptionally well to overcome today's problems and the roadblocks to meeting the vision statements.

Visions of the Organization

What's the difference between a *vision* and a *mission* for an organization? Or with the *purpose* of the organization? Many organizations go through an agonizing process of trying to determine the distinction based on vague definitions offered by consultants and academics. Could it be *vision and mission?*

Instead of trying to split the "definitional hairs," I have found that a useful perspective can be gained by utilizing the approach of linking your organization's internal efforts with the external world in which you compete and serve customers. In this regard, having a compelling vision of the future of the market and the industry is another vital ingredient in a business plan. By "vision" we mean a view of what the business will be like 10 to 20 years from now. It could be as simple as "an affordable, easy-to-use personal computer on everyone's desk" or "news available immediately from anywhere in the world." The winners tend to be able to express an energizing picture of the future in terms of market presence and customer benefits and have enough reality to it to make it aggressively believable. The losers tend to lack any vision and exist from day to day reacting to the market and the leads of other competitors.

Mission

The mission statement is essential to linking the organization with its vision of the future. Some organizations call this their "purpose statement" or the central reason why they are in business. A good mission statement will require leadership and be externally focused with customers in mind and will serve as a motivating "to be" or "to do." Either type of mission statement can be effective, so the academics' or consultants' arguments over which one is correct are unnecessary.

An example of a "to be" mission from Boeing is:

- Our long-range mission is to be the number one aerospace company in the world, and among the premier industrial firms, as measured by quality, profitability and growth.

An example of a "to do" mission from McDonald's is:

- To satisfy the world's appetite for good food, well-served, at a price people can afford.

Winners make their missions short, clear, and compelling, while losers will have missions focused on shareholder value or some other noncustomer, noncompetitive emphasis. In addition, winners embody a strong leadership emphasis in their planning, as reported in a study of almost 300 companies in Ernst & Young's "American Competitiveness Study." This emphasis on being a leader is essential to both the development of strategy and to motivating the organization's people, customers and suppliers by focusing on being a winner and not a follower.

Values

Values can be defined as the deeply ingrained operating rules or guiding beliefs of an organization. Some may see them as the specific cultural attributes that drive behavior. The winning organizations set out to create a specific culture and operating style to further define their strategic change and focus.

Owens/Corning Fiberglas uses "guiding principles" in place of "values." Their guiding principles are:

- Customers are the focus of everything we do.
- People are the source of our competitive strength.
- Involvement and teamwork is our method of operation.
- Continuous improvement is essential to our success.
- Open, two-way communication is essential to the improvement process and our mission.
- Suppliers are team members.
- Profitability is the ultimate measure of our efficiency in serving our customer's needs.

Call them basic beliefs, guiding principles, operating rules. Call them what you will. The important thing is that they must be defined, and the organization must live up to them, for they surely are the "Stakeholders' Bill of Rights."

In contrast, losing organizations tend not to have explicit values or "hollow" values, or perfectly stated values that no one operates by.

Strategic Focus

Organizations which have been successful, based in part on good planning (there's still the need to execute, which will be discussed later as well as the element of good fortune), know that they must provide the organization and its employees a road map to help "translate" the vision and mission into "things people can do." The next most critical element of what's in a good plan is the strategic focus of the organization in terms of "how" it will compete. Over the past several years there has been a lot of leading-edge thinking and many definitions put forth of how these areas of emphasis might be viewed and what they might be called, including:

- *Core Competencies.* The technologies and production skills that underlie an organization's products or services (e.g., Sony's skill at miniaturization).
- *Core Capabilities.* The business processes that visibly provide value to the customer (e.g., Honda's dealer management processes).
- *Strategic Excellence Positions.* Unique and distinctive capabilities that are valued by the customer and provide a basis for competitive advantage (e.g., Avon's distribution system).

These leading definitions are actually quite compatible, and debate over which is proper is not time particularly well spent for most organizations. Regardless of the definitions applied, what is common among all the successful organizations is their ability to identify those four or five key areas of strategic focus that are characterized by the following:

- Customers value the benefits that the focus provides.
- Concentration of resources toward being the absolute best in your chosen areas of emphasis will enable you to excel.
- Excellence in these areas will be difficult for competitors to imitate.
- These areas of focus are your organization's capabilities or what you're really good at, not outcome measures like market share, profit margin, etc.

The clarity provided by having a few key goals to focus on can help set the foundation for dramatic improvements in business results. For example, years ago, Hewlett-Packard established the lowering of their product failure rate as one of their key business goals. They also coupled this goal with a very clear objective or measurement, with the specific target being a tenfold improvement in results. Their continuing success in a number of rapidly changing markets speaks for itself.

Critical Success Factors—Obstacles to Success

Planning with a leadership emphasis requires a focus much beyond where you are today. Pushing your vision into the future often requires "thinking out of the box" or an unconstrained strategic perspective. The winning planning processes link this back into today's reality, by specifically focusing on the obstacles to success or things that we can affect and know would prevent our implementation of the plan. These obstacles can range from the lack of sufficient funding to excessive organizational layers. The point here is to better link the vision to today's starting reality, and being honest enough with yourself to highlight and correct the obstacles. This often creates several additional strategies to be incorporated into the plan. Winning organizations will attack this with a positive attitude, and all the others will use it to tell themselves they can't do something. Many organizations will translate the obstacles to success (things in your control which prevent you from successfully implementing the plan) into *critical success factors* (things that must go right in order to succeed). This positive transition can help set the proper winning tone on the challenges of moving forward.

■ Establishing Expectations (Measurements)

One of the major purposes of a business plan is to define what management and the stockholders expect from the organization's performance over the next 5 to 10 years, then to communicate how success will be measured. The outputs that are used to communicate these expectations are:

- *Business Objectives.* Business objectives are used to define what the organization wishes to accomplish over the next 5 to 10 years.

- *Performance Goals.* Performance goals are used to quantify the results that will be obtained if the business objectives are satisfactorily met.

How often have you heard the saying, "Anything worth having is worth working for!"? So it goes with business or strategic planning, but how can we tell if our efforts are paying off? We can surely see sales, profits and cash flow, all tangible measures of current business results. Both winners and losers utilize these very traditional measures. However, the winners tend to also track some very simple measures or objectives related to their progress in developing long-term competitive strength in their chosen areas of strategic focus. For example, if unsurpassed service is one of the areas of strategic focus, the management of the winners would have regular reporting on a few simple measures of unsurpassed service. These measures might even include things like the number of customer referrals, in addition to measures like customer problem resolution cycle times and the customer retention rate. This linking of very visible, regularly communicated measures to the strategic focus of the organization is crucial to the transition process required to make that vision become a reality. Losers tend to only rely on traditional financial measures to gauge their success.

Another reason why selecting and defining measures is important is their value in informing the organization of the pace of change and implementation required. Organizations which expect quantum leaps in performance against measures will need to prioritize and focus their resources in support of these stretch targets. Winners exhibit a keen flexibility and an intuitive sense of how far to stretch, and seldom set the easy-to-beat or absolutely unrealistic targets that losers tend to use. As we saw with Hewlett-Packard when John Young set aggressive targets (e.g., 10 times improvement in a critical business process within 10 years), they will use this to substantially reprioritize their resources and to create a compelling challenge for their organization to rally around.

Business Objectives

Business objectives set the direction for the organization over a period of time. They are a well-publicized set of objectives that provide management and the employees with information related to what the organization wants to accomplish in the next 5 to 10 years.

At the beginning of the 1980s, IBM released the following objectives that it planned to accomplish during a 10-year period.

- To grow with the industry.
- To exhibit product leadership across our entire product line. To excel in technology, value, and quality.
- To be the most effective in everything we do. To be the low-cost producer, the low-cost seller, the low-cost administrator.
- To sustain our profitability, which funds our growth.

At first, IBM measured themselves based upon these business objectives, reporting their progress to the world through their *Think* magazine. As IBM progressed through the 1980s, they lost sight of these business objectives and stopped reporting progress or lack of progress against these objectives. The lack of progress led to the problems they had and are having in the 1990s. Good business objectives not followed, measured and reported, are just scraps of paper that do no one any good.

Business objectives should be very aggressive. They should set a challenge for the total organization. No one feels good about making an easy objective, but we all feel great when we accomplish something that even we thought was out of our reach.

Motorola set "stretch" objectives for itself. William J. Weisz, chief operating officer of Motorola, explained,

> In 1981, we developed as one of the top ten goals of the company, the Five-Year, Tenfold Improvement Program. This means that no matter what operation you're in, no matter what your present level of quality performance, whether you are a service organization or a manufacturing area, it is our goal to have you improve that level by an order of magnitude in five years.

By 1986, Bill Weisz was talking about another tenfold improvement, but accomplishing it in just three years. The result of these aggressive objectives has increased Motorola's markets in Japan and the United States, and resulted in their winning the Malcolm Baldrige National Quality Award.

When a president of an organization was told by one of his vice presidents that there was no way they could improve 10 times in just five years doing business as they are—maybe 10 to 20 percent per year, but not 200 percent per year—the president answered, "You are right. You can't improve that much doing business as you *are*. You got the message!"

Compare this aggressive type of thinking to the objectives set by the U.S. government. On February 25, 1986, then President of the United States Ronald Reagan released Executive Order 12552, stating:

> There is hereby established a government-wide program to improve the quality, timeliness, and efficiency of services provided by the federal government. The goal of the program shall be to improve the quality and timeliness of service to the public, and to achieve a 20 percent productivity increase in appropriate functions by 1992. Each executive department and agency will be responsible for contributing to the achievement of this goal.

The government's objective was a five-year time interval to improve 20 percent. That's the type of management direction that will lead an organization into bankruptcy. For only an organization that can print its own money can exist when its debt is 50 percent of the Gross National Product.

Business objectives are developed and published by top management. They are subject to change, as the business climate changes, and as specific objectives are met.

Performance Goals

Performance goals can take the form of short- and long-range targets that support the business objectives. They should be quantifiable, measurable, and time-related. (Example: Increase sales at a minimum rate of X percent per year from 1995 to 2010, with an average annual growth rate of Y percent.) A typical long-range performance goal would be to decrease the cost of maintaining customer-purchased equipment at a minimum rate of 10 percent per year for the next five years; or to correct 99.7 percent of all customer problems with one service call per customer over the next 24 months. Each year, a set of short-range goals should be generated by first-line and middle management, directly tied into their budgets. These goals should be reviewed and approved by top management, to be sure they support the business objectives and are aggressive enough.

> *Note: Goals have two key ingredients. First, they specifically state the target for improvement; and second, they give the time interval in which the improvement will be accomplished.*

■ Defining Actions

Another purpose of a business plan that drives the organization's change process is to define the actions that will be taken to implement the plan over the next five years. It is designed to focus the resources of the organization in line with its expectations. The outputs that are used to communicate these actions are:

- *Strategies.* Strategies define the approaches that will be used to meet the performance goals.

- *Tactics.* Tactics define how the strategies will be implemented. They explain how the strategies will be accomplished.
- *Budgets.* Budgets provide the resources required to implement the tactics.
- *Performance Plans.* Performance plans are contracts between management and the employees that define the employees' roles in accomplishing the tactics, and the budget limitations that the employees have placed upon them.

Strategies

Once clear targets have been identified, a set of strategies must be decided on to further the organization's efforts. Strategies are defined here as specific programs, initiatives, and decisions which will require resources allocated to them. They can range from the development of strategic alliances to developing and conducting special in-house training for customer service. These operating strategies are usually very specific since the good plans are very clear and focused. Occasionally there may be some strategies that are critical and yet don't get openly shared, particularly since they may be extremely sensitive and widespread knowledge of them would put you at a competitive disadvantage (e.g., mergers and acquisitions and dispositions of certain business units). Once the strategies are identified, then lead responsibilities should also be assigned in order to clarify the implementation steps and create accountability in the organization.

The strategies document the approach that will be used to meet the performance goals. They are generated by middle management and approved by top management. Every effort should be made to keep the strategies up to date, without making major changes. Drastic changes in the strategies upset the organization and require a major expenditure of resources in reacting to those changes. Major changes can also result in the termination of projects that are only partially completed and/or have not become totally effective. It should be apparent that many strategies are generated by many different functions, supporting the business objectives.

Tactics

Tactics are the "how-to's." They are the actions that are planned to be taken or that are being taken to meet the performance goals. Tactics are generated by employees and first-line management, and are approved by middle and top management. They are updated at least once a year and change frequently, based upon experience and business needs. Employees in the first-line departments are encouraged to participate in the preparation of the tactics, since they will eventually be responsible for implementing them.

Budgets

Management has been making budgets since the beginning of time. Many managers spend more time preparing and defending budgets than they do providing career counseling to the employees reporting to them. Too many managers prepare their budgets by adding X percent for inflation, plus enough more to cover any new programs that are coming down the road. That's the type of reasoning that has caused our organizations to use 50 percent more resources than necessary.

Budgets should always be based on the tasks that are to be performed, as defined by the business plan. All other activities should be justified on an individual basis. In all cases, a productivity improvement projection should be incorporated. In most cases, the productivity improvement index should reduce employee costs by a conservative 10 percent. Budgets should be very specific for the next 12 months, and in detail for the following two years. This helps align each individual department with the tactics and strategies agreed to by management.

Performance Plans—Linking the Plan to Individual Goals

One of the key challenges facing any management team is the design and implementation of a process to link strategic business goals to individual performance. This is perhaps one the of most vital and yet most misunderstood processes in any organization. There have been a variety of methods to accomplish this and they do certainly have some common features such as simplicity, meaningful job content, and an enlightened management. In some organization-wide efforts, the improvement process has often taken on the role of breaking the plan down into the lowest level of detail as it gets deployed. In these situations, all departments will create a mini-business plan with a mission, goals and strategies linked into the organization's overall business plan. In managing the various departments, individual performance goals would then be set (usually mutually) with a much more direct connection to the business plan. Reward systems are then aligned with these individual performance goals.

Many organizations are using a concept called "Quality Policy Deployment" to help align the organization's business plan with the individual performance plan. In many of these organizations, every individual will have a card that is kept with them that states the organization's critical success factors for the specific year, and on the reverse side, the improvements that the employee will be making to support these critical success factors.

■ What's in an Effective Planning Process?

Good strategic thinking doesn't really have a definite starting and ending point, particularly in today's everchanging business environment. As pointed out earlier, the winners really try to gain a significant advantage by focusing on being a leader and creating part of the change in their industry as opposed to always reacting to it. In this regard, defining an appropriate planning process is critical. Today there are as many planning processes as there are planning formats and tools. Some are very long, elaborate affairs, and some are quicker-hitting events. The length is influenced by factors such as the current position of the organization (are they the leader or are they in trouble), the style of the management team, and time and scheduling. However, the most successful planning processes have many attributes in common. This even holds true independent of the form the plan takes, although they are essential to the success of developing the plan content discussed earlier. These attributes can be summarized as follows:

- In-depth consideration of customers, competitors and capabilities
- Involvement of implementors as planners
- Identification of obstacles to success and related critical success factors
- Testing against resource availability (time, money, and people)
- Development of specific communications plans
- Consensus on how the plan will be used in the implementation efforts

Customers, Competitors, and Capabilities

Just like there are many different forms that a business plan can take, there are a like number of planning processes and supporting analytical tools or frameworks. We have all seen the "how to" page-by-page workbooks related to business planning and the various environmental and competitive assessment frameworks; e.g., SWOT (Strengths, Weaknesses, Opportunities, and Threats) Analysis, or the Five Forces Assessment for industry analysis which focuses on understanding customers/buyers, suppliers, current competitors, new competitor entrants, and substitute products. All these tools are helpful and their contribution can be significant, particularly if they help to organize and expand your thinking. Regardless of which one you use, just remember that a good planning process will incorporate a serious effort to really understand your *customers, competitors, and capabilities* in the context of looking objectively at the future. Only then can you truly identify what will make you distinctive and valuable from the market's viewpoint and scrutiny.

With respect to customers, winning organizations continually seek new insights into their future product and service needs, buying criteria, and satisfaction levels. Losing organizations look at customers as "someone to sell to now." The planning process often incorporates primary (direct with customers) and secondary (independent third party studies) market research. The winners go even further and often find ways to organize and utilize the market knowledge of their sales force or find other means to get direct customer input, such as through field visits or market research.

Competitor analysis in many planning processes is often relegated to getting statistics on their financial performance. In many cases the losers don't even know much about their competitors. This lack of true understanding of your competitors' strengths, capabilities, motivations, and intentions is a severe handicap in strategic planning. The winning organizations often do more innovative things, including exhaustive analysis of all public communications by a competitor and doing internal role playing as if they were the competitors. This level of effort is deemed to be required in order to truly get competitive insights into the opportunities for the future.

One of the biggest challenges in the planning process is to do an honest, objective assessment of your organization's capabilities. Can you answer "What are you really good at?" Even if you could, would you be able to match this against customer needs to help determine if there is sufficient value in it? This assessment, if done objectively as the winners do it, can provide excellent insight into your competitive strengths and weaknesses.

Implementors as Planners

The winners know that even the best plan, not implemented, is useless. They work very hard to develop a useful planning process which requires that the management team responsible for its execution is involved in the development of the plan. This not only taps into their experience but it improves both the level of buy-in and the understanding of the organization's direction and initiatives. Although consensus is desired, the winners know that unanimity is seldom possible, but understanding and support is. In contrast, the losers don't involve line management and prepare the plan in a vacuum. They run the risk of both making poor strategic decisions and having poor execution. No wonder they're the losers!

Testing Against Resource Availability

One very important dimension of the potential obstacles deserves highlighting here. In many instances, strategic or business plans are created without specific consideration of how they link into the resource or financial and organizational planning efforts of the organization. At a minimum, the better planning efforts will include a related financial assessment or projection, not just of the desired results (which is always more fun and exciting to dream about) but the level and timing of the financial resources required (which often brings it all back to the practical side of planning). The same is true for assessing and identifying the adequacy of the people resources in terms of skill and amount of resources. These efforts do not always have to result in a demand for more resources, but can result in a much smarter redeployment of resources and/or the "freeing up" of resources through the elimination of unnecessary efforts and initiatives.

One particular organization had decided that they had to move aggressively into a direct sales and account relationship program, given the needs of their customers and the leadership position they wanted to establish in their market. Upon bemoaning their lack of resources to do this, one of the planning participants simply asked if the millions of dollars being spent on advertising was necessary, given the new approach to sales. After a bit of discussion, the team determined that a great deal of resources could indeed be redirected toward the implementation of this shift in strategy.

Communications

A planning process would be woefully inadequate without considering the communications plan for sharing your strategy with your organization and any external constituencies like your customers and suppliers. While certain aspects of strategy are indeed proprietary (e.g., specific R&D or mergers and acquisitions), most elements cannot be kept a secret only for management to know and for everyone else to guess at. How can you excite the organization if you don't tell them? Broadening the communications is usually not a problem once you realize that superior execution is a definite strategic advantage, and that the winner between two organizations with the identical strategy is the one that outexecutes the other.

This emphasis on communications and creating a widespread understanding of the strategic plan by people inside and outside the organization was recently proven to be one of the most universal "best practices" of successful organizations in the "International Quality Study" conducted by Ernst & Young and the American Quality Foundation. This study found that in particular, increasing middle management's understanding of the strategic plan had a very positive impact on the profits of the organization and its industry position.

Consensus on Implementation

Winning organizations work very hard to use the results of their business planning efforts in managing their implementation efforts. They always decide during the planning process how they will use the plan in managing their implementation efforts. A commitment to consistently reviewing the plan and its targeted measures is normally a hallmark of the best organizations. This very explicit effort goes a long way toward eliminating the dreaded "dusty business plan on a shelf" usually found in losing organizations. The losing organizations just give "lip service" and excuses for not implementing the plan and meeting the targeted measures.

■ How Do You Use a Business Plan?

A sound business plan provides an organization and its people both focus and freedom. Focus comes from the clarity of your customers and their needs, an understanding of the most critical strategic initiatives you must pursue in order to be a competitive leader, and consensus on the appropriate measures and strategies. The freedom really comes from the autonomy individuals will have during the implementation process and the new implementation ideas which will surface from throughout the organization. This unleashing of the energy of an organization is one of the most powerful potential outcomes of an effective planning and ongoing strategic management effort.

As soon as a plan is completed it has the chance of becoming outdated. In addition, the positive commitment gained during successful planning efforts can also erode during the initial stages of implementation. Managing to the plan and adapting it as you move forward is essential to realizing the opportunities your business has. Some of the most critical features of successful strategic management include:

- Visible leadership commitment

- Regular reporting and sharing of results
- Widespread sharing of the plan throughout the organization
- Well-understood linkages between individual performance and business success
- Adaptation from experience

Visible Leadership Commitment

While this point is often noted in discussions regarding new projects, markets, etc., it is perhaps even more vital here from an impact perspective since an organization's business plan should have a lot of meaning to all employees.

Regular Sharing of Results

Winning organizations tend to share the results of their efforts with their employees. What is normally shared at these organizations not only includes the basic financial results but also the key measurements of their strategic emphasis.

Widespread Sharing of the Plan

As noted earlier in the discussion, developing a communications plan is an essential part of the planning process. One of the more controversial areas is deciding on what gets shared broadly and what gets more limited distribution. As a general rule, it is just impractical to ask people throughout the organization to do the truly great things that winning organizations do if you don't share your vision, mission, and strategic priorities with them.

■ Adaptation from Experience

Smart management teams know that while having a road map is vital to success, experience is also a good teacher. This is especially true when it comes to implementing ideas and experience. The winning organizations have found that by having regular forums on both the results and progress, they will be able to share the lessons learned during implementation as well as new ideas on how to do things better. During these sessions, management teams have the delicate task of balancing between "staying the course" and modifying some of the initial strategies. There are no rules for this, but winning organizations know that you can't really determine if an effort is successful until you've given it a full commitment and learned from the results. This level of discipline serves them well when they consider possible changes required in the plan's strategies.

■ Summary

Here is a quick summary of what the winners and losers do (or don't do) in their business planning efforts. The survivors fall in the middle and they often try to emulate the winners, but without any real conviction.

	Winners	Survivors	Losers
The business plan			
	Market focus	Clear and targeted	Widespread
	Vision	External and compelling	None
	Mission	Customer-focused and leading	Internal
	Strategic focus	Limited, high-impact areas	Unfocused
	Measurements	Simple, linked to strategy	Financial only
	Strategies	Critical initiatives	Broad list of to do's
	Values	Explicit and real	Hollow
The planning process			
	Customers, competitors, and capabilities	In-depth knowledge	Superficial
	Implementors	Involved in planning	Top down only
	Obstacles to success	Considered and solved	Ignored
	Resource availability	Priorities set	Constraint only
	Communications	Throughout organization	Limited
	Implementation	Consensus achieved	Unaddressed
Using the plan			
	Leadership commitment	Visible	Counter-productive
	Reporting on results	Regular	Inconsistent
	Sharing the plan	Widespread	Limited
	Individual goals	Linked with plan	No linkage
	Adaptation from experience	Plan as guide	Plan ignored

By following the planning lessons learned from the winning organizations, you can help yourself become or remain a leader in your industry. While good business planning does not guarantee success, it goes a long way toward ensuring a common focus throughout the organization, thereby maximizing utilization of resources and total performance. Plan for success or rush into failure. The choice is yours.

The nice thing about not having a business plan is that you can't be off course if you don't know where you are going.   DR. H. JAMES HARRINGTON

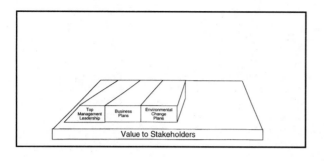

Value to Stakeholders

3

Environmental Change Plans: Best Practices for Improvement Planning and Implementation

Dr. H. James Harrington
Principal, Ernst & Young

Mark B. Hefner
National Director, Ernst & Young

and

C. Keith Cox
Staff Consultant, Ernst & Young

The lack of good long-range improvement plans has wasted billions of dollars. Today more and more organizations are applying the five Ps—Proper Planning Prevents Poor Performance.

DR. H. JAMES. HARRINGTON

■ Introduction

Everyone is talking about the need for a cultural change, but we believe that focusing on it is the wrong answer to today's problems and does not prepare most organizations to enter the twenty-first century. Culture is defined as your background, your history, your heritage, your religion, your beliefs. Most organizations want to hold on to their culture and, in fact, are worried about losing it. America should be proud of its culture. It is a culture rich in imagination, hard work, caring, risk taking, and accomplishment. It is a culture that made us the richest, most powerful, most productive nation in the world. Our culture is not the problem. It is the personality of today's population that is the problem. We talk about "workaholics" like "work" is the worst four-letter word in the English language. People work overtime begrudgingly if they are notified 72 hours in advance, and, if not, they refuse. It is the personality of today's work force and our children that needs to be changed. Personality is defined as an individual or group's impact on other individuals or groups. We need to change the personality of our people before we lose the culture that our forefathers worked so hard to create. It is the personality of our key managers that dictates the personality of the total organization. When a new CEO is appointed, the total organization adapts to his or her personality. If he or she is a basketball fan, you would be surprised at how many people know last night's basketball scores.

We cannot go back to what we used to do, for this old world has changed. The amount of information available to the individual doubles every five years. According to Richard Worman in "Information Anxiety," the weekday edition of *The New York Times* contains more information than the average person was likely to come across in a lifetime in seventeenth century England.

How do we create a change in the personality of our people? We do that by changing the environment that they live in. Our personality is molded by an ongoing series of environmental impacts that we are subjected to. It starts at birth and ends with our last heartbeat. The biggest impact occurs during the formative years of a child's life—the period before they enter school.

Let's focus on what we can do to influence and change the environment that impacts the personality of today's work force. What the organization needs is to develop a plan that will change the environmental factors that impact the personality of the employees, placing special emphasis on the management team. If we sustain a positive change in the personality of the organization for a long enough period of time (about five years), we will change the organization's culture.

■ Business Plans versus Environmental Change Plans

There is a big difference between a business plan and an environmental change plan. The business plan sets the product and service strategy for the organization—the markets that they hope to penetrate, the new products that will be introduced, the production strategy, etc. It is a plan that directs and guides the business. The business plan is primarily directed at meeting the needs of only two of the stake-holders associated with the organization—the customer and the stockholder. It is a plan that is primarily focused on the external opportunities. The environmental change plan, on the other hand, is an internally focused plan that is designed to transform the environment within the organization to change its personality (be-havioral characteristics) to be in line with the business plan. It takes into considera-tion the needs of all of the organization's stakeholders from an improvement standpoint. The environmental change plan defines the transformation in the busi-ness personality of the organization. It provides an orderly passage from one state or condition to another. The environmental change plan supports the business plan so the two of them, although different in content and intent, must be kept in close harmony.

■ Why Do You Need an Environmental Change Plan?

"Why does my organization need to develop an environmental change plan to improve? I know a lot of problems that we can start working on right now. In fact, we are already working on them." Don't stop working on your problems. You cannot afford to stop putting out the fires that your present organization is fueling, but as long as you continue to do what you have been doing, you will continue to get the results that you have been getting. Unfortunately, probably your competition is not content with their situation and they are changing, so if you don't change they will improve their competitive position. In fact, the twenty-first century will be dominated by those organizations that improve the most and can change the fastest.

Most organizations, in their hunger to improve their relative performance, have embraced many different improvement tools. It seems like each time someone went to another conference they came back with another improvement tool. Our research shows that more than 400 different improvement tools exist that will provide a positive impact on the environment. Each of these tools work under the right conditions. Many solve the same type of problems. The following is a list of 10 of the more than 44 different tools used to improve the environmental category called Management Leadership and Support.

1. Management Self Audits
2. New Performance Standards
3. Improvement Policy
4. Improvement Visions
5. Annual Strategic Improvement Plans
6. Leadership Skills Development
7. Self-Managed Work Teams
8. Responsibility Charting

9. 7-S Model

10. Risk/Opportunity Management

No organization can afford or effectively utilize all 400 improvement tools. Many of the tools overlap in approach and the problems that they solve. Many of them are not applicable or have little impact upon your organization.

Just as individuals differ, organizations differ in many ways. They have different management personalities, customers, products, cultures, locations, profits, quality levels, productivity, technologies, and core competencies.

Add to this complexity, the fact that winning, surviving and losing organizations have to do a very different set of things to improve. It becomes readily apparent that there is no one approach to improvement that is correct for all organizations or even for different sites within an organization.

■ What Creates Your Organization's Culture?

The organization's culture is created over a long period of time as a result of the way management implements the organization's basic beliefs (discussed in Chap. 2), combined with the way the employees react to management's stimulation.

There is a great deal of confusion about which term or terms to use: *values, beliefs,* or *principles.* I don't care what you call them, but every organization should have a set of statements that communicate to management and the employees an understanding of what the organization's culture is based upon. These statements provide direction to management that governs their performance. To the employee, they provide a promise of conditions that the organization is built upon.

Ford Motor Company's guiding principles are:

- Quality comes first.

- Customers are the focus of everything we do.

- Continuous improvement is essential to our success.

- Employee involvement is our way of life.

- Dealers and suppliers are our partners.

- Integrity is never compromised.

The following is a quote from IBM's Managers Manual:

An organization, like an individual, must be built on a bedrock of sound beliefs if it is to survive and succeed. It must stand by these beliefs in conducting its business. Every manager must live by these beliefs in making decisions and in taking action.

The U.S. Government has had a set of basic principles since the beginning. They are called the "Bill of Rights" and they have helped guide the United States for more than 200 years.

These fundamental statements that we will call basic beliefs are the things that attract new employees to your organization. They define the rules that the organization will not compromise. They are the employees' "bill of rights" so they should be changed infrequently, and then only when they have become obsolete because of social and/or external environment. I worked for IBM for 40 years and their basic beliefs remained constant throughout that time period.

Management's job is to live every business minute complying with the organization's basic beliefs. This is the first obligation of every manager. No employee should accept a management position unless they have already lived up to these basic beliefs, and no manager should be left in a management role unless they live up to these basic beliefs. Employees have a responsibility to work in support of the basic beliefs, refusing to compromise them. Management who ask their employees to bend the basic beliefs are not doing their job, and this condition should be brought to the attention of upper management.

■ Assessment of Today's Personalities

In developing our improvement plan we need to define the "as is" status of the organization related to:

- Competitive position
- Core competencies
- Core capabilities
- Basic beliefs
- Customer satisfaction
- Employee satisfaction
- Quality management systems
- Successful and unsuccessful improvement activities
- The organization's commitment to improve
- Major new programs

To do a complete assessment can be very time consuming and expensive, but when you take a look at the alternatives it usually is less expensive in the long run. If the organization is relatively advanced, most of this data should be readily available within the organization. If this is the case, it is best to take the time to do a comprehensive analysis and define any voids in the "as is" organizational structure. The advanced organizations will typically already have benchmarked their competitors' products, defined and compared their core capabilities and competencies, calculated the customer satisfaction index, hold focus groups with customers, conduct employee opinion surveys, and have a business plan that is actively being followed.

If this is not the case, a minimum analysis may be the right answer for your organization rather than completing a more thorough analysis in preparation for the three-year plan. This allows the organization to start the improvement process and modify the plan as additional data is made available. At a minimum, the assessment would include the following:

- Review of customer-related data
- Review of competitor-related data
- Review of the business plan and critical success factors
- Review of upper management measurements and performance
- Review of the organization's measurements and performance
- Review of quality systems audit data
- Review of past and planned-for improvement activities
- Review of grievances
- Private interviews with all top management to identify potential improvement opportunities and/or problems.
- Focus group meetings with middle management, first-level management, and employees to define the present personality of the organization, compliance to business principles, and to identify roadblocks to change.
- Perform a quality management systems audit
- Conduct an improvement needs survey

This type of assessment will provide a good definition of the "as is" status. It will also identify many improvement opportunities that should be included in the three-year improvement plan. In addition, the focus groups and the surveys will identify the differences in management and employee priorities. Some surveys provide a measure of the organization's dedication to improvement that can be used to decide the probability of successfully implementing the improvement process. We find that this approach is very useful and greatly decreases the chance of failure.

It is very important that the data collected during the private interviews, focus groups and surveys be kept absolutely confidential. This is one of the best reasons for using a third party to do the assessment. The third party also provides an unprejudiced view of the organization which is needed if the true "as is" picture is to be defined. When individuals live in a problem situation for a long period of time, they become used to it and begin to consider it normal, rather than the exception that it really is. Do not rely only on the people who created the organization's personality and problems to define improvement opportunities.

Once the assessment is complete, the results should be reported back to management and, at a minimum, to every employee that took part in the assessment. If you ask anyone for their suggestions and/or to evaluate present status, you have an obligation to get back with them to show them how their data was used and reported to upper management.

■ How Do You Change an Organization's Personality?

Change is difficult under the very best circumstances. Western nations' change of activities has been primarily driven out of fear—fear of losing market share, fear of losing jobs, fear of not making enough profit to keep stock prices up, and fear of failing. Eastern nations, on the other hand, are changing because they see an opportunity. It is too bad that most U.S. organizations had to get into trouble before they got serious about improvement. For example,

- Xerox and General Motors lost more than 30 percent of their market.

- IBM began posting record-breaking financial losses.

- The U.S. government got so far in debt that there seemed to be no way out.

All organizations need to have an improvement plan, not just the losers. It's too bad that our leaders have been so slow in realizing this. It has to be discouraging to the nation as a whole when the U.S. government is just starting to talk about eliminating a quarter of a million jobs by improving its effectiveness, when the potential has been there since the early 1970s. Even Japan is looking at increased focus on improvement as corporations start downsizing. (Example: Toshiba eliminating 5000 jobs, and Fuji eliminating 6000 jobs.)

What needs to be done to make improvement possible and long lasting? What needs to be done to bring about a change in the way we think, the way we talk, and the way we act? The following is an effective model for change.

1. Everyone must feel that change is necessary.
2. There has to be a common vision of how the change will affect the organization's environment.
3. Everyone should feel ownership in the improvement plan.
4. Management must change first and be the model for the rest of the organization.
5. Barriers must be broken down and removed by management.
6. The impact of the change must be openly communicated to each of the stakeholders.
7. Everyone needs to be trained to perform well in the new environment.
8. A measurement and feedback system needs to be put in place.
9. A risk-taking environment must be provided.
10. Directing and coaching must give way to leading.
11. Desired behaviors must be rewarded.

■ Organized Labor Involvement

It is strongly recommended that if some of the employees in the organization are represented by organized labor that the union leaders get involved as early as possible. We recommend that the appropriate union leaders became active partners in developing the environmental vision statements and the plan that will transform the organization. This will help to align the organized labor's goal with the environmental vision statements. Early involvement of these key people in the planning process often slows down the process a little, but in the long run much time will be saved as the plan will be much better and much more effectively implemented.

■ Establish Environmental Vision Statements

Management has control over relatively few things. They do not control the economy, their customer, their competition, their suppliers, government regulations, the stock market, etc. The only things that management can change are the environmental processes that they control. If you want to bring about change in the organization, the environmental processes within the organization that impact the desired results must be changed.

The Executive Improvement Team (EIT) should develop the visions. This EIT should be made up of the highest officials in the organization (example: president and all vice presidents) plus the key union leaders. The EIT should be limited to about 8 to 14 people.

The EIT now needs to define the five to ten environmental processes that have the most impact upon the organization's performance. Typical processes that impact organizational performance are:

1. Measurement processes

2. Training processes

3. Management processes

4. Customer partnership processes

5. Supplier partnership processes

6. Business processes

7. Production processes

8. Employee processes

9. R&D processes

10. After-sales service processes

The Executive Improvement Team (EIT) should set aside two or three days to define the key environmental processes and develop a set of preliminary visions of how these processes will evolve during the next five years. To accomplish this, the meetings should be held off-site in a sterile environment where no one is interrupted by phone calls. The time during the day discussing visions is important, but equally important are the evening activities where the executive team gets to socialize and informally interact with each other.

The EIT should define the key environmental processes that impact the organization's performance by brainstorming to make a comprehensive list, and then consolidating it and prioritizing it down to a maximum of 10. Once the key environmental processes are defined, the EIT should review the "as is" analysis and define the present state of each of the processes. Then they should discuss each process to define if it needs to be improved over the next five years. If the process needs to be improved (and most of them will), the EIT should define how it should change.

The EIT should think beyond the present boundaries and define the future desired state. After the EIT has brainstormed a list of phrases that best represents the future state of the specific environmental processes they are discussing, they should prepare a vision statement for the specific environmental process. This five-year vision statement should clearly define the desired way the organization is operating five years from now.

Jack Welch, CEO of General Electric, says, "Leaders—and take everyone from Roosevelt to Churchill to Reagan, inspire people with clear visions of how things can be done better. Some managers, on the other hand, muddle things with pointless complexity and detail. They acquaint it with sophistication, with sounding smarter than anyone else. They inspire no one." Father Theodore Hesburgh, former President of Notre Dame University, stated, "The very essence of leadership is that you have to have a vision. It's got to be a vision you articulate clearly and forcefully on every occasion. You cannot blow an uncertain trumpet."

A typical example of a five-year vision statement for management support and leadership is:

> Management fosters an environment of open communication where opinions and suggestions are encouraged and valued: visions, plans, and priorities are shared throughout the organization.
>
> Management provides the necessary time, tools, and training for employees, which enables everyone to contribute their personal best toward the mission of the organization.
>
> Teamwork is stressed: decision-making is accomplished at the lowest appropriate level. Bidirectional feedback occurs on an ongoing basis to measure results and provide input for a continuous improvement process.

Stakeholders' Involvement with Vision Statements

The EIT develops the preliminary vision statements. These statements reflect the way management interprets the data they have and the way they picture the evolution of the organization's environment. But management is only a small part of the people who are affected by these vision statements. There are three more stakeholders who also need to influence these vision statements. They are: (1) the customer, (2) the employee, and (3) the suppliers.

Each of the executives should take the preliminary vision statements back to their organization and hold a series of focus group meetings with their direct reports, first-line managers, and employees. Each focus group should review all the vision statements to determine:

1. Is this the type of environment you want to live in?
2. Is this different from today's environment?
3. Do you understand the vision statement and what each word means?
4. How could it be improved?
5. Do you think it is achievable?
6. What would keep us from achieving it?

Often these sessions are kicked off with the person who did the assessment reviewing the findings. This helps everyone obtain a better picture of the "as-is" condition.

Flip charts should be used to record all comments, negative or positive. This is an effective way to document the discussion and come to common agreement on key issues. Somehow things look different in print.

Procurement should ask the major suppliers to attend a focus group meeting where all the vision statements are reviewed, but most of the supplier focus group time would be directed at the supplier partnership vision statement. Marketing should do the same thing for their major customers, with particular emphasis focused on present customer partner status and on the customer partnership vision statement. With both the suppliers and the customers, it is better to review the vision statement with too many, rather than too few.

This is one of the most exciting parts of the whole improvement process. In most organizations, it is the first time that management has ever asked the employees what type of environment they would like to spend their lives in.

Preparing the Final Vision Statements

When the results of the focus groups are available, a second meeting of the EIT is held to develop the final vision statements. At this meeting, the executives present their teams' inputs and represent their teams' views of the desired future. Big arguments break out over small points; i.e., which synonym to use, which adjective to use, where the comma goes, etc. After some agonizing debate, a final group of vision statements is agreed to. Each word in each statement has a common meaning to the members of the EIT.

The outcome from this meeting is a new, final group of vision statements. In our experience, most of the final vision statements are very different from the preliminary vision statements. This difference is very important. Even though an individual employee did not have his or her suggestion included, he or she is able to see that the executives changed their visions after they talked with the employees. It becomes very obvious that management is listening and really caring about the employees' opinions.

Organizations tap the full potential of their employees when meaningful, common visions are created jointly. Management's job is to promote these agreed-to, common visions. "Promote" does not mean to only talk about them or to support them. It means to live them, to sell them, to be enthusiastic about them. Confusion reigns supreme when management talks and writes one message, but acts and lives another. If management cannot live and act the visions, for heaven's sake, don't talk or write them.

There is one bank in Arizona whose management felt strongly enough about their vision statements to the point that they rented billboards on the route their employees traveled to and from work. On these billboards, the bank's mission and visions were posted. The message was not there for their customers. The customers would see the results of these visions. They were there for the employees. It showed the employees that management was firmly committed and that with the employees' help, they could bring about the major environmental changes that the vision pictured.

■ Setting Performance Improvement Goals

The executive team should now define how it will measure success for the improvement effort. Every one of the executives has expectations of what should be accomplished in their function and the organization as a whole. To guide the improvement planning cycle, the executive team needs to focus on setting goals for only a few critical organizational measurements. For example:

- Return on investment
- Customer satisfaction
- Response time
- Value added per employee
- Market share
- Error-free performance
- Dollars saved
- Morale index

We suggest that the executive team select three to six organizational measurements and set yearly goals for them that will be used to design the improvement process around. The ultimate design of the improvement process will be greatly impacted based upon how aggressive these improvement performance goals are.

■ Desired Behavior and Habit Patterns

The start of the personality change to the organization is developing a set of vision statements. If they are worthwhile and are embraced by management and employees alike, then the individual's feelings and thought patterns will begin to change. If the organization and the individuals involved are rewarded personally and socially as these new feelings are embraced, over time they will transform into normal behavior and/or habit patterns. For example, if part of your management support vision statement was to "empower your employees at all levels," this part of the vision could first be reflected in your employees as they begin to feel that they don't have to get management approval to take action on unplanned events. They would begin to gain confidence that they could make the right decisions in many cases without management's help. With time and continuing management support, they will begin to feel confident that they will not be hurt because they make a decision, and their behavior patterns will change. Now more and more, they will take the needed action, often telling management after the fact about the problem and how it was handled. They will start to come to management, explaining how they are going to correct the problem instead of asking management how to solve it. Positively reinforced behavior and actions become habits. At this point in time, these special patterns become a natural pattern. "It's just the way we do things around here; it's nothing special."

For every vision statement, a list of habits and behaviors that would exist in the organization if the vision was realized should be prepared. To accomplish this, the EIT may decide to focus on key words or phrases in the vision statements, or the vision statements as a whole. Typical key words or phrases that might be included in your vision statements are:

- Empowered employees
- Customer-driven
- Process focus
- Streamlined operations
- Quality first
- Technology-driven

Using "empowered employees" as a key phrase, the following is a list of some of the behavior and habit patterns that would be observed in an empowered work force:

- Self-managed work teams are used effectively.
- Wild ideas are encouraged and discussed.
- Unsolicited recommendations and suggestions are often turned in.
- Business information is readily available to all employees.
- Management defines results expected, not how to get them.
- Decisions are made more quickly and at lower levels.
- There is less second-guessing.
- People define their work process and time schedules.

Now the EIT needs to select key behavior and habit patterns and establish a way of measuring how they are changing within the organization. For example:

- Self-managed work teams could be measured by the percentage of people that are part of these teams.

- The degree that wild ideas that are encouraged and discussed could be measured by reviewing brainstorming lists to determine what percentage of the items stretch the imagination.

- How often unsolicited recommendations and suggestions are turned in can be measured by the number of performance improvement ideas and suggestions that are turned in per eligible employee.

Using this type of thought pattern generates a very extensive list of behavior and habit pattern measurements, many of which are not being measured in most organizations. The EIT should include many of them when creating their TIM vision behavioral measurement plan.

■ Three-Year Improvement Plans

Organizations that want to eliminate the piecemeal, flavor-of-the-month approach to improvement are stepping back and looking at all their improvement options before committing a course of action. It takes time up-front, but it saves total cycle time, cost and effort over a three-year period. In addition, it produces much better results. Properly designed, it will create an organization that is creatively bringing out the best each employee has to offer. Work becomes a rewarding, enjoyable, exciting experience. The environment promotes a team spirit without taking away the individual's sense of accomplishment, achievement, and self-esteem. The excitement of belonging and achievement creates an electrifying air that snaps like lightning bolts between management and employees alike, breathing new life into the entire organization.

David T. Kearns, chief executive officer of Xerox, stated "Our primary motivation in applying for the National Quality Award was to find out how we could improve. Sure, we wanted to win. But we wanted to learn even more. We spent the last year using the award process to identify areas in which we can improve. And we're using what we learned to kick off a second five-year plan for quality improvement at Xerox worldwide."

Yes, a multiyear improvement plan lies at the heart of every successful improvement activity, and not just for the large corporation. We reviewed Globe Metallurgical's "QEC Continuous Improvement Plan." Globe was the first small business to win the coveted Malcolm Baldrige National Quality Award. Globe's continuous improvement plan contained 96 different objectives, with multiprojects for most of the objectives, with target completion dates distributed over a two-year period.

Factors Impacting the Three-Year Improvement Plan

An organization must consider many factors before finalizing the three-year improvement plan. It can be divided into two categories: (1) impacting factors, and (2) influencing data. *Impacting factors* are things like the organization's mission, values, performance goals, business plans, etc. *Influencing data* are things like customer feedback, opinion surveys, poor-quality cost, competitive performance, etc.

Certainly the environmental influencing data can have a major impact on the final three-year improvement plan. Some of the things included in these considerations are: technologies, standards, desired pace of change, competitive environment, etc.

The proliferation of improvement tools has certainly increased the complexity of the improvement planning cycle. Crosby's 14 Steps, Deming's 14 Points, Feigenbaum's 10 Benchmarks of Quality Success, plus hundreds and hundreds of others, many similar, but still all different. The result? There are more than 400 different improvement tools now available for the organization to select from.

Difference Between Planning and Problem Solving

We find there is a lot of confusion between planning and problem solving. North Americans are good problem solvers, but they hate to plan. Planning sessions continuously flow over into problem-solving sessions. It is important to separate planning from problem solving if the three-year plan is to be completed in an expeditious manner. Planning is upper management's responsibility. Problem solving is the responsibility of middle management, first-level management, and the employees. Look at the difference between the two:

Planning	Problem solving
Define direction	Define solutions
Identify change areas	Implement changes
Assign resources	Use resources
Identify needed action	Take action
Highlight symptoms	Find root causes
Look at big picture	Focus on single issues
Short-term cycle	Long-term cycle

Individual Environmental Improvement Three-Year Plans

The EIT should look at each vision statement and develop a plan to transform the organization's environment over the next three years in keeping with the vision statement. This environmental improvement plan must provide a logical transition from the "as is" state to the desired future state as defined by the vision. Transition is defined as an "orderly passage from one state, condition, or action to another." An effectively planned-for transition:

- Is not abrupt
- Does not create morale problems

- Does not have schedule slippage
- Is not uncontrolled
- Is not unplanned for
- Does not create customer complaints
- Accomplishes desired results without rework

The EIT will then generate a list of today's problems related to the environmental process that is being planned for and a list of roadblocks that will impede the change of state. When this is complete, the tools that impact the environmental process under study will be reviewed. For example, as mentioned earlier in this chapter, in the management support and leadership category, about 44 of the 400 improvement tools are directly applicable in helping bring about this transformation.

The appropriate improvement tools and the list of problems and roadblocks are then analyzed to determine which tools are used to correct which problems. In many cases, different tools are effective on the same problems. The EIT must study these interrelationships to determine which tool provides the best combined results in their particular environment. Once the EIT has selected the appropriate tools, an implementation plan for each tool will be prepared and an individual assigned the responsibility for ensuring the plan is implemented. At this point in the planning process, priorities are not given to individual tools unless there is some type of interdependency.

After the individual environmental improvement plans are completed, the EIT should then review the performance improvement goals to identify which measurements are impacted by the specific environmental change plan. The EIT should then evaluate how much improvement the specific environmental change plan will have related to the affected performance improvement goals. Although a number of environmental change plans can impact one measurement, the total sum of the impacts should add up to at least the minimum goal for each measurement. If this is not the case, the individual environmental improvement plans need to be improved.

The EIT should also evaluate each of the specific environmental change plans to ensure that they will be conducive to the "desired behavior and habit patterns" previously developed. If the individual environmental improvement plans do not meet this test, they will need to be modified.

■ Combined Three-Year Plan

When the EIT has completed developing the individual three-year improvement plans, they are now ready to combine the plans and prioritize activities. There are a number of things that should be considered when the individual plans are combined, not the least of which are the performance improvement goals that were developed earlier. The executive team should review the individual plans to define which activities impact each of the performance improvement measurements and schedule the activities so that the performance goals will be met.

Other things that should be considered in scheduling the activities in the three-year plan are:

- Available resources
- Other activities going on within the affected area
- Holidays and vacations periods
- Seasonal and/or new product workload fluctuations
- Interdependencies
- Organized labor involvement
- Change management timing

If, when all the constraints are considered, the improvement plan will not bring about the desired improvements as defined in the performance improvement goals, then either the goals or the plan should be modified.

The biggest single mistake most organizations make is in trying to implement the improvement process too fast. Great care should be taken to balance the improvement effort and resources required with the other activities going on within the organization. Most organizations want to overcommit to the improvement effort at this point in time. In fact, it is much better to be conservative during the first year rather than too aggressive. It is our experience that organizations already feel that they have a workload that is 110 percent of their work force capability.

When the improvement process starts, there is an increase in workload. In some areas, the increased workload can go up as much as 30 percent. The total workload typically does not drop down to below its original value for 12 to 18 months. To offset this short-time, peak workload, consultants and temporary employees should be used. As employees are freed up from their present jobs because of the improvements made in the process, they should be re deployed to replace the consultants and temporary personnel.

A major portion of the three-year improvement plan is the supporting change management plan that will help ensure the smooth implementation of the individual process changes. The whole concept of change management will be discussed in detail later in this chapter. Special focus is being placed on it because it is such an important part of the three-year improvement plan and it is often overlooked.

■ Rolling 90-Day Improvement Action Plan

Now that the organization has agreed on a combined three-year improvement plan, it is time to put theory into action. The rolling 90-day improvement action plan is used to provide the organization with an agreed-to, short-range schedule for implementing the combined three-year improvement plan. This schedule will be divided into weekly segments, but specific target dates are often added to the plan. (Example: The EIT will meet on the first and third Tuesday of every month from 9 AM to 12 noon; or, The final report is due February 3.)

To accomplish this, any activity that starts during the first three months of the combined three-year plan should have a detailed, day-by-day implementation plan developed for it. This plan should be prepared by the individual assigned the responsibility for that activity by the EIT. These plans are then combined into a rolling 90-day improvement action plan.

■ Making the Improvement Process Work

The problem that we have had in the West is not what we do; it's how we implement it. Almost all the improvement tools that individual organizations have tried are good under the right conditions. Unfortunately, in most cases, we have done the right thing, but did it very ineffectively, thereby minimizing the potential gains from the activity. Literally billions of dollars have been wasted in the West training employees to do things that they are not using today. The key to successful implementation of the improvement plan is an excellent change management process. To ensure that the improvement activities are successful, the organizational change management activities must be an integral part of the three-year improvement plan.

■ Organizational Change Management

Only 5 percent of the organizations in the West truly excel. Their secret is not what they do, but how they do it. They are the ones that manage the change process.
                        DR. H. JAMES HARRINGTON

A critical component of an integrated three-year improvement plan is a structured and disciplined process for managing and implementing change. The adoption of a TIM philosophy will create a great deal of organizational change that will have a major impact on organizational members' current beliefs, behaviors, knowledge, and expectations. To make this challenge even more daunting, the changes brought about by TIM will impact people who are probably already overwhelmed with the increasing acceleration of change in their professional and personal lives. Therefore, all organizational members must realize that organizational change can, and must be managed. Change cannot be viewed as a one-time event or a passing phase. Change must be seen as the manageable process which it is. (See Fig. 3.1.)

For TIM to bring about real, sustainable business improvements, it is imperative that managers at all levels of the organization have the ability and willingness to deal with the tough issues associated with implementing major change. They must be capable of guiding their organization safely through the change process. This involves convincing people to leave the comfort of the Current State, move through the turbulent, new way of doing things (the Transition State), to arrive at what may be an unclear, distant Future State. Specifically, these three states are defined as follows:

1. *Current State.* The status quo or the established patterns of expectations. The normal routine an organization follows before the implementation of TIM.

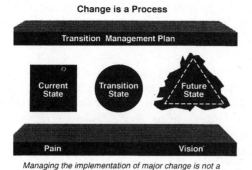

Change is a Process

Transition Management Plan

Current State

Transition State

Future State

Pain   Vision

Managing the implementation of major change is not a mystery; it is a manageable process.

Figure 3.1. The process of change.

2. *Transition State.* The point in the change process where people break away from the status quo. They no longer behave as they have done in the past, yet they still have not thoroughly established the "new way" of operating. The transition state begins when TIM solutions disrupt individuals' expectations and they must start to change the way they work.

3. *Future State.* The point where change initiatives are implemented and integrated with the behavior patterns that are required by the change. TIM goals and objectives have been achieved.

The focus of Organizational Change Management (OCM) implementation methods is on the transition between these various states. The journey from the Current State to the Future State can be long and perilous, and if not properly managed with appropriate strategies and tactics, it can be disastrous. Each major improvement effort in the three-year improvement plan will undertake this journey. It is for this reason that OCM must be part of each of the individual improvement plans.

Over the last 20 years, we have been involved in an ongoing field research project, focused entirely on change management. The primary purpose of this project has been to identify the distinguishing characteristics of those managers/organizations who were highly successful at implementing major change, and those who were not. As a result, we have been able to identify some of the "best practices," as well as common "pitfalls" of implementing major change projects. Prior to an in-depth look at some of the more critical best practices of Organizational Change Management (OCM), it is important for us to first set the proper context from which the best practices were identified.

■ Working Definition of OCM

What is the definition of successful implementation of major organizational change? In order to determine which organizations were successful at implementing major change projects and those that were not, it was necessary to develop an operating definition of "success." The definition created is as follows: *"Successful implementation equals achievement of the stated human and technical and business objectives on time and within budget."* Granted this is a very restrictive definition of success, but in order to truly identify the "best practices" of what it takes to make change happen successfully, it was critical to identify those actions that were taken by organizations that achieved this definition of success, and to identify those actions that resulted in a less than acceptable (i.e., unsuccessful) change project. The philosophy of Total Improvement Management and the identification of improvement projects will bring little value to the organization unless fully implemented. The remainder of this chapter will be dedicated to highlighting those "best practices" which an organization should follow to increase the likelihood of successfully implementing improvement projects. It will also discuss the most common pitfalls which would cause implementation failures.

■ OCM Best Practices

Determining When a Change Is Major

We can apply our first "best practice" to any improvement project which is being implemented, by first identifying the pitfalls. Many organizations have a tendency to assume that every change or improvement project requires the same level of implementation effort. In essence, they tend to repeat their past implementation history: they budget for cost and time requirements for both the technical and human objectives as if all change projects were the same. The "best practice" that should be applied here deals with accurately determining when an improvement project is going to be a major change for the people impacted within the organization. If it is a major change, then it is worth some special implementation effort and some special allocation of implementation resources. There are some guidelines, in terms of when or how to determine if a major project needs special implementation effort.

Factors to Consider

Essentially, there are three factors to consider:

1. *Is the change a major change for the people in the organization (human impact)?* A major change is any change that produces a significant disruption of an individual's normal expectation patterns. In order for management to determine if a change initiative is considered major, 14 specific factors should be examined. These factors that disrupt expectation patterns are:

- Amount
- Scope
- Transferability
- Time
- Predictability
- Ability

- Willingness
- Values
- Emotions
- Knowledge
- Behaviors
- Logistics
- Economics
- Politics

One, or any combination of these factors, can cause a change to be considered major in the eyes of the targets. Management must have a handle on the way their employees are perceiving even what seems like the most insignificant of changes.

2. *Is there a high cost of implementation failure*? What is the price associated with failing to implement a specific improvement project (cost of failure)? It is imperative for management to understand the consequences of failing to successfully implement any change. Not only will resources be wasted on a problem that is not solved or an opportunity that is not exploited, but there may be other implications such as morale suffering, job security threatened, and the organization losing confidence in its leadership.

3. *What are the risks that certain human factors could cause implementation failure (resistance)*? Questions that need to be answered include: Is senior management truly committed to this project, how resistant will the organization be, and does this change "fit" with our culture? Once again, ignoring any of these human factors can cause a project to fail. Later in this chapter there are specific "best practices" that address a majority of these factors.

These three factors must be considered for every improvement project identified. Senior management must be able to recognize these business imperatives, which require a dedicated effort in managing the human and technical objectives in order to achieve success. Therefore, in order to leverage this "best practice," each improvement project resulting from the adoption of a TIM philosophy should be assessed with regard to these three factors. Hence, an accurate determination can be made of the level of disruption change causes the organization, and how much time, effort, money, and resources will be required to ensure successful implementation.

Building Resolve to Manage Change

Next, we can discuss pitfalls and "best practices" with regard to building the resolve and commitment necessary, not only to initiate an improvement project, but also to sustain that project all the way through to completion. One of the common pitfalls that we have seen many organizations make is strong, zealous initiation of improvement projects, only to have them flounder from lack of resolve to sustain the project through to completion. Obviously, then, the "best practice" in this case, is to build the necessary commitment to sustain the change with senior and middle management, thereby enabling the organization to manage the change process over time.

Achieving "informed commitment" at the beginning of a project is one of the main issues in any change project. There is a basic formula that can be applied which addresses the perceived cost of change versus the perceived cost of maintaining the status quo. As long as people perceive the change as being more costly than maintaining the status quo, it is extremely unlikely that the resolve to sustain the change process has been built. The initiator of the change must move to increase people's perception of the high cost of maintaining the status quo and decrease their perception of the cost of the change, so that people recognize that even though the change may be expensive and frightening, maintaining the status quo is no longer viable and is, in fact, more costly. This process is referred to as "pain management."

"Pain management" is the process of consciously surfacing, orchestrating, and communicating certain information in order to generate the appropriate awareness of the pain associated with maintaining the status quo compared to the pain resulting from implementing the change. The "pain" the initiator is dealing with is not actual physical pain. Rather, change-related pain refers to the level of dissatisfaction a person experiences when his or her goals are not being met or are not expected to be met because of the status quo. This pain occurs when people are paying or will pay the price for an unresolved problem or missing a key opportunity. Change-related pain can fall into one of two categories, "current pain" and "anticipated pain."

Current pain revolves around an organization's reaction to an immediate crisis or opportunity, while *anticipated pain* takes a look into the future, predicting probable problems or opportunities. It is very crucial that management understands where their organization is located on this continuum of current vs. anticipated pain. This understanding enables management to better time the "resolve to change." This resolve/commitment, which must be built and sustained, can occur during either the current or anticipated time frames. If this attempt to build resolve is formed too early, it won't be sustained; if it's formed too late, it won't matter. Management has a wide variety of pain management techniques from which it can choose. Some of these techniques being used by *Fortune 500* companies include: cost/benefit analysis, industry benchmarking, industry trend analysis, and force-field analysis, among many others. When this process has been accepted by senior and middle management, a critical mass of pain associated with the status quo has been established, and the resolve to sustain the change process has also been established. It is only then that management can begin to manage change as a process, instead of an event.

Any project that results from the Total Improvement Management philosophy will, by necessity, cause change in an organization. The application of this "best practice" is critical in beginning to mobilize support and understanding for the reasons for change, to help them let go of the status quo and move forward to a very difficult state, known as the "Transition State." Managing people through the Transition State to project completion requires resolve not only to initiate change, but to sustain it over time, with management continually communicating the necessity for change and supporting the actions required to bring it about.

■ Identify and Orchestrate Key Roles

Identification of key roles in the change process is vital to successful implementation of change projects. Many scholars have studied the field of change management and have formulated different names for the five key roles. The names that we will use are as follows:

1. *Initiating Sponsor* is the individual or group with the power to initiate or legitimize the change for all the affected people in the organization.

2. *Sustaining Sponsor* is the individual or group with the political, logistic, and economic proximity to the people who actually have to change. Often we talk about initiating sponsors as senior management, and sustaining sponsors as middle management, but that's not necessarily the case. Often sponsors can be someone in the organization who has no real line power, but has significant influence power as a result of relationships with the people affected by the change, past successes of the individual, knowledge or power.

3. *Change Agent* is the individual or group with responsibility for implementing the change. They are given this responsibility by the sponsors. Agents do not have the power to legitimize change. They do not have the power to motivate the members of the organization to change, but they certainly have the responsibility for making it happen. They must depend on and leverage sponsorship when necessary.

4. *Change Target* is the individual or group who must actually change. There really is nothing derogative associated with the word "target." In fact, it's really more of an indication of where the resources which are allocated to any specific project must be focused to achieve successful change.

5. *Change Advocate* is an individual or group who want to achieve change, but who lack sponsorship. Their role is to advise, influence and lobby support for change.

Identifying the members of an organization who must fulfill these roles, and then orchestrating them throughout the change process is a "best practice" which organizations can leverage to greatly increase their likelihood of success with any specific improvement project. Once these roles are identified, management should maneuver those key roles to optimize each of them throughout the change process in order to achieve successful implementation. To be effective in that task, management must understand the intricacies of each role, how they interact with each other, and how they work in an organization. The first thing that needs to be understood is that in all major change projects, key roles will overlap. When this occurs, the individual(s) should always be treated as a "target" first (i.e., surface pain, manage resistance, build commitment). An example of this would be if a divisional president would sponsor Management By Objectives (MBO) for his division. He might also serve as a change agent by promoting MBO's use to mid-level mangers, and could be considered a target if he uses MBO himself.

Impact of Change on the Organization

Another important point management must understand is the interaction of these key roles in the three most basic organizational structures. These three basic structures in most organizations are: linear, triangular (e.g., staff), and square (e.g., levels of business). (See Fig. 3.2.)

All of these structures are usually found in an organization and can be useful; however, each one can become dysfunctional. The linear structure is the simplest to understand. A sponsor delegates implementation responsibility to a change agent who implements down to the target. The triangular structure is more complicated because the change agent and target report to the same sponsor, but the target does not report to the change agent. What tends to happen in this situation is that the change agent often will try to use his or her own legitimizing power to implement the change. The target, however, usually knows who "the boss" is, so the sponsor-change agent relationship must be clarified. The appropriate action in this situation is for the sponsor to play "watch my lips" with the target to ensure he or she understands that the change is the sponsor's idea and no one else's.

The final basic organizational structure, the square, can also be very dysfunctional. The problem here is that a sponsor, or a sponsor's change agent, will try to implement a change on a second sponsor's employees/targets. What sponsor No. 1 is usually unaware of is that these targets will rarely respond to change directives unless those directives are received by those who control the consequence

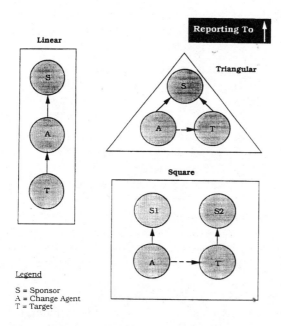

Figure 3.2. Key roles three basic organizational structures.

management. The proper solution is for sponsor No. 1 to become an advocate to sponsor No. 2 in order to bring him or her on board with the change initiative. If this tactic fails, appealing to a higher authority to intercede is necessary.

Management needs to comprehend that these types of problems involved with the triangular and square structures are associated with a high risk implementation failure. Therefore, with a solid understanding of what the key roles are and how they affect change in an organization, sponsors, both initiating and sustaining, will help to identify these roles and fill them. It is then up to management to establish an optimum performance level and orchestrate and manage that performance.

Build Sponsorship to Support and Sustain Change Objectives

One of the common pitfalls associated with implementing major change among the sponsorship group is an assumption that once the decision has been made and communicated regarding what must be implemented, no further involvement on their part is necessary. Top management tends to ignore the importance of the other key roles, relying instead on employee compliance with orders from above. They are often dismayed to find that after six months, their directives still are ignored. In fact, if major change is involved, sponsor involvement and commitment is most important. The "best practice" that we have helped organizations apply to successful change initiatives is the accurate and timely identification of initiating and sustaining sponsors, exactly what their roles entail, and installing an architected, proactive approach to building the sponsorship necessary at each level to support and sustain the change through to completion. This process has been referred to as the building of "cascading sponsorship."

"Cascading sponsorship" is an effective way to eliminate the corporate "black hole." "Black holes" are those places in organizations where change decisions enter but are never heard from again. These typically occur when there is a manager who does not sponsor the change and, therefore, the targets beneath him or her do not adopt the change. There is little initiating sponsors can do to maintain the change at lower levels of the organization because they do not have the logistical, economical, or political proximity to the targets. The result is that change cannot succeed if there is not a network of sustaining sponsorship that maintains the integrity of the implementation as it moves down through all levels of the organization; hence, cascading sponsorship.

Essentially, the way cascading sponsorship works is by starting with the initiating sponsor and working down through the different levels, specific to any improvement project. Sponsors prepare the change agents to fulfill their roles, giving them the necessary skills to manage not only the technical aspects of the engagement, but the human aspects as well. The success of any improvement project starts at the top and ultimately rests on the shoulders of the sponsors. Sponsorship is the most critical risk factor in any change project. To have an effective network of sponsors, organizations implementing major change need to adhere to five critical rules. Those rules are:

- Sponsorship is critical to successful change, so all sponsors must demonstrate, publicly and privately, unsurpassable commitment.

- Weak sponsors must be educated or replaced, or failure is inevitable.

- Sponsorship cannot be delegated to change agents.

- Initiating and sustaining sponsors must never attempt to fulfill each other's functions. Initiating sponsors are the only ones who can start the change process, and sustaining sponsors are the only ones who can maintain it.

- Cascading sponsorship must be established and maintained.

Sponsors must also be sure not just to speak rhetoric about being committed to the change; they must openly demonstrate it. Sponsors need to develop a clear vision of the future state at the strategic and tactical level, and surface the appropriate amount of pain to move toward that vision. Sponsors need to become educated to understand the effect the change will have on the organization, and empathize with what the targets are being asked to change about the way they operate. Sponsors not only need to be prepared to allocate necessary organizational resources (e.g., time, money, people, etc.) for successful implementation, but they must also be prepared to personally pay the price for success. Finally, sponsors need to develop a reward structure that recognizes those who facilitate the implementation process and discourages those who attempt to inhibit the acceptance of change.

Preparing Change Agents in the Required Skills

We are living in a turbulent environment, where change is accelerating dramatically in three ways: volume, speed, and complexity. This means that we can no longer manage as we have in the past. A possible pitfall in this area is that sponsors, often incorrectly, assume that those identified as change agents and advocates possess the skills necessary to successfully deal with human as well as technical implementation problems.

In today's unstable environment, it is necessary to have finely honed skills for managing and implementing technical and human change. These skills are in very high demand, and special training is often necessary. The "best practice" for organizations is to build the capacity to manage change. To accomplish this objective, the change agents must come into the engagement with a different perspective. Part of the development of these skills is also a bit of a shift in mind-set.

When you look at agents historically, you will note that they tended to have a more of a technical expertise mentality. Their primary focus and objective was to be sure that the change, whatever it might be, was technically sound and that if people weren't able to use it, then it wasn't the agents' fault. That mind-set must shift to a mind-set that follows the idea of facilitators of change. The mind-set that says change agents need to be responsible for not only managing the technical aspects of the change, but also the human aspects of the change. This mind-set also dictates that the change agent should focus on the process as well as the content issues, with technology being designed to accommodate human interests, needs and values.

Truly effective change agents should be skilled in a complex combination of characteristics that can be brought to bear on a given change project. Successful change agents must have the ability to work within the parameters set by the sponsor and understand the psychological dynamics regarding how individuals and organizations can modify their operations. Change agents must optimize their performance by placing emphasis on the technical aspects, and especially, the human aspects of the change. Change agents must be skilled in dealing with resistance. To understand that resistance, it is critical for change agents to be able to identify, relate to, and respect the targets' and sponsors' diverse frames of reference. Change agents must also be the "cheerleaders." They must constantly strive to build commitment and synergy among targets and sponsors, while at the same time being aware of and utilizing power dynamics and influence techniques in a manner that reflects a capacity to achieve results in an ethical way. The bottom line is that a change agent must be professional, setting aside a personal agenda for the good of the change.

Success ultimately is judged by achievement of both the human and the technical objectives. That mind shift needs to occur if change agents are going to more effectively manage the human aspects of implementing major change projects. The change management skills within the organization are critical to the success of any change project. This change of perspective must occur if change agents are going to successfully manage the implementation of change projects. However, even the most skilled change agents in the world cannot successfully implement major change by themselves. The other roles in the change process, specifically the advocates, must have their skills finely honed and be prepared to use them.

Preparing Advocates in the Required Skills

To be a successful advocate, as with change agents, a special mind-set is required. Successful advocates must focus their attention on the sponsor(s) of the change and position the change as being beneficial to the sponsor(s). Advocates must seek the approval of those in power and avoid at all costs wasting time/energy with people who can't say "yes." Effective advocates are results oriented and are willing to accept nothing less than successful change. They are unwilling to simply adjust to the unacceptable status quo. Advocates must be confrontative and assertive and when faced with poor or declining sponsorship, they must be prepared to either educate the sponsor, replace the sponsor, or ultimately prepare to miss deadlines and projected budgets. With this mindset in place, an advocate simply needs to follow five basic steps to be successful. These steps are:

1. Precisely define the change that needs to occur and how success would be measured.

2. Identify the key targets that must accommodate the change.

3. With each key target or target group identified, determine the initiating and sustaining sponsor(s) that must support the change.

4. Evaluate the current level of sponsor commitment.

5. Develop "pain management" strategies to increase and gain the appropriate sponsor commitment level.

When everyone in the change process knows their role(s) and has the skills necessary to fulfill those roles, successful implementation will be completed on time and within budget. Without the correct perspective and skill base, especially for change agents and advocates, the chance for successful implementation is nearly eliminated.

Enable Sponsors, Agents, Targets, and Advocates to Work as a Synergistic Team

The next "best practice" which we can discuss, deals with the idea of synergy—building synergistic work environments and synergistic work teams. Synergy is a very important concept when implementing change projects. Synergy occurs when two or more people, working together, produce more than the sum of their individual efforts. Much has been said about empowerment, participative management, cross functional teams—all of which are very good ideas and necessary, but none are likely to be successful without a basic synergistic environment. A common pitfall is that management promotes the idea of synergistic output and synergistic teams, yet most fail to achieve them. The advised "best practice" is to enable sponsors, agents, and targets to work effectively as a synergistic team throughout the change process.

Integral to synergy is allowing people to work in a synergistic environment. Synergistic environments are open, there is no fear, there is five-way communication, and people in those environments really do feel as if they can have some influence over the outcome of any specific project or business issue. To really build this kind of environment and this kind of teamwork, it is necessary to first meet the two prerequisites of synergistic work teams:

1. There needs to be a very powerful common goal shared by these sponsors, agents, targets, and advocates for the change.

2. Goal achievement requires a recognition of interdependency: sponsors, agents, targets, and advocates must recognize that the goal cannot be achieved without working together.

Therefore the "best practice" here is primarily to focus on making sure that those prerequisites exist. Once the existing prerequisites are confirmed, a team and a process can be built so that people can really capture the potential synergy. With the prerequisites for synergy met, a group or organization can begin its journey through the four phases of the synergistic process and team development. It is important to note that all teams must go through this process—there are no short cuts. However, the length of time a team has already been in existence can affect the duration of time it spends in each phase of the process.

These synergistic relationships are generated through a four-phase process. The four phases are:

I.   Interacting

II. Appreciative Understanding

III. Integrating

IV. Implementing

Each phase is interdependent on the others, and individuals or groups on an implementation team must demonstrate the ability and willingness to operate to the characteristics associated with each phase.

Interacting.   For people to work together effectively and synergistically, they first must interact with one another. If this interaction is going to be meaningful, people must communicate effectively. At first, this task is not easy and one usually filled with conflict. This is what is referred to as a group "storming." What happens is that the inevitable misunderstandings individuals are bound to have go unresolved. This causes anger, frustration, blaming, suspicion, alienation, hostility, and possibly withdrawal. In an attempt to stop this destructive cycle and a total breakdown of the team development process, teams must move on to group "norming." Here, the group decides on some basic ground rules as to how it is going to operate.

Appreciative Understanding.   As important as effective communication is for successful change, something more must occur. Group members in a synergistic team effort must value and utilize the diversity that exists among the members. This is a continuation of the "norming" process that occurs in team development. Valuing a different point of view can be difficult for individuals because of the emphasis our culture places on rational, linear, left-hemispheric thinking processes that encourage critical analyses. This thought process can produce an "I'm right, you're wrong" attitude, where as synergy dictates that people should support each other and look for the merit in another's viewpoint.

Integrating.   Even though a team has passed through the first two phases of the synergistic process, it is not yet sufficient to produce synergistic outcomes. Synergy is the result of communicating, valuing, and merging separate, diverse viewpoints. Once again, accomplishing this integration is extremely difficult because our culture does not teach and reward the skills needed. For team members to work through the "norming" process of team development and move on to "performing" the final phase, they need to develop the skills necessary to make integration possible. Specifically, team members need to:

- Tolerate ambiguity and be persistent in the struggle for new possibilities.

- Modify their own views, beliefs, and behavior to support the team.

- Generate creative ways of merging diverse perspectives into new, mutually supported alternatives.
- Identify issues, concepts, etc., that cannot or should not be integrated.

When teams can successfully complete these tasks, they are one step away from synergistic results.

Implementing.   Even the best plans and solutions are useless unless they are fully implemented. The bottom line for synergy must be the successful implementation of change initiatives. The culmination of synergistic events should be well thought-out, change-oriented action plans. The final phase in the synergy process is designed to build on all the momentum that the previous phases have built and to direct that energy into completing the task at hand. It is at this point in the team development process that teams actually begin to "perform." The key to success in this phase is basic management skills. As individuals' capacity to grow beyond themselves (synergy) is increased, it must be managed as any other valuable resource would.

Most implementation problems are the result of nonsynergistic behavior. This behavior can be attributed to human nature and bad habits. Fortunately, if a team follows the guidelines developed to create synergy and effective team development, successful implementation will be the result.

■ View Resistance as a Natural Reaction That Must Be Expected and Managed

"Resistance" is any opposition to a shift in the status quo and is a common response to change. Resistance occurs because people are control-oriented and when their environment is disrupted, they perceive that they have lost the ability to control their lives. This resistance will begin as soon as major change is initiated and can be expressed overtly or covertly. The amount of resistance generated will vary from person to person because each individual has their own unique frame of reference that influences how they view the change. An individual's frame of reference is comprised from his or her values, emotions, knowledge, and behavior. Organizational resistance usually takes the form of a key sponsor not supporting the change because of his or her frame of reference. Although similar to individual frame of reference, organizational frame of reference is comprised of logistics, economics, and politics.

One of the common pitfalls with regard to resistance is that sponsors who are driving change tend to think of resistance as inexplicable. They view resistance as a mysterious force that affects people. They think that resistance is avoidable, believing that if it does occur, it is really a result of somebody's failure. Typical responses are: "What's wrong with that person?; What's wrong with that group?; Why won't they support our change effort? There must be something wrong with those people." In fact, that perspective on resistance is a major barrier to successful change. Typical perspectives that can become major barriers to change include:

- An unclear vision that causes confusion.
- A poor implementation history.
- No consequence management to accompany the change.
- Too little time to implement the change.
- Lack of synergy.

When these implications are understood, management will see that they cause disruption to people's expectations, producing resistance. The "best practice" applicable to resistance is to first view the phenomenon as a natural and understandable human reaction to disruption, and, that as a result, that we respond to resistance as an inevitable part of managing a major change. The greater the change, and the more disruptive it is to the status quo, the stronger the resistance will be. This is true not only of changes that are perceived as negative, but also of changes that are perceived as positive. The reason is that any change, positive or negative, brings about unknown implications. Resistance must be expected, budgeted and planned for, and managed for change objectives to be successful. To effectively manage resistance, one must first understand the reasons for resistance.

To add perspective to the process of managing resistance to change, explore options available when managing resistance. Expect resistance and manage it through a prevention approach or through a healing approach. It is important to recognize that there is a price associated with managing resistance to change. Resistance will always be the companion of major disruptive change, regardless of whether people view it as positive or negative. Acceptance of this fact implies the recognition and acceptance of the price tag. There is no choice in whether or not you pay for resistance if your change project is going to be successful. The question is: How are you going to pay? There is a choice between paying for prevention, which incorporates planning and allocating resources in advance for managing resistance, building plans to overcome the phenomenon, and building commitment to change objectives; versus healing payments, which are the later maintenance costs associated with changes which were forced on targets.

People accept change, not because they believe in it, but because they must. In order to ensure compliance, there is a later and greater cost to pay. Included in a compliance system is the burden to the organization of hiring additional supervisors and managers to ensure that employees are doing what they are told. Resistance which is managed by healing is manifested in higher turnover, lower productivity, low morale, pessimism and distrust of management. These costs can be significant. Resistance always has a price tag. A conscious decision must be made whether that invoice will be paid in advance or after the fact.

With a clear understanding of resistance and how it can be paid for, there are a number of actions management can take to manage it. The first step that must be taken is to determine if the resistance is an ability deficiency or a willingness deficiency. If it is an ability deficiency, management should identify the additional knowledge and skills that are required and provide the appropriate education and training.

If it is a willingness deficiency, management should proceed as follows: First, the reason for the change should be effectively communicated to the targets so that they understand why they are being asked to change. Second, the targets should be evaluated to see if they have accepted the change. Third, any inconsistencies with what is necessary to motivate people in the direction of the change objectives (i.e., clear vision, committed sponsorship) should be identified. Next, the current rewards, recognition, performance measures, and compensation should be analyzed. Then plans for rewards, recognition, performance measures, and compensation that will support the change objectives should be developed. Finally, the new consequence management system should be communicated, implemented, and enforced.

Although the prescriptions to manage resistance appear to be "cut and dried," one must remember that the introduction of major change and the resistance that follows it are fueled by very emotional responses. However, these emotional responses fall into predictable patterns that have been researched, and like resistance, can be managed. The first pattern, the emotional response to a positively perceived change, has five stages an individual goes through. Those stages are:

1. Uninformed optimism

2. Informed pessimism

3. Hopeful realism

4. Informed optimism

5. Completion

Individuals move from believing the change at hand will be easy, to a point where pessimism sets in as they realize the change will be considerably more complicated than first anticipated. It is at this point that individuals may choose to "check out," either publicly or privately. Checking out refers to dissatisfaction and a withdrawal of one's investment in the change. During this period, change agents should increase opportunities for public expression of pessimism. Attempts to suppress negative reactions only lead to increased target frustration and private "checking out." Once an individual has passed through informed pessimism, they begin to "see the light at the end of the tunnel." They now have renewed confidence in the ability to complete the change and integrate it into the normal work routine.

The second pattern, the emotional response to a negatively perceived change, has eight stages a target must pass through. Those stage are:

1. Stability

2. Immobilization

3. Denial

4. Anger

5. Bargaining

6. Depression

7. Testing

8. Acceptance

The first four stages of this model are a natural progression. The change is introduced and the target is in disbelief. He or she then refuses to believe that the change is really going to happen. When the target finally realizes the change cannot be stopped, a period of anger sets in, followed by bargaining. It is at the bargaining stage where change agents must be cautious. Targets will make every attempt to delay the change or get back to the status quo. The change agent must use confrontative reality testing (i.e., the clear and firm affirmation of the change). The target will inevitably get depressed but as time goes on, they will learn to live with the change, eventually accepting it.

With actions in place to manage resistance and the emotional response to change, additional follow-up activities need to be initiated to ensure successful implementation. Management should be sure to continually provide the targets with information to keep them informed of the progress of the change. Management needs to mark the ending of the change and celebrate the successes. The past status quo should be treated with respect and never denigrated. Most importantly, the change needs to be reinforced and its implementation cycle should be minimized.

■ Recognize the Levels of Commitment That Are Required

Commitment is the cornerstone of successful change implementation, but it is wrought with pitfalls and obstacles. One of the most common pitfalls observed here is underestimating the initial level of commitment necessary for the change to be successful. In looking at any specific change project, there needs to be some level of commitment at some level of the organization for it to be successful. Many changes however, require very little commitment, while others require a great deal of commitment at every level of the organization. The "best practice" is to recognize the initial level of commitment required from key people in order for the change to succeed. Once that level of commitment is identified, proactively pursue building that commitment throughout the organization at every level necessary. An operating axiom requires (1) knowledge of the level of commitment required to achieve your change objectives, and (2) only ask for the level of commitment necessary to achieve your change.

The commitment model that management should work from consists of three developmental phases (see Fig. 3.3). Each phase consists of specific and identifiable stages that cannot be avoided. At each stage in the model, there will be reasons why resistance is generated. Change agents should use the appropriate actions at each stage to build commitment.

The first phase, *Preparation*, forms the foundation of either support or resistance to change. The first stage in the commitment model is the *Contact Stage*. Here, the target will either become aware of the change or they will be unaware because contact was not perceived by the targets. Once contact has been acknowledged, the

Phase 1. Preparation

Stage 1. Contact

Stage 2. Awareness

Phase 2. Acceptance

Stage 3. Understanding

Stage 4. Positive Perception

Phase 3. Commitment

Stage 5. Installation

Stage 6. Adoption

Stage 7. Institutionalization

Stage 8. Internalization

Figure 3.3. The commitment model.

targets will move to the *Awareness Stage* where they realize that their jobs will be modified. This does not mean that people understand the full impact of the change. If understanding is achieved, the targets move to the next phase. If not, they are left in confusion and progress is halted.

Phase two, *Acceptance*, marks people's passage over what is called the disposition threshold. The targets' new awareness enables them to make decisions about accepting or rejecting the change. The third stage, *Understanding*, is critical. Targets now become more familiar with the ramifications of the change and begin to form an opinion about it. Negative perceptions will lead to resistance. Once targets accept and support the change, they will move to the next stage, *Positive Perception*. At this point, people must decide if they are going to support implementation. If the targets refuse to support the change, then resistance is increased, and change agents must raise pain and manage the resistance. If a positive perception of the change is developed, the person has reached the point of action (installation). A formal decision to install and use the change is made. The target has now reached the *Commitment Phase* and passed the commitment threshold.

In the next stage, *Installation*, targets demonstrate truly committed action by installing the change. However, this decision does not mean the maximum level of commitment has been reached. In fact, the possibility of reversibility should be stressed. If the change is not aborted and long-term usage commences, the *Adoption Stage* has begun. During this time the change is still in a trial period despite the enormous amount of commitment it took to reach this point. If, after extensive usage, the change is successful, the target will reach the seventh stage of the commitment model, the *Institutionalization Stage*. The change has been around a long time and become part of the everyday routine and corporate culture. Reversibility is no longer a possibility at this point. It takes time to institutionalize a change and the organizational structure should be aligned accordingly. At this level all the targets regard the change as standard operating procedure, yet, this does not necessarily mean they have to believe in the change.

The final stage of the commitment model is just that, *Internalization*. Here, the targets have reached maximum commitment because the change reflects their own beliefs as well as those of the organization. Once the level of commitment has been decided, management should use communication to assist in reaching that level. However, when using communications to build commitment, the following considerations should be taken by management:

- Who must receive the message (internal/external)?
- What do the targets need to know?
- What is the desired likely response?
- How should the message be delivered (media)? How often?
- Who should deliver the message?

Once these questions have been answered, achieving the desired level of commitment will be easier. The one thing that management must know is that building commitment is a time-consuming, expensive process and the investment should only be made to build a level of commitment necessary for the change to be successful—no more and no less. All changes do not need to go through the internalization stage.

■ Understand the Strategic Importance of the Organization's Personality and Culture

Adoption of a Total Improvement Management philosophy will, in and of itself, have implications. For some organizations it may be a minor cultural modification, but for most, the adoption of a TIM philosophy will be a major organizational personality and cultural shift. In either case, the current organization's personality and culture are a huge issue that must be addressed for change projects to be implemented successfully. Because organizational personality and culture are difficult to understand, hard to measure and manage, they are relatively easy to ignore. Commonly, organizations ignore them or do not treat them as key variables when implementing a major change initiative. Obviously the "best practice" is just the opposite. Senior management must understand the strategic importance of the overall organizational personality and culture to the change initiative, and work hard to understand and manage the impact they have on the successful implementation of improvement projects.

Corporate culture is the basic pattern of shared beliefs, behaviors, norms, values, and expectations acquired over a long period of time by members of an organization. Organizational personalities reflect the way the present management team is operating. If an improvement project or a change initiative is consistent with that set of behaviors, beliefs, norms, values, and expectations, then the organization's personality and culture are actually enablers or facilitators of that change. On the other hand, a change project may be fundamentally counter to the organization's personality and culture, making acceptance of the change much more difficult.

One thing we are very clear about is that whenever there is a discrepancy between change in culture and existing culture, existing culture wins. So, to apply this "best practice" to any change initiative, we need to understand whether organizational personality and culture are enablers or barriers to the change. If they are barriers, we must identify why they are barriers, what the existing barriers are, and proactively modify the change or modify the organization's personality, or some combination of both, for change objectives to be successfully met. There are only three options available:

Option 1: Modify the change to be more consistent with the organization's personality and culture.

Option 2: Modify or change the organization's personality to be more consistent with the achievement of the change objectives.

Option 3: Ignore Options 1 and 2 and plan the change initiative to take significantly longer and cost significantly more than what you may have originally budgeted. (This is not really an option.)

■ Summary

The piecemeal approach to improvement usually produces results, but not the best results. To become a winner, or to stay a winner, organizations need to define how they want the organization's environment to evolve over the next five years, by preparing a series of vision statements that define the future desired state. Once their direction is defined, they can design the improvement process that will uniquely meet their transformation needs. Organizations can no longer react to the latest improvement fad. They must consider all of the options available to them, then patiently implement them over a period of time.

It is a fact that in today's turbulent environment, where demands for change are continually accelerating, there will be losers, survivors, and winners. Throughout this book you will receive many ideas on identifying what improvements your organization can make to be more successful. However, these improvements will not bring substantial value to your organization unless they can be implemented successfully. To implement those solutions, your organization must be able to manage change. It is your organization's ability to manage and implement change that will determine if you will be a loser, survivor, or winner. The most critical issue of managing change is the ability to manage the people who must change, helping them to become more resilient and adaptable in the process.

Managing change is no longer a luxury or a means to achieve a competitive advantage. It is a necessity. The future success and survival of an organization will depend on how well change decisions can be implemented.

Now let's look at how different types of organizations implement their improvement process.

Tool Selection

Losers: Always look for the latest fad.

Survivors: Select one guru approach and hold to it.

Winners: Look at all options and their impact upon the organization.

Planning for Improvement

Losers: Plan to implement the tool of the day.

Survivors: Use a standard, proven approach that other organizations have had success with.

Winners: Use a highly customized plan.

Payback on Improvement

Losers: Do not measure return on investments. Measurements are focused on activities, not on results.

Survivors: Long-term payback. Primary emphasis still on activities.

Winners: The improvement process must pay for itself as it goes along. Measurement systems are established early in the process to measure return on investment.

Use of Visions

Losers: Top management develops an overall future general vision statement that describes how the organization will be perceived in the future.

Survivors: Top management develops an overall general vision statement. Management then widely communicates and discusses this vision statement with the employees.

Winners: A series of vision statements that relate to the organization's internal environment are developed jointly by management and employees.

Purpose of the Improvement Plan

Losers: Reduce customer complaints; used as a marketing tool.

Survivors: To stay competitive.

Winners: To improve the organization's value as viewed by all its stakeholders.

Approach to Planning and Implementation

Losers: Are neither proficient in solution identification or solution implementation skills. The only thing that saves the loser from poor decision-making is their inability to adjust during the implementation stage.

Survivors: Are not highly proficient in both solution identification and solution implementation skills. They may be competent in one or the other, but not both. A survivor's change project usually fails as a result of effectively implementing the wrong solution or ineffectively implementing the right solution.

Winners: Understand that the identification of future state improvements is only half the battle, and knowing how to improve is just as important as what to improve. Winners know that the success of a change project depends on the relationship between their decision-making skills used to identify the solution, and their implementation skills used to make it work. They are highly proficient in both skill sets and realize that there is only one alternative: the successful implementation of a correct solution that generates support and commitment from those affected.

Best OCM Practices

Losers: Are not aware of the "best practices" and "pitfalls" described in this chapter. They do not understand the change process and are unable to manage change.

Survivors: Have little knowledge of the "best practices" and "pitfalls" described in this chapter. They do not thoroughly understand the change process or the strategies and tactics used to manage it. Therefore, survivors are unable to achieve a competitive advantage through effective implementation.

Winners: Are aware of the "best practices" described in this chapter and use them as an approach to manage the implementation of major change. They are also well aware of the "pitfalls" and make every effort to stay away from them.

Results Focus

Losers: Fail to achieve change initiatives. More likely than not, initiatives will be abandoned.

Survivors: Rarely accomplish what they have intended. If they do manage to achieve their change objectives, it is usually only after consuming more money and time than they had originally anticipated.

Winners: Focus on successful implementation. They will only accept change initiatives being completed on time and within budget.

You can only be the best when you have tapped the full
potential of your people and they are working for a
common vision.   ANONYMOUS

■ References

1. Ernst & Young Technical Report TR 93.004-1 HJH
2. Ernst & Young Technical Report TR 93.004-2

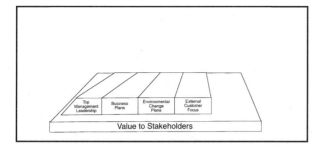

4

External Customer Focus: Best Practices for Outstanding Customer Relationships

An organization without an external customer is an organization that is a blight on humanity.
DR. H. JAMES HARRINGTON

■ Introduction

Since the very beginning, mankind's progress has been largely driven by the need to provide service to an external customer. Sure there are times when we are our own customer. For example, when we read a light novel, we satisfy our personal need for relaxation but in the longer view, the break from the hectic pace of life should reinvigorate us so that we can provide better service to an external customer. Individuals or organizations that exist just to service themselves are a blight on mankind because they consume resources but give back nothing.

■ Why Single Out External Customers?

When you look at the stakeholders of an organization, it is hard to say which is most important, and although all of them are required for the organization to function, the external customer and/or consumer plays a unique role in the total process. They are the starting and ending points for the process. An organization is normally formed to fulfill potential customers' needs and the process is complete when these needs are fulfilled. It is thus fitting that the customer is one of the stakeholders that is considered early in the improvement process.

■ Today's Customers

As the world transformed from a seller's market to a buyer's market in the late 1970s, the importance of the customer came into focus. Sure, customers were always important, but in the last half of the twentieth century, our transportation and communication systems advanced so rapidly that our customers suddenly have many options available to them that they never had before. With these options our customers' buying habits took on a whole new level of sophistication. Today's customers are looking at life cycle costs, not just purchase price. Quality has given way to perceived value. The customers today are far smarter than they were 20 years ago because they are armed with data that they never had before. Magazines like *Consumers Report* provide customers with facts that used to be concealed in the vaults of big corporations. Customers will no longer pay for the mistakes that the organization makes.

■ Customer Focus

Customer satisfaction and customer partnerships have become a hot management issue in most organizations lately. Management has suddenly realized the importance of the customer in the business equation. This has resulted in a number of major changes in management practices at all levels. Customer satisfaction is no longer a sales and marketing issue only. To some degree, it touches all parts of the organization. It has been proven over and over again that it is easier and more profitable to do business with loyal, satisfied customers than to relentlessly pursue new customers. If through a relentless, energetic sales campaign you grow your customer base by 10 percent but lose 15 percent of your present customer base, you're going downhill very fast and paying a lot to do it.

This increased focus on the existing customer has spun off many new projects and slogans. For example, IBM uses "customer-driven quality." Another uses "customer satisfaction—our first priority." The 1990s can be labeled as "the decade of the customer." Many organizations brag about being customer-driven. As Ford Motors' chairman, Donald E. Petersen, put it, "If we are not customer-driven, our cars won't be either."

In the 1980s the customer was crowned king, and our organizations lined up to serve him. Competition to service the customer has become so fierce that new performance standards are being developed every day. One afternoon, Premier Industrial Corporation got a call from Caterpillar informing them that their production line was down because a $10 relay had failed. Premier's replacement part was located in a warehouse in Los Angeles. By 10:30 PM that evening, Caterpillar's line was up and running again. Service like that costs Premier a lot of money, but it pays off. Premier receives as much as 50 percent more for its parts, and its return-on-equity is 28 percent. Their chairman, Morton L. Mandel, says, "To us, customer service is the main event."

Pepsi Cola Company's president, Roger A. Enrico, puts it this way, "If you are totally customer-focused and you deliver the services your customer wants, everything else follows."[1]

Customer-Related Measurements

Measurements are key to an improvement effort. Management measures what they are interested in. Unfortunately, most organizations' management systems have been designed to measure the customer interface from the suppliers' standpoint (example: profit, return-on-assets, percentage of market, etc.). Although this serves the organization's purpose, it does not serve the external customers' needs. In fact, they are often adversely affected. What we need to do is to step out of our jobs and put on the customer's shoes, then ask ourselves what is important to the customer as it relates to our organization.

To do this, *start by making a list of every time the customer comes in contact with the organization.* Define who is contacted, why they are contacted, and how they are contacted. Then ask the question, "Which contacts are very important and which are unimportant?" Double-check the ones that you classified as unimportant to be sure that if there was a problem with them, it would not have a negative impact upon the organization's reputation. For example, when you go into a shop and open the door, you come in contact with the door handle. Is that important? Well, if it comes off in your customer's hand, what impression of the store and its products is created in the potential customer's mind? Yes, the door handle *is* important. Remember, there is nothing so small that you can't lose a customer over it. Once you have defined what's very important, develop a set of measurements for each of them that reflect the customer's impression. When you do this, things like packaging, instructions, cleanliness, and ease of use become key measurements, along with things like reliability, safety, and competitive pricing.

Using the Right Words

Words are important. Organizations that are customer-focused talk and think differently than other organizations. Many organizations talk customer focus, but have labeled the customer as an inferior being, group, or organization. The word "customer" has taken on a bad meaning. Employees in poor-performing organizations talk about their customers like this:

- The customer won't know the difference.
- The majority of the customers believe what they are told.
- The customer will follow the crowd.
- Customers, oh yes, they're the idiots that buy this stuff.

Obviously this is an organization that has poor customer satisfaction because the word "customer" itself is thought of as being inferior. If you increase the organization's focus on the customer, customer satisfaction will get worse, not better. If your organization defines the word "customer" as an inferior being, you need to change this definition or use another word before starting your process to improve your customer relationships. Some organizations talk about the person receiving their output as their client, their partner, consumer, etc. There are many words that can be used to replace the word customer but if you don't want to change the word, your other choice is to change the meaning. Few of us stop to think that the chair we sit on to eat dinner each night was bought for us by our customers, that the car we drive to work was bought for us by our customers, etc. It is the customer who, in reality, is making our mortgage payments. When you think of them that way, they're pretty great people and all of us (with very few exceptions) need them.

Management needs to help the employees see the customer in a new light. An organization that has a high level of customer satisfaction thinks of its customers in the following ways:

- Without customers, I wouldn't have anything.
- This customer is the most important thing right now.
- Every customer presents a new challenge that I have an opportunity to solve.
- When I have satisfied customers, I have job security.

How can you find out how your organization defines a customer? That's simple—ask them. Take a tape recorder and interview people from manufacturing, marketing, sales, engineering—from every area. Conduct interviews one-on-one and in focus groups. Be sure that you put the employees in a situation where they feel the information will not be held against them. Often, an external consultant can be used to ensure anonymity.

Not only is the definition important, but organizations that provide great customer satisfaction talk differently. The pronouns that the interface personnel use (i.e., sales and marketing) directly reflect their customer relationships. Customers who feel that you are close to them are most satisfied with the relationship. Look at the way the following words are used by your customer interface team. Count their frequency of use and make an effort to use the words that do not distance the relationship.

Pronouns	Distance
I and we	Very close relationship; the user is part of the "in" group.
You and your	Familiar relationship; the user is not part of the group.
He/she/it	Unfamiliar relationship; the user is not part of the group.
They	Distant relationship; the user is not part of the group.

The organizations should strive to increase the frequency that the customer contact group uses "I, we, you, and your." Even better, limit the use of "you" and "your" and replace them with the person's or group's name. Dr. Charles Cleveland, a semiotician who has worked with organizations like 3M, American Express, Hallmark Cards, and Eli Lilly & Co., points out that you can tell the difference between organizations that have a high level of customer satisfaction and ones that have a low level of customer satisfaction by counting the frequency that certain words are used. He points out that the words used more frequently by organizations with high levels of customer satisfaction are words like: *the, yes, was, people, service,* and *again.* Words used more by organizations with poor levels of customer satisfaction are: *they, a, not, are, them,* and *no.*

Dr. Cleveland points out that based upon his experience, there is a five-part rule of good customer satisfaction. Organizations with high levels of customer satisfaction spend four parts discussing "what we can do for them" and one part discussing "what they can do for us." Organizations with poor customer satisfaction focus two parts on "what we do for them" and three parts on "what they do for us."

It is interesting to note that organizations with poor customer satisfaction use the word "you" 13 percent more than the organizations with good customer satisfaction. The reason that the organizations with good customer satisfaction use the word "you" less is because they have replaced it with the person's or organization's name.

I am always impressed when I check into a hotel and in my room is a terry cloth robe laid out on the bed waiting for me. Then I am insulted to read a little note on the robe that states, "This is our robe, not yours. We are letting you use it. If you steal it, we will send the FBI after you. Your picture will be posted in every post office around the world and never fear, we will, I repeat, *we will* get you." Well, maybe the message is not quite that strong, but many of them mean the same thing. For example, a hotel in Boston provides robes with tags that read, "This robe is the property of the hotel. Do not remove from the room. Missing robes will be charged to you at $75 each after room inventory upon check out." Compare that message to the following, "Robes provided for your comfort." Inside the tag, "Robes are available for purchase through our concierge—Hotel Crescent Court." The message is the same in both cases, but the customer's impression is very different. The words you and your team use and write have a major impact on your customer's satisfaction, both directly and indirectly.

■ Customer Perception

There are more than three times as many customers lost due to poor service than as a result of poor product performance. Your customers today look at your company from the total organization standpoint, not just based upon the output it provides. They weigh many inputs before making a buying decision. They look at your advertising. They read about you in articles. Their friends tell them about their experiences, and they have their own personal experiences with your organization. From all this data, they develop a perception of the organization. This perception may or may not be factual, but it is the only thing that is important to your organization. To close a sale, the customer must perceive your output as their best value, no matter what your facts say. To prove the point of the importance of customer perception, let's look at the auto industry. The Mitsubishi Eclipse and the Plymouth Laser are exactly the same car, manufactured in Illinois, but the Eclipse outsells the Laser by over 100 percent. Understanding the customer's perception is crucial to a successful organization. It is not enough to provide excellent output—the customer must perceive it as excellent. Forget about the facts; perception is all that counts.

Today's customer perception of an organization's performance can be classified into four major categories.

1. *Poor performers.* Most of these organizations went out of business in the 1980s and early 1990s, but there are still some organizations that are not meeting requirements. All the organizations that do not meet requirements will go out of business in the 1990s.

2. *Good performers.* These are organizations that receive very few complaints, and very few compliments. They meet the customer's requirements all the time. Most of these good organizations will go out of business by the end of the 1990s.

3. *Better performers.* These organizations meet the customer's needs and expectations most of the time. These organizations will survive but will lose market share and struggle through the 1990s.

4. *Best performers.* These organizations meet the customer's needs and expectations all the time. In addition, they delight the customer by providing surprisingly better products and service that set new customer standards and fulfill their desires. These are the organizations that will grow market share and profits, exploding into the twenty-first century as the leaders.

■ Needs versus Expectations versus Desires

Customers' needs and expectations are often very different. In most cases, needs are much easier to satisfy than expectations. Customers tend to communicate and prepare their specifications based upon their needs, but they measure the organization's performance based upon their expectations. For example, when you ask a person what they need in a hotel room they will tell you a bed, shower, toilet, and maybe climate control, but what they expect in addition is clean towels, clean bedding, a TV, phone, fax machine, and a hotel dining room.

There is one category that is even more demanding than needs and expectations. I call it the "desire level." In a hotel, the desire level includes things like a pool, exercise room, in-room jacuzzi, well-stocked refrigerator, mirrors on the ceiling so you can shave in bed, etc. Normally, customer specifications are written at the needs level. Dimensions are indicated on prints with tolerances of plus or minus some amount. What customers need is all parts to be within the tolerance band. What they really expect is to see the parts evenly distributed on each side of the midpoint and well within the tolerance band. What they really desire is to have each part exactly on the center dimension.

You need to understand your customers and be sure you understand their needs, expectations, and their desires. At a very minimum, you need to meet their needs and expectations. If you want to be the preferred supplier, you need to meet all three.

◼ When the Customer Remembers Your Name

The question, "Is the customer king?" is still up for debate, but the need for supplying the customer with superb products and services is not debatable. The customer remembers your organization's name under two conditions and two conditions only.

Condition 1. Your name is remembered when you provide poor products or service to the customer. You know what I'm talking about. When I leave a restaurant with my wife after a bad meal, Marguerite turns to me and says, "Wow, that was bad!" I respond, "Yes, the ice cream was hot and the soup was cold." Marguerite turns to look one more time at the restaurant's name saying, "Tom's Fine Food, ha! We'll have to remember not to come back here again."

Condition 2. Your name will be remembered when you provide extremely good value. In this case, my wife turns to me saying, "That was great! The sherbet just before the main course set the stage for that delicious lobster tail, and it was only $23. Can you believe that? Let's go back inside and get a matchbook. I don't want to forget this restaurant's name. We will bring Bob and Betty Maas here next week when they come to visit us."

When you just meet your customers' requirements, you don't become the preferred supplier. Your customer can be lured away for a few pennies. They won't go out of their way to patronize your organization. They don't complain, but they most likely won't return if they have a choice.

Surprisingly good customer service is only achieved when each employee feels personally in charge of customer satisfaction. What makes surprisingly good service quality? Surprisingly good people. To get surprisingly good people you must select employees who are customer-directed and provide them with superb training.

The truth of the matter is that you can never have surprisingly good service without a superior effort, and superior effort is driven by:

▪ Trust in management

▪ A shared vision and missions

▪ Shared values of what is important to everyone

People relate and react to the values set by management. A bus driver was asked why he was speeding past lines of people, smiling and waving his hands at them, but not stopping to pick them up. He replied, "It is impossible for me to keep my time schedule if I have to stop for passengers." Sure, customers value bus lines that are on schedule, but this example obviously proves that there are other things that they value more. American Express' quality statement expresses it better than I can.

Their standard is, "Not just better than our competition or in line with our customer's expectations—but noticeably superior to the competition and above our customers' expectations."[2]

■ Customer Satisfaction

The level of customer satisfaction is directly proportional to the difference between your perceived performance (not actual performance) and the customer's expectations (not needs). In today's relationships, customer expectations are continuously increasing. Performance that was outstanding yesterday, just meets requirements today, and will be inadequate tomorrow.

Customer satisfaction must be one of the primary considerations in our strategic planning process. The International Quality Study conducted by Ernst & Young and The American Quality Foundation reported: "The percentage of businesses indicating the importance of customer satisfaction in the strategic planning process as secondary or less."

Canada	19 percent
Germany	27 percent
Japan	5 percent
United States	22 percent

"Between 20 to 30 percent of the businesses in Canada, Germany, and the United States relegate customer satisfaction to a minor priority in the planning process."

Customer satisfaction is all the rage in Japan today. Organizations like NEC, Hitachi, and Matsushita have established departments dedicated to improving customer satisfaction. In other fields like securities, restaurants, and insurance, Japanese leaders recognize that quality defined as product reliability is not enough. J.D. Power, III is generally regarded as the father of customer satisfaction by the Japanese. It is interesting to note that seven of the top 10 customer satisfaction rated organizations are service industry organizations. This is an unusually high number compared to the performance of American service organizations.

The keys to customer satisfaction are:

1. Superior products
2. Superior sales and delivery staff
3. Superior after-sales service

If your organization has error-free products and your sales force does not set unrealistic expectations for the products, after-sales service should not be required. It is only when you fail at one or both of the first two that after-sales service comes into play, hoping to save the situation and the customer.

Marketing and sales are the heart of customer satisfaction because they are the organization's key interface with the customer. Although this could lead an organization to believe that only marketing and sales should be in contact with the customer, this would be a strategic error, but one that many organizations make. Yes, sales and marketing are the primary interface and should serve as the customer's ombudsman within the organization, but all functions need to have an interface with the external customer. For example: Production control needs to understand when, where, and how the customer needs the product delivered. Engineering needs to understand how the customer will use the product and have the customer involved in preparing the specifications. It is very important that we make it easy for the customer to do business with the organization. They should not have to go through a list of names to obtain the information they need. Today, more and more marketing and sales functions are being combined so that the customer has only one contact within the organization. This practice is proving to be very beneficial.

Although every part of the organization can and does impact the customer's opinion of your organization, we will focus this part of the book on the primary contact interface—sales and marketing.

■ Marketing's Impact on the External Customer

The basic marketing concepts could be defined as ways to influence customer buying habits, and these concepts have not changed in the last 100 years. Sales' basic concepts, on the other hand, have undergone some very basic changes since World War II, from quick, one-time sales to a relentless pursuit of customer satisfaction, thereby ensuring future sales.

Basically, marketing's primary rule is to work with future, potential, and present customers to identify their future needs, and to nurture and romance the customers to the point that they are excited about buying the product or service the organization offers. We believe that, with the exception of the COB and the president, it is the marketing functions in most U.S. organizations that are accountable for the organizations' poor performance for the following reasons:

1. Poor forecasting

2. Inadequate follow-through

3. Lack of accountability

Poor Forecasting

Marketing's primary responsibility is to define future products, target cost structures, relevant market windows, and volumes. In most cases, marketing has met these obligations, but the high failure rate of new products is a testimonial to the inadequacy of their work. Frequently, errors in volume projections cause large quantities of equipment to be under-utilized or sales opportunities to be lost due to lack of production capabilities. In the late 1980s, Campbell Soup Co. received many awards for the large number of very innovative new products that they developed. The only problem was that very few were financial successes. Chicken noodle soup and tomato soup were paying the bills for the mistakes the marketing and development functions were making. When their current CEO took charge, one of the first things he did was cut back on the number of new products that were being announced each year, and profits began to soar.

Not only have most marketing departments had a poor record at projecting customer requirements, but many of them do not even forecast key supporting requirements. Marketing departments need to forecast not just product requirements, but also sales, delivery, and after-sales service requirements. The marketing forecast and product description should include all of the support performance requirements that impact customer satisfaction. They should also include a customer satisfaction index that would be achieved if their product description was met.

It is time for the marketing profession to make a major breakthrough in its performance. There is far more room for improvement in sales and marketing than any other function I know of. A typical example is quality function deployment (QFD), which more correctly should have been called market forecast development. In most organizations, QFD is championed by development engineering or quality assurance, when in reality it should have been marketing that originated and implemented the concept. QFD should be a mandated course for every marketing major.

The methods used to define new products should be different based upon how the organization is presently performing.

Losers: For organizations that are losing the competitive battle, they need to get very close to their customers. To improve, it is imperative that you get out, meet with your customers, understand exactly what their needs and requirements are, and develop products that fulfill these needs. Primary input for new product development comes from straightforward customer requests and customer-focus groups. Visiting customers and seeking feedback from current customers also provide a positive impact. Don't just rely on customer surveys.

Survivors: For organizations that are holding their own, heeding customer requests is also important, but other data sources also come into the picture. Internal market research survey techniques using the mail and personal contact provide much of the input to the development cycle. Supplier suggestions are also very helpful in identifying ideas for new products and services.

Winners: For organizations who fall into this classification, direct customer feedback is useful, but external market research is the source of most new ideas. These organizations must look beyond customer requirements as customers currently perceive them, and anticipate total new opportunities that are developing in the marketplace. This is the real challenge for the marketing function.

Inadequate Follow-Through

The trademark of most marketing functions is to prepare the new product requirement and throw it over the wall to engineering. As a result, the released product specifications bear little or no resemblance to what the marketing group submitted.

Of course, this is obviously wrong, but most organizations do not go back and compare the product characteristic requirements developed by marketing to the new product specifications generated by development engineering. Marketing, as the customer's ombudsman, needs to be held accountable for ensuring that engineering-released specifications are in keeping with their inputs and have not violated any of the key assumptions that marketing made originally. The marketing department must be responsible and be held accountable for ensuring that the released specification, at a minimum, meets the requirements set forth in their input.

Lack of Accountability

No one takes marketing forecasts seriously. Everyone thinks marketing is an art, not a science—I don't agree. We spend millions of dollars every year in universities teaching marketing to our future leaders. Marketing subjects have been in the university curriculum longer than quality subjects. Marketing personnel and departments should be evaluated on the following criteria:

1. How complete is their input to engineering?
2. How similar is the released specification to the product requirements defined by marketing?
3. How close is the customer satisfaction level to that projected by marketing?
4. Sales volumes for the first 6-, 12-, and 18-month periods.

These four measurements should be benchmarked and a plan developed to reach the benchmark level in two years. Programs should be implemented to improve these four measurements by a factor of 10 in five years. Marketing personnel salaries and bonuses should be directly related to how well they do in defining future markets.

■ Sales and Delivery Staff Impact on the External Customer

Marketing should not be given all of the credit or all of the blame for the relationships that exist with our external customers and/or consumers. At the other end of the product cycle is the sales and delivery staff. A good sales team can offset poor product performance, but a great product cannot offset a poorly performing sales team.

Remember the point made earlier: Three times the number of customers are lost due to poor service than poor products. Sales performance in most organizations has not improved as much as it could because it is hard to measure. Most organizations measure how satisfied their customers are, not how dissatisfied the ones they lost are. We need to consider how much the organization loses each time a customer stops buying from us. If we are a manufacturer of automobiles, customer loyalty represents a $140,000 life-long investment. Why would you argue over a $40 repair bill? In banks, each customer represents $900 a year profit, or $45,000 over an average lifetime. Appliance brand loyalty is worth $160 per year. In a supermarket, a typical customer represents $4400 in profit.

The sales interface is key to customer satisfaction. No matter how you cut it, each customer represents a new challenge to the salesperson. Each customer has unique needs and unique problems. Not only is each customer unique, but their personality traits change each time you deal with them. A person who is normally even-tempered can blow up over a seemingly minor issue, making it impossible for them to save face and resume doing business with your organization. Often these explosions are not justified based upon the salesperson's actions, but the customer had been preconditioned as a result of some problem at home or an unpleasant meeting with his or her manager, or even someone that cut him or her off in the parking lot. The job of the salesperson is very difficult, even when they are provided with an outstanding product, because the combination of personalities, moods and expectations make up an almost endless set of different challenges.

How do you ensure that your sales and delivery personnel are prepared to meet these challenges? The answer to this rests in understanding what the customer values in a salesperson. They are:

1. Courteous individuals who are interested in and like people.

2. Knowledgeable, informed individuals.

3. Availability.

4. Reliability—People they can depend upon to perform an individual service the same way each time.

5. Empowered individuals—People who can make decisions without escalating it up the management ladder.

Sales Force Measurements

Salespeople are often measured on the wrong thing. Bonuses are based on sales volume only, when they should also be based on the following:

- Percentage of time with customers
- Percentage of time with potential customers
- Percentage of customer contacts that result in sales
- Level of customer satisfaction
- Percentage of lost customers

The sales team needs to be motivated and competitive, but most of all, they need to have a high level of personal satisfaction with their jobs. The old saying, "Give me a person who is in debt and I will show you a great salesperson" is no longer true. There is a very different work attitude shown by the sales force that works as a team and is serviced well by their support personnel, than the old type of sales force that is always working to get around the "system."

Old Type of Sales Force	New Type of Sales Force
Does what needs to be done	Knows the whole business
Know the product	Knows the customer's business
Every sale is unique	Every sale is based upon the standard process, modified to the customer
Works hard to look good	Makes the organization look good. Takes pride in the organization
Resists any type of measurements	Uses measurements to benchmark activities
Monetarily motivated	Performance motivated
Sells the products	Sells solutions
Gets new customers	Keeps the proven customers
Changes jobs often	Long-term commitment
Sells the product even if it is not the best answer for the customer	Helps the customer find the best answer even if it is another organization's product
Finds ways to beat the system	Changes systems that are bottlenecks to progress
Heavy-handed selling	Haggle-free buying
Measured by number of new sales	Measured by customer retention

■ Other External Customer Contacts

The customer is impacted by many other people besides marketing and sales. Everyone that comes in contact with the customer needs to be a salesperson for the organization. Customer satisfaction and the reputation of the organization are impacted by every contact the customer makes with the organization. At the consulting firm of Harrington, Hurd & Rieker, the young lady that answered the phone had extensive training in all of the organization's consulting methodologies. As a result, when she was unable to locate the consultant that the client wanted to talk with, she was in a position to collect the information that the consultant needed to respond to the situation. Very often, she was able to provide the client with the correct answer immediately. She also developed a personal interest in all of the major clients and could say, "Good morning, Mr. Abbott," by recognizing his voice when he said, "I'd like to speak to Walter Hurd, please." It's great when your clients volunteer good comments like "She makes me feel like I am your most important client."

Everyone who comes into contact with your customer needs to be prepared for that contact. This includes the bellhop, the accountant, the floor sweeper, the secretary, and yes, even the president. They all need to have customer relations training and need to understand the products and/or services that are being provided.

Every Nissan Infiniti dealer's employees, including clerks and receptionists, attend a six-day boot camp. They learn how to treat customers as "HONORED GUESTS." On the "Firebird Raceway" they drive the Infiniti and competitors' cars like the Lexus 400, Mercedes 300E, and BMW 535. Ken Petty, the class instructor, says, "The receptionist probably talks to more customers than any other person in the showroom." Is it any wonder that J.D. Power and Associates' research reveals that Infiniti and Lexus both get the highest ratings for customer satisfaction, in spite of Lexus receiving much higher ratings for car quality. When it gets right down to it, all of your people either directly service the external customer or service those that do. Keeping this in mind, we all have impact on external customer satisfaction, because the people who service the external customer cannot provide excellent service unless they receive excellent service from the rest of us.

■ The Customer Satisfaction Process

How do you get high customer satisfaction ratings? The 1980s were driven by customers who wanted high-quality, luxury products and services at any price, but the long recession that occurred in the early 1990s has changed customer expectations. Today they are looking more and more at value. The way to provide high value to your customer, and, as a result, get high customer satisfaction ratings is:

1. Define new products and services based upon customer inputs and undefined needs.

2. Deliver products that have outstanding quality, durability, and performance.

3. Select customer interface personnel who enjoy people, and train them so that they are technically competent.

4. Aggressively seek out suggestions from your customers and your employees.

5. Provide a wide range of products and/or services that are priced lower and perform better than your competition.

6. React quickly and nondefensively in handling complaints.

7. Look for things and trends that could cause future problems and correct them before they become complaints.

8. Be sure that all employees at all levels receive feedback from the external customer and that some of the employees from each area are given a chance to meet with the external customer.

■ Customer Data

How do you measure customer satisfaction? Well, the best way to measure your customer's satisfaction with the organization is by reviewing your change in market share. If you are providing the best value to your customer and reaching your potential customers, your market share will continuously grow, and if customers are selecting your organization in preference to another, you probably have a high level of customer satisfaction.

But market share does not tell you how you are performing today. It tells you how well you performed yesterday. When you measure a downturn in your market share, it probably is too late to do much about it. As a result, we need a much faster measurement. In most organizations today, management wants to be proactive, defining weaknesses in their products and services before they become a critical factor. This is where customer surveys come into play.

Information is the critical foundation that a strategic plan is based upon. Without the comprehensive capturing, understanding, operationalizing, and utilization of internal and external intelligence, an organization can never become world-class. An external customer data collection and analysis system needs to be defined that will allow the organization to understand what is really important to the customer, how these needs are changing, and measure the positive and negative gap between the organization, its competitors, and its customers' expectations.

A thorough external customer data system includes three parts.

1. Baseline studies
2. Monitoring studies
3. Long-term studies

Baseline Studies

Baseline studies are directed at collecting data and analyzing it to provide insight into key issues, developing trends, new opportunities, changing technologies, competitive trends, etc. They are the foundation of the strategic plan. Baseline studies include current customers, your competitors' customers, and potential future customers. They answer questions like:

- How should we use our resources to provide the best performance?
- What do we need to do?
- What is the competition doing and what direction are they moving in?
- What strategic alliances should we have?
- Where are we now?

Monitoring Studies

Monitoring studies are designed to collect and analyze current trends. They are keys to the continuous improvement process. They measure the ultimate success or failure of the organization's change process. Typical data sources are:

- Field reports

- Customer surveys
- Benchmarking
- Lost-customer studies
- Complaint analysis

Part of the monitoring studies has to be related to delivery of the organization's products and services. Measurements should take place right after the delivery process. Follow-up data should be collected during the first 90 days after delivery and again about 12 months after the product/service was delivered to the external customer. For example, about 90 days after I purchased a Lincoln, my wife received a dozen red roses from the salesman, asking for her evaluation of the sales process, the car, and the service organization.

Monitoring studies tell you how the organization is performing, how things are changing, and if the competitive gap is changing.

Long-Term Studies

Long-term studies often utilize monitoring and baseline study data as well as special data collected for the individual study. Long-term studies are used to help make critical business decisions. Typically, they focus on long-term changes in:

- ROI
- Organizational performance
- Organizational effectiveness
- Organizational mobility
- Changing customer perspectives

Long-term studies answer questions like:

- How can the organization make better decisions based upon fact?
- What is the payback from different alternatives?
- What did the organization achieve?
- How can we react better to the changing environment?

■ External Customer Data Systems

The one voice you cannot turn your back on is the voice of the customer. In fact, you need to amplify the very slightest murmur into a deafening roar that sets the corrective action system in motion. As Tom Peters, the author of "In Search of Excellence" put it, "Listening to customers must become everyone's business. With most competitors moving ever faster, the race will go to those who listen (and respond) most intently." At IBM in the 1970s, we used to say that if a customer had a problem, we would blacken the skies with planes getting to the customer to correct it. In truth, if we had really been sensitive to the voice of the customer, the problem would have been solved by a phone call three days earlier without upsetting the customer.

The following are the four phases to managing the customer database.

Phase 1—Understanding the external customer. During this phase the organization collects and analyzes direct customer input. This is accomplished through round tables, focus groups, user groups, and customer interviews.

Phase 2—Developing customer satisfaction measurements. Measurement systems should include a set of core questions that are used year after year, and special operational issues that reflect the present data requirements. Data is collected using personal contact, mail surveys, phone surveys, benchmarking, data service firms, etc.

Phase 3—Organizing and reporting. Many sources of data are now consolidated in order to be analyzed and prioritized. Data is segregated into different classifications—those that require immediate corrective action, and those that will be used in trend analysis. The data system should focus on exception reporting using statistical trend analysis. Both positive and negative trends should be analyzed to correct negative trends and to insure that the root cause of the positive trends are analyzed, understood, and institutionalized.

Phase 4—Improvement. This phase uses the problem-solving cycle to define root causes, identify appropriate corrective action, and monitor the corrective action to verify that it was effective. The status of each improvement opportunity should be tracked as part of the data system described in Phase 3. Be sure that the improvement process provides direct, rapid feedback to the customer that identified the problem.

■ External Customer Satisfaction Measurements

The voice of the customer must be fed back to all who come in contact with the external customer on a very personal basis. The most probable customer contact points are:

- Marketing
- Product engineering (technical solutions to customer problems)
- Sales
- Delivery
- Administration
- After-sales service

For each of these contact points, a series of measurements needs to be developed to evaluate the contact from the external customer's standpoint. It is extremely important that we manage the external customer's experience. For example, typical questions we should ask about administrative contacts could be:

- Was the billing process on time and was it accurate?
- Were there any unexpected costs?
- How responsive was the organization to your problem?
- Was it easy to obtain information?

In addition, data needs to be collected on the total customer experience. For example:

- Overall satisfaction
- Would the customer buy from the organization again?
- Would the customer recommend the organization to friends?
- How does the organization compare to its competition?
- How well does the organization understand the customer's business and their problems?

IBM Rochester reports that an increase of 1 percent in their customer satisfaction index equals a revenue opportunity of $257 million.

■ Customer Complaint Handling

Too many people view customer complaints as a necessary but nonproductive part of their business. In truth, every complaint should be treated as a gem. It is an opportunity for the organization to improve. You should look at every complaint as an opportunity to save a potential lost customer.

The Consumer Affairs Department reports that only four out of every 100 dissatisfied customers ever complain. The other 96 just walk away and start to look for a new source of the product or service. These 96 percent of your customers are either too nice, or don't want to take the time and effort to complain. They are called "the silent switchers." Of the 4 percent that give you a second chance, 95 percent will continue to do business with you if you quickly respond and resolve their problem. What is quick? Well, there is no one right answer. It depends upon the product. As a rule of thumb, if the product is under $100, quick is less than 10 minutes. If it's over $100, quick could be as much as three days.

These figures are for general products or services. Individual products vary a great deal from the averages. IBM Rochester looked at customers that were purchasing mid-size computers and found that 84 percent of the customers that had no problems would buy from IBM again, and 91 percent would recommend IBM to other people. Of the customers that had a problem and were completely satisfied with the way it was resolved, 91 percent would buy from IBM again and 94 percent would recommend IBM to others.

If your organization does not respond quickly to customer complaints, between one-third to one-half of them will find a new source for the service and/or product you provide. In addition, they turn out to be your competitors' best advertisement.

An unhappy customer will tell his or her story to an average of nine other people, and 13 percent will tell it to more than 20 people. If you can get the dissatisfied customer to share his or her concerns with you, you have a chance of saving the customer. There are three factors that impact the customers' desire to share their concerns with you. They are:

1. How easy it is to complain.
2. Their belief that something will be done about it.
3. The impact that the dissatisfaction has on the customer.

Make it easy for your customers to express their disappointments and/or concerns about your service and/or products. Ask them for their feedback while they are receiving your output so you can correct any problems on the spot. For those who are embarrassed to tell you, provide them with an evaluation form and a pen. Use follow-up phone calls to get their impression after they have used your output. A phone call makes it easier for some people to share their displeasure because they don't have to look you in the eye. Ask for their suggestions (not their complaints) on how the organization could be better.

I have an item that symbolizes how many customers feel about your complaint departments. It is a hand grenade with a number 1 on the pin. The sign above the hand grenade reads, "Complaint Department—Please Take a Number." Too many complaint departments view themselves in just that way. They look at the customer as another problem, not as another person that they are there to service. Eliminate your complaint departments and replace them with a customer service department. What you call it makes a big difference in the way your employees react to the customer.

Suggestions versus Complaints

Complaints are perceived as negative and, in most cases, they are. Suggestions, on the other hand, are ideas designed to help the organization. As beneficial as it is to react to complaints, proactive organizations welcome and reward suggestions that come from customers. Many organizations, like First Chicago Bank, bring customers into their performance reviews to solicit their comments and ideas. Roger S. Penske, former race car driver, invited 40 customers (independent distributorships) to the Detroit Diesel Corporation warehouses in Canton, Ohio. The purpose of this visit was to have the customers understand how the warehouses that they interface with operate, and make suggestions on how they could be improved. These customers provided about 250 different suggestions that helped cut engine part delivery time from five to three days. Emergency orders now take less than 24 hours. As a result of the increase in customer focus by Detroit Diesel, Penske was able to make a major turnaround in the company within a 24-month period. It went from a company that had been losing money for years to one that made $21 million profit, and increased its market share from 3 to almost 6 percent.

■ Getting and Staying Close to Customers

The key is to get close and stay close to your customers, to build up a personal bond with them so that your organization understands their needs, expectations, desires, and moods. This understanding needs to penetrate the entire organization at all levels.

Top management needs to get out from behind those big mahogany desks and start talking to customers, waiting on them, understanding them, and understanding the difficulties that face the front-line troops. Detroit Diesel requires all managers to call or visit four customers a day. Xerox executives spend one day a month taking complaints. At Hyatt Hotels, all executives, including the president, put time in as bellhops. American Airlines executives spend time behind the counter issuing tickets and making seat assignments.

The employees behind the scenes also need to be exposed to the customer. Management has built up walls between most of their employees and the external customer, buffering them from a feeling of responsibility to these important individuals. All too often the people that produce the output never face the person that is going to consume it. In these cases it is easy to lose touch with reality and the customer. Why give that little extra effort if the results are absorbed into the total process so that they are not recognized? It is for this reason that employees have to be put in touch with the external customer. This can be accomplished in a number of ways:

1. Employees making sales calls with salespeople.

2. Employees calling customers directly to find out how they like the product that the employee produced. For example, General Motors has made use of this technique.

3. Job rotation to an interface assignment.

4. Employees hosting site visits.

5. Employees following up on complaints.

6. Performance data feedback from the customer's environment.

Because it is often impossible to have every employee contact customers every two or three months, it is important that the employees that do have the opportunity share their experience with the other employees. Customer-related comments from a coworker have much more impact than when the same data is provided by management.

Are you close to your customer? Are you doing the following?

1. Do customer phone calls take priority?

2. Do you have a customer contact plan?

3. Are you living up to your contact plan?

4. During your contacts, do you:
   - Solicit their opinions and suggestions?

- Understand how their expectations are changing?
- Understand what is important today and what today's priorities are?
- Define potential future needs?
- Use a check sheet to be sure you collect the required data.
- Review action taken on previous suggestions and problems?
- Provide feedback on suggestions and problems within three days?
- Collect and analyze the check sheets to identify early trends?

5. Do you follow up on each lost customer to define the root cause for them changing suppliers?

6. Do you listen to your competitors' customers?

7. Do you have a personal list of your key customers and their phone numbers?

8. Do you have a way to identify trends in buying patterns and take action to determine the cause of negative trends?

Losing organizations normally can answer yes to up to five of the eight questions. Your organization is a survivor if it can answer yes to all eight questions. Of course, just doing these eight activities does not make a winner out of an organization. What makes the big difference is how well they are executed.

Designing for Customer Satisfaction

In the 1970s, we designed products to make a profit. In this process, the cost of the product drives the selling price. If the selling price is too high, we go back and redesign the product and the process. This turns out to be a very costly and time-consuming way of developing new products.

In the 1990s, the more advanced organizations are designing from the customer satisfaction standpoint. Figure 4.1 shows this process.

In the new product development model, primary emphasis is placed upon what the customer will pay for the product. The product is designed around this target cost. Subdividing this cost to the component level drives designers and engineers from all areas of the organization and its suppliers to work together, making the proper tradeoffs required to meet customer expectations. This reduces product cycle time and maximizes product performance levels.

Developing Strategic Customer Partnerships

Being all things to all people is difficult, and trying to meet this standard often can lead to disastrous results. More and more organizations are beginning to realize that it is best to select a few core customers with whom they will establish very close personal relationships. These core customer relationships are called "strategic customer partnerships" or "formal customer partnerships." A partnership is a relation-

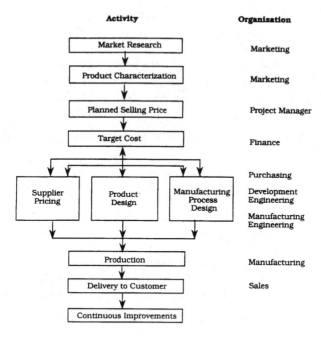

Figure 4.1. Design process for winning organizations.

ship between two or more individuals or organizations that is mutually beneficial to all parties.

What are the advantages of forming strategic partnerships?

- To identify business issues for high-impact customers and qualify sales opportunities (apply opportunity analysis methodologies).
- To gain insight into future business trends.
- To develop closer working relationships with high-impact customers at the executive level.
- To align development activities to the customer's development cycle, resulting in increased sales revenue and sales team productivity.
- To improve the success rate of new products.

The degree of strategic partnering will vary a great deal based upon the product or service the supplier provides. If the supplier is a management consulting firm, the partnership relationship can be very comprehensive and have a major impact on both organizations' future. If the supplier is a component supplier, it could be limited to cooperation during the product development cycle and an annual supplier commitment to improve component performance in the customer's environment. In return, the customer grants the supplier preferred supplier status. For example, a customer partnership between a materials distributor and a tool and die shop resulted in the distributor being given the responsibility for managing the customer's warehouse.

Let's look at a situation where a close, long-term strategic partnership is desirable. In this case, the steps to develop a customer partnership are:

1. Select the target organization.
2. Meet with the organization's executives to get their agreement to collaborate.
3. Understand the customer's strategic direction.
4. Define opportunities for joint cooperation.
5. Present proposal to customer executives.
6. Sign a partnership agreement.

In today's environment, customers around the world are looking for suppliers that want to establish a partnership relationship with them. They want their suppliers to understand their business and their competitive issues. Many of these customers are talking about empowering their employees, but the truth of the matter is that more and more of the buying decisions are being made at the senior executive level as they select organizations to partner with. In this kind of environment, customers are looking for a different kind of salesperson, one that uses consultive selling methods (providing solutions to the customer's problems), not one that sells items. The salesperson needs to focus not only on the quality of his or her sales approach, but also on the effectiveness of the approach.

Competition in today's marketplace is fierce, and if you're going to win, you have to have a sales team that is performing value-added service as viewed by your customers. For this to be accomplished they should:

- Target strategic customers.
- Make sales calls at the highest level.
- Develop partnership relationships.
- Differentiate your products and services.
- Sell solutions to the customer's problems.
- Demonstrate value in the relationship.

Selecting Target Organizations

A partnership represents a major commitment on the part of the supplier and the customer. Partnership agreements should not be entered into lightly. What do you look for when you are trying to identify a potential customer partner?

- Is the organization one of the top revenue drivers for your organization?
- Is the organization a technical leader that will push the state of the art in your organization's products and/or services?

- Will the added profit offset the cost?
- Is there a potential of greatly increasing your revenue from the customer?
- Do you believe that they would consider a partnership relationship?

Based upon your answers to these questions, select five to ten potential customer partnerships to target. Now collect all the information you can on these organizations. Get copies of their yearly reports and their Dunn and Bradstreet reports. Find out what their values are and their critical success factors. Review the new products they have released in the past five years and how successful these products were. Get information on how these organizations use your products and your competitors' products. Analyze this data and be sure that you still want to develop a partnership with each organization. This step will usually reduce your list by about 50 percent.

Meeting with Potential Customer Partners

Now that you have a set of potential customer partners, your job is to convince them that the partnership would represent a valuable asset to them. It will take time and effort for your partners to acquaint your organization with their business and its critical success factors, so they must feel there is value in investing their time and effort in preparing your organization for a long-term relationship. To offset these expenses, your organization needs to be able to provide your partners with creative solutions for their problems that they could not conceive themselves for the same cost.

Evaluate your organization to determine how it could have a positive impact on your potential customer partner's performance in the following areas.

Direct value	Integrated value	Strategic value
• Increased revenue	• Improved service level	• Improved processes
• Increased sales	• Increased productivity	• Reduced cycle time
• Cost reduction	• Increased market share	• Improved quality
• Rework reduction	• Reduced variation	• New markets
• Cost avoidance		• New products

If the analysis reveals a beneficial relationship for both parties, contact your potential partners and set up a meeting at the executive level to review your proposal to form a partnership. When setting up the meeting, work through your normal contact. Do not go over his or her head to set up the meeting or you will create a potential problem.

During this meeting, point out how this partnership will help the potential partner, using the different value impacts. Ask the customer if they know of other value impacts that the partnership would have. Be sure you demonstrate that you understand their business problems and their critical success factors (if you have been able to obtain this information). You should also share your critical success factors at this meeting. Discuss what your organization hopes to gain from the partnership. Explain that you want to invest additional resources to be sure that your team better understands their priorities and how you can help the potential partner. If the customer agrees, ask that a customer sponsor be assigned to work with you to develop the relationship.

Understanding the Customer Partner's Direction

To understand the partner's direction, you will need to collect information about the organization through direct contact with key personnel, supplemented by focus group meetings. The purpose of this activity is to:

- *Determine organizational goals.* What are upper management's visions for the organization; what is its central mission?

- *Understand the organization's critical success factors.* What things must be accomplished for the partner to meet his or her goals?

- *Identify organizational obstacles.* What are the things that are or can be in the way of achieving the critical success factors?

- *Predict the impacts.* What are the things that could happen if the obstacles are not overcome?

- *Find potential solutions.* In what ways can your organization help the partner overcome these obstacles?

Defining Opportunities for Joint Cooperation

Based upon your study of the customer's direction, develop a plan that will support the customer's goals and provide unique value-added to the customer. Present the benefits of the plan to the customer and verify these values from the customer's standpoint. Use value impact analysis methodology to help identify opportunities. Develop a value matrix to show how the relationship will impact the customer.

To prepare a value matrix, rank each solution's impact on the partner's critical success factors and estimate the time frame for obtaining the value. The customer's sponsor should be very involved in these activities and should concur with the proposal and the estimated benefits. When the sponsor feels comfortable with the proposal, you are ready to move on to the next phase.

Presenting Proposal to the Customer's Executive Team

The sponsor should now arrange for a presentation to the customer's executive team. The presentation should be very carefully prepared since it will be the hallmark of future activities. In preparing the presentation, think about the following:

- What level of commitment do you need to proceed?
- What is the value-added from the customer's standpoint?
- What is the value-added from your organization's standpoint?
- Identify specific actions required from the customer, along with estimates of resources needed.
- Focus on priority opportunities.
- Define the next step in the process.
- Be sure to have time for closure.

The presentation should include benefits to both parties and a working relationship that will be required to implement the proposal. The output from this phase is often a signed partnership agreement that defines the relationships and potential future cooperations.

■ Summary

All organizations need to have an overriding objective of becoming the preferred supplier to all of their customers and potential customers. When this state is reached, all of the stakeholders are rewarded.

Let's look at the different ways three different types of organizations think about their external customer and the external customer's perception of them.

How marketing and sales personnel think about the external customer

Losers: They are herds of people that need to be manipulated. They don't know what they want. They are complainers.

Survivors: We need to find a way to make them want our products or service. Our sales campaign needs to be directed at getting them to sign on the bottom line fast.

Winners: We need to understand their problems and provide them with solutions that delight them.

How employees think about the external customer

Losers: They don't impact me and I never see them.

Survivors: Marketing and sales need to get more of them for us.

Winners: We need to do better for them. When I foul up, it hurts our external customers.

How management thinks about the external customer

Losers: We need to find more new customers.

Survivors: We need to get a bigger share of the market. We will try to keep the old customers, but our major effort needs to be directed at getting new customers.

Winners: We need to improve the way we service them and predict their future needs so that we can provide the product or service when they need it. Our old loyal customers come first. We can't afford to lose even one.

How the customer thinks about the organization

Losers: They're OK. I guess we get what we pay for and nothing more.

Survivors: They're good organizations. They meet my needs.

Winners: They're great! They went out of their way to service me. I'll recommend them to anyone.

Remember, if it costs x dollars to keep a current customer, it costs $5x$ dollars to get a new customer, and $12x$ dollars to win a lost customer back. The key to keeping your customers is to manage the quality of the total customer experience.

■ References

1. *Business Week*, March 12, 1990, p. 88.
2. Raymond J. Larking, Vice President, American Express.

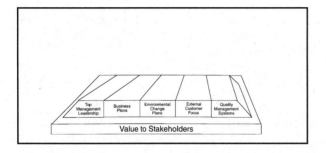

5

Quality Management Systems: ISO 9000 and More

Ralph Ott
Manager, Ernst & Young

*A Quality Management System will not win the game
for you but the lack of it can cause you to lose the game.*
DR. H. JAMES HARRINGTON

■ Introduction

Many management systems are required to run an organization. There are financial systems, personnel systems, production control systems, etc. These are all very important and must be addressed as part of a total improvement process. Three management systems, however, can be considered "Improvement Management Systems." They are the:

- Safety Assurance Management System
- Environmental Assurance Management System
- Quality Management System (QMS)

These three systems are crucial to a successful improvement process and serve as basic building blocks in the "Total Improvement Management (TIM)" methodology. Although all three are critical, due to the large scope of the subjects, this chapter will focus only on the Quality Management System. This does not mean, however, that the other two can be ignored when you design and implement TIM.

We are focusing on the QMS because it is a key to effective improvement processes and because it focuses on customer satisfaction, an important goal for virtually all companies. Without a good QMS, improvement activities are unstable. It is like building a big, beautiful castle on sand with no foundation. It will look good for awhile, but will soon collapse under its own weight. The QMS is a prerequisite for an effective improvement process. Once you have it in place, you can start your continuous improvement activities.

■ What Is a Quality Management System?

Quality Management Systems (QMS) can take many forms. Therefore it is better to define Quality Management Systems in terms of their function or intent rather than in terms of how they are implemented. The intent of a Quality Management System is simply: *To ensure that an organization consistently meets customer requirements.*

Quality Management Systems define *how* organizations operate to consistently meet customer requirements. Clearly, there are many ways that organizations can achieve this goal, and so the systems are specific to the organizations which implement them. There are some common characteristics, however, in most Quality Management Systems:

- They cover a broad scope of activities in the organization. Quality is defined in broad terms and includes not only product performance characteristics, but also service characteristics (e.g., delivery, support, etc.) that customers demand.

- Since *consistency* of results is paramount, Quality Management Systems focus on consistency of the work process. This often includes some level of documentation to standardize work.

- Quality Management Systems emphasize prevention of errors rather than reliance on error detection and reaction.

- Recognizing that not many systems will be 100 percent effective at prevention, there is also emphasis on corrective action of problems that are encountered. Quality Management Systems are, in this sense, "closed loop" systems. They include detection, feedback and correction.

- Finally, most Quality Management Systems include elements of measurement to increase their effectiveness and/or identify problems.

While these elements are common to many Quality Management Systems, they will take different forms for individual organizations. As one would imagine, the needs of organizations the size of Xerox and IBM can be quite different from those of smaller firms like NetFrame or Solectron. Other factors like markets, technology, and organization strategy also place different demands on Quality Management Systems, forcing them to take a variety of forms for different organizations.

Benefits of a QMS

There are some obvious benefits to an organization which can consistently meet or exceed its customer requirements. These include:

- Improved customer satisfaction
- Increased customer confidence
- Improved market reputation
- Improved market share

From an organization's internal perspective, some of the benefits are:

- Reduced rework
- Lower costs
- Lower inventory
- Less employee frustration associated with rework and recurring problems

While these benefits are obvious, let's take a closer look at the underlying value of Quality Management Systems and what they will and will not do for an organization. Quality Management Systems are not the answer to uncompetitive lead times in your semiconductor plant or weakening brand awareness for your toothpaste line. Quality Management Systems act as a foundation for improvement efforts and as the mortar to hold past improvements together.

When I have asked the leaders of organizations in various industries to rate, on a scale of 1 to 5 (where 1 very few and 5 very many), how many organizations in their environments are trying to improve their operations and financial position, I have gotten consistent answers in the 4 to 5 range. When I have asked these same people to rate, on the same scale, how many organizations in their environments are actually realizing improvements in their financial positions, I have gotten consistent answers in the 1 to 3 range. The results beg the question, "Why do only few organizations reap the rewards of improvement efforts?"

There are many reasons why this gap may exist. First of all, when we speak of improvement activities, people generally think about operational improvements, such as improving execution of a given marketing strategy. Unfortunately, the perfect execution of a poor marketing strategy will yield poor results. Let's consider IBM's Market-Driven Quality (MDQ) efforts and its recent problems. Most outside observers agree that IBM is a quality-focused organization. It put into place Quality Management Systems well before its MDQ efforts. Was an MDQ program enough? For anyone who owns IBM stock, it was not. A strong QMS did not prevent problems for IBM because the innovation in information technology moved the market from mainframe technology to networked workstations faster than IBM's assets would allow it to move. Quality Management Systems generally focus on the execution of strategies. They do not explicitly address strategy development. Some consider this a shortcoming of Quality Management Systems. What is most important, however, is to realize that this limitation exists and to act accordingly.

Because they drive toward consistent results through consistent process execution, Quality Management Systems provide the means for sustaining the improvement activities gain. In doing so, they are in essence, a prerequisite for effective improvement. Quality Management Systems create consistent performance, whether good or bad. In doing so, they provide a sound basis for managing the process.

Quality Management Systems provide a way to institutionalize an organization's best practices, to turn best practices in one area of an organization into everyday practices throughout the entire organization. They also provide for better control of the business, much like financial management systems. This control is especially important in high growth organizations, where it is likely that work processes change rapidly. Often in these environments, because of the high degree of change, results also change rapidly but not always for the better. A QMS allows an organization to understand what its current practices are and make decisions to change them consciously. The understanding of the current processes is an important step in determining opportunities for improvement. Implementing a QMS will help an organization make changes and internalize the quality process at every level of the organization, improving its overall business processes and reducing errors and their costly consequences. Cuts in an organization's waste and cycle time will reflect favorably on the bottom line. The requirements of a QMS are especially helpful in realizing these benefits because they are customer-oriented.

In any environment, consistent process execution is the key to consistent process results. This is what Quality Management Systems address and is perhaps their biggest benefit. This drives organizational performance and results in improved customer satisfaction and efficiency.

The ISO 9000 series does not define the best Quality
Management System, but it is an excellent starting
point and provides a strong foundation to build upon.
DR. H. JAMES HARRINGTON

■ The Development of Quality Management Systems

With the simple concepts and obvious benefits previously described, it should come as no surprise that Quality Management Systems are not a new idea. Organizations have pursued improved customer satisfaction and internal efficiencies, in one way or another, for as long as businesses have existed. In fact, in the days of the Egyptian Pharaohs, there was an extensive and formal quality system related to burials. "The Book of the Dead" described the methods by which burials should be performed and specified exact procedures for burial. The successful application of the standard was proven by the mark of the Superintendent of the Necropolis.

The more formal and widespread adoption of quality management concepts, however, began during Word War II. Since then, organizations in many industries have incorporated Quality Management Systems into their own operations and/or have required their suppliers to implement Quality Management Systems. With this movement, organizations developed guidelines for themselves or for their suppliers to follow. These organizations would then audit their operations or suppliers' operations to ensure compliance with the guidelines.

There are now a variety of Quality Management Systems in place around the world. They range from systems developed by a plant for its suppliers (IBM San Jose Supplier Manual) or by an entire organization for its suppliers (Ford's Q1), to systems developed to regulate an industry (Department of Defense's MIL Q 9858A), and international standards (ISO 9000 series).

The Department of Defense developed MIL-Q-9858A as a guideline for all defense contractors to follow. DOD contractors are audited to the MIL-Q-9858A specifications. Ford Motor Company developed the Q1 program which specifies guidelines for Ford suppliers, to which those suppliers are audited. The Q1 program has been very successful, and many organizations have modeled quality system requirements after this program.

Every organization has a QMS, formal or informal. Outside of the formal ones mentioned above, a company might have an informal, unwritten policy that "we never ship product X until John has taken a look at it" among many other informal procedures. Whether formal or informal, the quality system is only effective if it *ensures that an organization consistently meets customer requirements.*

With all the publicity the quality movement received during the 1980s, the marketplace moved quality from a differentiator to a requisite for market entry. Because of this, customers have had to become intimately familiar with their suppliers' Quality Management Systems. This has been especially strong in cases where organizations have included their suppliers' components in their products. If you include a battery from organization X in your new car model you had better be able to vouch for the quality of the battery. If the car doesn't start, *your* new car model's reputation will suffer, not necessarily that of your supplier.

The result of the 1980s movement was increased importance of an old World War II concept. In most organizations, the supplier quality engineer began visiting supplier factories to certify their quality systems. The natural side effect of this movement was that many *different* guidelines were developed. Organizations would each develop guidelines for Quality Management Systems which their suppliers had to comply with. Each organization would ask its suppliers to comply with its guidelines and audit the supplier to ensure that it did comply. Looking from the suppliers' viewpoints, each supplier was receiving different requirements from different customers and having to comply with them all and be audited by them all. Imagine this picture when there are 50 customers and 50 suppliers, or even 100 or 1000. The result is a huge transaction cost for the world economy and an enormous barrier to exchange of goods or services when different standards are used. In essence, this is the situation that has been building for decades.

There are really two issues here. First of all, with so many different standards, organizations would have to expend significant resources to comply with all those standards imposed by their customers. Secondly, even if organizations agreed to common standards, this would reduce the preparation necessary to achieve compliance with them, but would not necessarily reduce the number of audits performed and the time and effort associated with them. Picture a small organization of 200 people with one quality manager that has 50 customers coming in each year to audit the QMS. Each audit lasts two days, and the quality manager and one quality engineer spend the two days hosting the customer. It is easy to see that both would spend 50 percent of their time on the QMS audit process, even if everything goes well.

As an analogy, think of the cost we collectively would incur if we had to spend two weeks checking the backgrounds of our new dentists or doctors. The collective effort would be astronomical. Board certification of these professionals lessens the need for each individual to check backgrounds. The quality profession felt that the same approach could be applied to reduce the effort and conflicting requirements that were being placed on organizations' supplier networks around the world. As a result, the International Standards Organization (ISO) brought together quality professionals from around the world to prepare a QMS specification that could be used internationally by all industries.

The International Standards Organization (ISO), based in Geneva, Switzerland, is an organization comprised of representatives from the standards-setting bodies of over 90 countries around the world. The American National Standards Institute (ANSI) represents the United States at the ISO. Representatives from many other organizations get involved on specific issues as well. The mission of the ISO is to facilitate the international exchange of goods and services. Understanding that varying requirements for Quality Management Systems presented a barrier to the efficient exchange of goods and services, the ISO initiated a technical committee (TC 176) to develop standards for Quality Management Systems. After several years of work, the ISO published the first version of the ISO 9000 series standards—standards for Quality Management Systems, in the mid-1980s.

The ISO 9000 standards are very generic standards for Quality Management Systems in that they can be applied to any organization in any industry. They are not written as industry, process or product specific; the interpretation and implementation of the standards will vary depending on the environment.

Along with the development of the ISO 9000 standards, a system for "certifying" (or "registering") organizations to the standards through a third-party audit system was developed. Since the standards are the same for all organizations in all industries, it is possible for independent bodies (called registrars or certifying bodies) to audit organizations to ensure compliance with the standards. The third-party audit system is designed to have one or two thorough audits conducted each year that will meet the requirements of all 50 customers.

The ISO 9000 standards and their associated third-party audit system were developed to address the cumulative cost issues mentioned above, and to address the associated barriers to trade.

ISO 9000 certification is not intended to replace a reputation for quality goods with a stamp that says "certified." It is meant to supplement it and remove some of the cost of investigation. It does not guarantee that it will provide you with a quality output. It does indicate that the organization has a system that meets the requirements of the ISO standard. Again, in terms of our analogy, ISO certification serves a similar purpose as professional certification. Offered two professionals, one with certification and one without, and everything else being equal, a consumer will generally choose the certified professional. This is not a guarantee, however, that the professional will never make a mistake or even that the professional will always perform better than the noncertified individual.

The ISO 9000 standards have been very successful, as evidenced by the adoption of the standards in a wide variety of industries worldwide. To the time of this writing, over 50 countries have adopted ISO 9000 as a national standard. Tens of thousands of organizational operations have been certified to the standards worldwide and thousands of others are currently pursuing certification. Manufacturing sites, software organizations and service organizations have all shown interest, though most of the activity has been in the manufacturing sector to date. Some European Community directives incorporate ISO 9000 systems certification as alternatives or additions to product certification requirements. The true measure of ISO 9000's success in the future will be based upon how many organizations are willing to drop their unique requirements in preference to specifying an ISO 9000 standard, and this has yet to be fully observed.

The initial success of the ISO 9000 standards at organizations like IBM, Hewlett Packard, and Sun Microsystems is a statement about the belief regarding their value. While there is not much hard data to date that demonstrates the value of the ISO 9000 standards, anecdotal evidence is strong and growing. But again, the most compelling evidence is in the standard's rate of and breadth of adoption. Although there are other more comprehensive Quality Management Systems, the ISO 9000 series provides a strong and widely accepted base to build your improvement process on, and work is underway to release an even more comprehensive update that will guide organizations in the twenty-first century.

■ The ISO 9000 Series Standards—An Overview

The ISO 9000 series is a set of five individual, but related, documents that define international standards for quality management systems. They were developed with the goal of documenting the quality management system elements to be implemented in an organization in order to maintain an effective QMS. The series does not specify the specific techniques or technology that should be employed in a QMS. ISO 8402 is a guide to the terminology and definitions used in the series. The ISO 9000 documents are:

- ISO 9000 is the road map for the series. Its purpose is to provide the user with guidelines for the selection and use of ISO 9001, 9002, 9003, and 9004.

- ISO 9004 sets guidelines for the implementation and auditing of the actual QMS.

- ISO 9001, 9002, and 9003 are quality system models for external quality assurance. These three models are actually successive subsets of each other. These are the documents which define the specific "contractual" requirements and which companies are actually audited against.

ISO 9001 is the most comprehensive standard in the series—covering design, manufacturing, installation, and servicing systems.

ISO 9002 covers production and installation. It does not cover design or service functions.

ISO 9003 covers only final inspection and test. ISO 9003 would apply to an organization which does not manufacture its products, buts sells products that have been purchased from other sources or facilities. ISO 9003 is limited in its guidance and may encourage a philosophy of "inspecting in quality" as opposed to "building quality in."

Whereas the standard is being applied here as a basis for an organization developing a quality management system and not external compliance, you should consider the most stringent standard, ISO 9001 (the applicable portions of it), with the guidance of ISO 9004. ISO 9004 represents the basic management code of practice upon which the rest of the series and a QMS are built. In the context of Total Improvement Management, you should aim to install a QMS for your own benefit, not to satisfy the need to be certified by a third party. The standards that apply for this purpose are ISO 9001 and 9004. The summary of each element of ISO 9001, shown in Fig. 5.1, was published in the June 1990 issue of *Quality Progress*, page 51.

Management responsibility. Requires that quality policy be defined, documented, and communicated throughout the organization; that responsibility regarding quality be clearly defined; that in-house resources are available for verification activities; that a management representative be appointed to ensure quality system requirements are being met; and that the management representative lead a management

Section

0	Introduction	4.9	Process control
1	Scope and field of application	4.10	Inspection and testing
	1.1 Scope	4.11	Inspection, measuring, and test equipment
	1.2 Field of application	4.12	Inspection and test status
2	References	4.13	Control of nonconforming product
3	Definitions	4.14	Corrective action
4	Quality system requirements	4.15	Handling, storage, packaging and delivery
	4.1 Management responsibility	4.16	Quality records
	4.2 Quality system	4.17	Internal quality audits
	4.3 Contract review	4.18	Training
	4.4 Design control	4.19	Servicing
	4.5 Document control	4.20	Statistical techniques
	4.6 Purchasing		
	4.7 Purchaser supplied product		
	4.8 Product identification and traceability		

Figure 5.1. Major headings in ISO-9001.

review periodically to ensure the continuing suitability and effectiveness of the quality system.

Quality system. Requires a quality system that meets the criteria of the applicable ISO 9000 series standard be established and maintained (documented as a quality system manual and implemented) as a means of ensuring that product conforms to requirements.

Contract review. Requires review of contracts to ensure requirements are adequately defined and to ensure the capability exists to meet the requirements.

Design control. Requires procedures for controlling and verifying product design to ensure that specified requirements are being met and to include procedures for design/development planning, design input/output, design verification, and design changes.

Document control. Requires establishing and maintaining procedures for controlling documentation through approval, issue, change and modification.

Purchasing. Requires that purchased product conform to specified requirements; ensured through subcontractor assessments, clear, and accurate purchasing data, and verification of purchased product.

Purchaser-supplied product. Requires procedures for verification, storage, and maintenance of purchaser-supplied product.

Product identification and traceability. Requires procedures for identifying product during all stages of production, delivery and installation, and individual product or batch-unique identification as needed.

Process control. Requires procedures to ensure that production and installation processes are carried out under controlled conditions, which include documentation, monitoring and control of suitable process and product characteristics, use of approved equipment, and criteria for workmanship.

Inspection and testing. Requires that procedures for inspection and test at receiving, in-process, and final stations be in place as documented in quality plan; must include maintenance of records and disposition of product.

Inspection, measuring, and test equipment. Requires procedures for selection, control, calibration, and maintenance of measuring and test equipment.

Inspection and test status. Requires that markings, stamps, or labels be affixed to product throughout production and installation to show conformance or nonconformance to tests and inspections.

Control of nonconforming p;roduct. Requires control of nonconforming product to ensure it is not inadvertently used; includes identification, segregation, and evaluation.

Corrective action. Requires procedures for investigating causes of nonconformance, taking action to rectify them, and creating controls to prevent future occurrence.

Handling, storage, packaging, and delivery. Requires procedures for handling, storage, packaging, and delivery of product.

Quality records. Requires procedures for identification, collection, indexing, filing, and storage of quality records.

Internal quality audits. Requires a system of internal audits to verify whether quality activities comply with requirements and to determine the effectiveness of the quality system.

Training. Requires procedures for identifying training needs and providing training for all personnel to meet those needs.

Servicing. Requires procedures for performing servicing as required by contract.

Statistical techniques. Requires procedures for identifying the use of statistical techniques in process, product, and service.

ISO 9000 Requirements Framework

Many business professionals, even those who have been through extensive ISO 9000 training courses have said to me, "I need to explain to someone what the [ISO 9000] requirements are all about, but it would take all day." It is easy to see how this perception can develop for two reasons. First of all, the generic language of the standards makes it difficult for people with little experience implementing such systems, to map the standards to a particular organization. This "mapping" is crucial not only for articulating what the requirements are, but for assessing an organization's current status relative to the standards and for planning an effective and efficient system implementation.

Secondly, if you read ISO 9001, the most comprehensive of the standards, from beginning to end, you read 20 requirements sections (4.1 through 4.20). When completed, it may seem as though there are 20 completely separate requirements without much linkage between them. This being the case, it may be difficult to get a sense for the value of implementing the standards. We have developed a simple framework which structures the requirements in an order which better shows the relationship of the various elements and thereby makes the logic and the benefits of implementation more obvious.

The ISO 9000 requirements framework has three basic parts.

- *The core system* is the most fundamental part of the management system. It includes the elements of the standard which are applied to all parts of the organization. These include Documentation Control, Training, Records, and Corrective Action.

- *The operating system* refers to the various parts of the value chain that are within the scope of the standards: Contract Review, Design, Purchasing, and so on. It also includes operations support elements such as Statistical Techniques, Handling, etc.

- *The management support and control system* refers to Management Responsibility and Internal Quality Audits. Management responsibility is specifically management's role in supporting the system, not to exclude management involvement from the other elements of the standard.

An in-depth discussion of the standard requirements is outside the scope of our purposes here. However, since the core system impacts all areas of the business, we will discuss it here in more detail. As already mentioned, the primary purpose of the standards is to help suppliers consistently meet customer requirements. One of the basic premises that ISO 9000 is built on is that consistency of results starts with consistency of process execution. This consistency can be achieved through documentation, training, automation, or some combination thereof. Two parts of the ISO 9001 requirements are related to Documentation Control (4.5) and Training (4.18), although the requirement for a documented system permeates the entire standard. In short, to achieve consistent process performance, *the way work is performed in all parts of the business should be planned with documentation and training*.

Following from the planning of work, the next step, obviously, is to execute the plan by performing work in accordance with the documentation and training. Next, there needs to be some mechanism in place to measure, or verify, if the process is working; if it is producing the intended results, both in terms of customer requirements and internal needs. The Records portion of the standard (4.16) addresses this need. Records are also referenced in various other parts of the standard. If records indicate that the system is not working properly, then there needs to be a mechanism in place for Corrective Action (4.14). Corrective action directed at the system ensures improvement of the documentation and training which did not produce adequate results.

The message this framework provides, is as simple as it sounds. In the ISO 9000 documentation, the relevant parts of the requirements we call the core system, are separated and not well-linked (they are found in sections 4.2, 4.5, 4.14, 4.16, 4.18). The reorganizing of the standard into the core system, operating system, and management support and control may provide you with a clearer understanding of the standard. You may have also noticed that the core system was put into the context of Shewhart's Plan-Do-Check-Act cycle. Regardless of the labels used to convey the message, the framework provides a meaningful interpretation of the standards *for any business*. This interpretation should prove useful in examining, improving, or implementing your QMS.

Many people today are articulating the interpretation of the standards as "Document what you do; do what you document." While this is essentially a good guideline, don't be lulled into thinking that it is a sufficient rule. When addressing the various elements of the standard, there is definitely a judgment necessary as to the adequacy of procedures. For example, if you document your contract review process, but it does not ensure a suitable means for your organization to have confidence in the delivery commitment you are making to a customer, the procedure is not adequate. Perfect execution of an inadequate procedure is inadequate.

It has been suggested by organizations that have pursued ISO 9000 certification that conformance to the documentation of processes is all that is needed. Some suggest that a key to success in the audit process is not developing detailed documentation so that auditors cannot observe inconsistencies. While some auditors of the standard might agree with the above premise, the documentation should be at a level that ensures the consistent performance of processes that will have an effect on the customer. We are using the ISO 9000 standard as a basis for building a

QMS that supports the organization's strategy; therefore, we should seek to apply the standard in a way that *ensures that an organization consistently meets customer requirements.*

Using the this framework to guide your implementation can provide added focus to your effort. The other beauty of the requirements framework is that the core system can be applied to departments or processes which are not directly addressed by the standard. One of the necessary components of a QMS is that it should be applied to all processes in the organization. For instance, you should use the same principles described in the core system to manage your finance, human resources, legal or *any* other processes, though it is not explicitly a requirement in the standards to do so. In some organizations, the capital allocation or personnel acquisition process is more important to the strategy and customer satisfaction than the manufacturing or order fulfillment process. Applying the framework will help you communicate the standard requirements to others and help you realize the intended benefits of ISO 9000 implementation.

■ ISO 9000 or QMS Implementation Tips and Traps

Align ISO 9000 Requirements with Your Customer Requirements

In designing your Quality Management System, start by asking, "What does my customer require?" Then ask, "What does my organization require?" Constantly review your system design to make sure your system meets both sets of requirements. Many organizations quickly find that their internal needs exceed their customers' requirements.

Many organizations implementing ISO 9000 standards today view the standards as a burden, as something they *have* to do. This unfortunate situation is undoubtedly due to the development and adoption history of the standards. For many organizations, ISO 9000 *has* been a "compliance" issue because of customer requirements, regulations or competitor certification. Some of these organizations view the ISO 9000 requirements as having little overlap with their customer requirements. Organizations with these views ask us to assist them to address that "small" portion of the ISO 9000 standards that they believe overlap with their customer requirements. These organizations want to get implementation over with as quickly and painlessly as possible. They are mostly interested in getting a certificate to show their customers. Unfortunately, they are missing out on a great opportunity.

Other organizations view ISO 9000 implementation as an opportunity and *want* to implement the standards. These organizations have a different view of the requirements. They see the ISO 9000 requirements overlapping directly with their customers' and their own requirements. This, after all, is the intent of the standards: to help organizations consistently meet customer requirements, their own internal requirements, and to maintain internal control. Any requirements which do not guide an organization toward these ends would not add value.

When an organization recognizes the importance of meeting or exceeding customer expectations and sees the ISO 9000 requirements as a direct link to this end, then the implementation is no longer seen as a burden. The organization aggressively pursues implementation with more focus and effort, and this results in a more effective and efficient total operating system.

Assess, Train, Implement

Organizations take two basic approaches toward ISO 9000 implementation. Some organizations start by sending people to educational courses and then have them return to lead the implementation. Then, after some time trying to implement a system, they have a second or third party assess their progress and make course corrections based on the assessment feedback. This approach, in general, does not work well unless the organization has selected very competent and experienced individuals to lead the implementation activities. Many organizations have contacted us because they have taken this route and had unacceptable results.

The alternative approach is to start with an assessment by an experienced third party, then receive training to the standards, and finally plan and implement the system. This approach has resulted in more effective and efficient implementations. The biggest difference between the two approaches is that the first includes a "course adjustment." Stated another way, it includes *rework*—exactly what the implementation of the standards tries to prevent. Approach 2 can minimize rework.

Many organizations have chosen Approach 1 because it delays third-party involvement. Conventional wisdom says, "Why should we have someone tell us that we have nothing in place?" The short answer is that early involvement of experienced third parties will save you effort in the long run, by preventing errors and simplifying implementation. Organizations that start implementation without early guidance are often surprised to find out that they have actually made their systems much more complex than they need to be. Good coaching can help you build a simple, effective, efficient system right from the start and avoid costly rework.

ISO-9000 Initial Assessment

"Why do an assessment when we have not yet implemented a system?" is a typical question. As we stated before, all organizations have quality management systems, either formal or informal. No organization can exist without one. The effectiveness and efficiency of these management systems make the difference between a loser or a survivor. There are three reasons why you would want to pursue an assessment before installing a new QMS. First, the assessments themselves can be an excellent educational experience regarding the standards. Just by listening to the kinds of questions assessors ask, the way they ask questions, and the level of detail they go into, your employees begin to understand the requirements of the standards, *specifically for your business*. Second, the assessment can focus on the adequacy of your informal (undocumented) systems. If your current systems are inadequate, just documenting them the way they are is not sufficient. Third, the assessment findings can be used to tailor training to your needs.

Documentation Hierarchy

As described in the requirements framework, documentation is an important part of the ISO 9000 system. A suitable structure for the documentation system greatly simplifies the task of implementing and maintaining a system. Figure 5.2 depicts the most common structure for QMS documentation.

The top level of the documentation hierarchy is the Quality Manual. The primary purposes of the Quality Manual are:

- To describe what elements of your organization are controlled by the system
- To describe the structure of the Quality system
- To demonstrate that all relevant elements of the ISO 9000 standard are addressed by your system or specifically state which elements are not addressed
- To state your organization's policies for the various elements of your system

The Quality Manual should link to the next level of your documentation hierarchy by referencing the relevant procedures for each part of the QMS described in the manual. The Quality Manual is *not* intended to be a collection of all your procedures. It should be a brief, stand-alone document which meets the above objectives.

The second tier of the documentation hierarchy is the Procedure level. Procedures provide a medium level of detail by focusing on *who does what* in your organization. They define the steps in your processes and the responsibilities for

Figure 5.2. Documentation hierarchy.

executing those steps. They are typically somewhat general, in that they can apply to multiple products, customers or suppliers. Procedures should reference the next level of detail (instructions) when needed.

The most detailed level of the documentation hierarchy is provided by Instructions. Instructions define *how* work is performed. They often provide specific information (e.g., product specific, supplier specific, customer specific, order specific, machine specific).

Another level described in the hierarchy, Records, provides evidence of the proper functioning of the Quality system and evidence of results of the Quality system. The use of Records should be referred to as appropriate, in the Procedures or Instructions. They are listed separately here because they are generally treated differently from Procedures or Instructions. Records tend to be event-driven. For example, your Instructions may specify that you record test results on Form Number XYZ. This form may be completed for each production run or each unit.

■ Procedures, Quality Manual, Procedures, Instructions

The most common question when beginning to develop documentation is, "Where do we start?" Do we start at the highest level (the Quality Manual) and work our way down, or do we start at the Instruction level and work our way up? We have found that the most effective way to document a system is to start in the middle, with Procedures, next develop the Quality Manual, then refine the Procedures and add Instructions. In reality, the various levels will be developed in parallel to some degree, but primarily in this order.

Procedures define processes by describing who does what. They describe the sequence of tasks and the responsibilities for those tasks. Procedures most closely match the way people think about their work. Therefore, they provide an easier mechanism for people to describe and document the organization's processes. Here we have found that flowcharting techniques, often used in process improvement approaches, can be used very effectively. In fact, we have seen clients use software-based flowcharting tools and databases to ease the documentation creation and maintenance processes. We have also found that organizations that have used these types of process documentation tools in reengineering or other process improvement efforts are able to use the documentation to support their QMS (example: *Work Draw—Business Process Redesign Tool Kit* by Edge Software, Inc., Pleasanton, CA). Other documentation methods, like the "play script method" also make documentation easier.

We recommended starting with Procedures, then developing the Quality Manual, and then refining the Procedures. The Quality Manual will describe the overall system and therefore, the relationships between the various elements. By developing this overview of the system relatively early on, you will be in a better position to ensure that the various elements of the system are well linked; that they create a cohesive picture. Use the development of the Quality Manual as an opportunity to test if you are capturing all the appropriate elements of the system in your Procedures and that they fit together in a logical way. In a sense, this will help you define the boundaries of your Procedures and also save rework later on.

System Structures—Three Generic Types

There are many right ways to organize an ISO 9000 management system. The structure of the management system refers to how the various parts of the system relate to each other; how the boundaries are defined for the procedures. This will become more clear shortly. There are three generic system structures. They are oriented around either:

- Standards
- Departments, Function, or Organizations
- Business Processes

In a standards-oriented system structure, your procedures would essentially follow the outline of the appropriate ISO 9000 standard. If ISO 9001 is the appropriate standard, for example, you would have a procedure covering Management Responsibility (section 4.1 of the standard), a procedure covering Contract Review (section 4.2 of the standard), and so on. In the simplest case, you would have 20 procedures in total. In actuality, you may wish to address some of the sections of the standard in more than one procedure. For example, you may wish to have a procedure addressing Receiving Inspection and Testing (section 4.10.1) and a separate procedure covering In-Process Inspection and Testing (section 4.10.2). In any case, the standard guides the way the procedures are organized.

In a department-oriented system structure, your organizational structure is the primary guide of system development. Your would likely have a procedure or multiple procedures for the Customer Service Department, a procedure or procedures for the Production Scheduling Department, and so on.

In a process-oriented system structure, your procedures would align with the natural boundaries of your work processes. For example, you might combine production elements because they occur in sequence as part of the same larger process. You might combine the activities associated with printing a pick list for the day's shipments with the procedure for final packaging and shipping.

Each of these system structures has its merits and may be appropriate for a given organization. Figure 5.3 summarizes the advantages and disadvantages of the three generic approaches. Remember that these are only three models for quality management system structures and that they have been presented as pure models (as applying to the entire system). It is also often useful to combine elements of these pure models to form an effective hybrid. One of the most useful hybrids that organizations employ is to utilize the standards-oriented structure for the Quality Manual with the process-oriented structure for procedures, to perhaps get the best of both worlds.

These three generic system structures provide useful models for organizing the quality management system. Whichever you choose, the process of examining the

Standard-Oriented	Department-Oriented	Process-Oriented
+ Easiest to ensure that all parts of the standard are addressed − May not fit what actually happens in your business	+ May be most natural for employees in some companies + Easy to define implementation tasks − Does not address cross-functional processes • Mostly used in large companies	+ Most useful for identifying improvements + Addresses inter-department links − Most effort to develop

Figure 5.3. Quality Management System structures.

pros and cons of each of them in your environment will be very beneficial in clarifying your vision of the quality management system.

Phased Implementation

Many articles, training materials, and consultants suggest an implementation plan that consists of four phases of training and planning, documentation development, then implementation of the procedures. This approach may sound simple and look good on paper, but it has one very practical flaw.

For most organizations just beginning to implement a system, there will be a fair initial effort to document current processes adequately to achieve consistency. While there is wide variation in the times organizations require to develop the documentation (four months to two years, depending on several variables), let's be optimistic and say it is a four-month effort. If the organization begins this document development in January, it will document some processes in January, some in February, March, and April, and then implement them in May. The problem is that if your organization is like many others, your processes are not remaining stagnant for four months. By the time May rolls around, some of the documents you produced in January or February may no longer be accurate and you will need to update them before implementing them. That cycle of updating may never end. It creates a moving target.

An alternative implementation plan can avoid the frustrations associated with moving targets as they relate to documentation development and implementation. The alternative is simply to implement the procedures as they are developed, one or two at a time.

There are other benefits of using this approach. It reduces system shock that can be associated with a one-day "cutover" to the use of the system. By slowly introducing the use of documented procedures to the various work areas, there can be an adaptation period which allows for more comfort with the system. It allows more time for people introducing the procedures to do any associated training, and ensures a higher level of comfort with the transition period. Also, phased implementation can provide you with feedback in the early stages that can be used to improve other procedures or future implementation activities. You can implement procedures that have the highest benefit to the organization first, without waiting for all the others to be completed. The phased implementation approach requires that the documents are developed in a specific order.

Implementation Order for the Phased Approach

As mentioned above, the phased approach to implementation carries with it some requirements for the order in which procedures should be implemented. Obviously, there is an opportunity to prioritize the implementation based on where the biggest benefit is perceived. For example, if your largest number of customer complaints pertains to invoicing errors, you may want to develop and implement procedures for contract review and invoicing early on.

In the absence of other priorities, we have found it most useful to begin the process with procedures which are closest to the customer (e.g., Contract Review, Design, Customer Returns, Customer Complaints).

The other priority which often compels organizations is their production or service delivery process, because it relates to many of the elements of the ISO 9000 standards. This logic does have merit, but there is also a drawback. The production and service delivery procedures will, in most organizations, affect the largest number of people, people that may quickly be disenchanted with a new system if it does not work perfectly at first. If your organization will be going through a learning period during the implementation, it may be better to learn while affecting fewer people.

In addition to these guidelines for prioritizing implementation, there is one absolute necessity. The Document Control system *absolutely must* be developed and functioning before any other procedure is implemented. This will be the mechanism to improve any procedures that are not optimal when introduced. If this improvement mechanism is not in place, then suggested changes will not be acted on promptly, the integrity of the system will suffer, and individuals may lose confidence in the system. Therefore, when using the phased approach to implementation, the Document Control procedures must be the very first to be documented and implemented.

The Change Mechanism—The Single Most Important Part of the System

The true test of the effectiveness of an implementation is if documentation and reality (how work is performed) remain in agreement or if they drift over time. Implementation does not mean that on one day, work will be performed in accordance with documentation and training. It means that on *every day*, from implementation until a different system is developed and documented, work will be performed in accordance with documentation and training. Anytime that reality deviates from the system documentation, there is a system failure.

The QMS will, and should, change. These changes are not necessarily driven by failures in the system, but more by the organization's continuous improvement process and response to its changing environment. In fact, management should become concerned if the QMS is not changing, because it may indicate one of two things:

1. The procedures are not in agreement with the way the organization is functioning.
2. The organization is not keeping pace with the changing environment.

This requires management action to bring the organization back on track.

The single most important part of the system to ensure effective implementation is the change control mechanism. When it is discovered that there is a better way to perform a process than the existing way, an effective change control mechanism ensures that reality and documentation remain in agreement. The change control mechanism is the *only* way to ensure system integrity. It is the mechanism for realizing continuous improvement. Its effective functioning is essential not only to the proper functioning of the entire system, but to the intent of the standards—consistently meeting customer requirements. There are two essential aspects to the effective functioning of the change control mechanism:

- That people understand the importance of using it
- That people understand how it works

Both conditions are necessary. Neither alone is sufficient. If people do not understand the importance of using the change control mechanism, then obviously they won't use it. But also, if people do understand the importance, but do not understand how the mechanism works, it will not be used. It is for these reasons that the Document Control procedures must be the first to be developed and implemented under the phased approach. The change control mechanism is an important part of the Document Control procedure.

Effective implementation of change control is largely a matter of mindset. Many people in traditional organizations think of "keeping the documentation consistent with how things are done." Let's take this concept down to the shop floor. If an operator finds a better way to produce a widget, the operator might say, "OK, I've found a better way. I'll produce the widgets differently. Then later on I'll change the documentation to reflect this." This is the beginning of a serious system failure.

The alternative mindset is "keeping how things are done consistent with procedures." Back to the shop floor example: With this mindset, the operator might say, "Ah ha. I've found a better way to produce widgets. I'll initiate a documentation change to reflect this better way so that the better way is consistently used in the future." This approach will keep system integrity intact and yield the same benefits as the first case above, but with greater consistency.

Some people think that change control is an inhibitor to continuous improvement. On the contrary, consistent performance is a prerequisite to continuous improvement. The implication is that change control is a crucial part of a quality management system and warrants very serious attention. A fast, effective change mechanism can be a true competitive advantage.

■ Why Do Organizations Get Certified?

With the many advantages that organizations reap from having good quality management systems, you would thank that management would require ISO 9000 registration because it improves the overall organization's performance. But this is not the case. In a survey done by Deloitte & Touche, they asked, *"What is the most important reason to attain ISO 9000 registration?"* The following is the result:

27.4%   Customer demands/expectations

21.8%   Quality benefits

15.6%   Market advantage

9.0%   E.C. Regulations

8.9%   Corporate mandate

8.9%   Part of larger strategy

Today there are more than 20,000 ISO 9000 certified organizations. In the United States at the beginning of 1994, there were 2250 organizations certified to ISO 9000 standards compared to just 222 two years earlier. Europe is ahead of the United States by almost 10 to 1, but that ratio will change drastically in the next five years. As Craig Verran, Assistant Vice-President for product marketing at Dun & Bradstreet puts it, *"They (U.S. organizations) view it as a very competitive thing. It's like an MBA degree, it doesn't make you a better person, but when you have a choice, you're going to go with the credentials."* (*New York Tribune*—January 30, 1994)

■ Summary

Quality Management Systems are a necessary condition for effective improvement. They define the processes that are necessary in order to continuously provide products and services that meet customer requirements. To achieve this goal, the processes must be effectively designed (appropriately structured), implemented and followed. Quality Management Systems ensure consistent execution of work processes and lead to consistent results. They provide a mechanism to sustain the gains of improvement activity and turn best practices into everyday practices. The ISO 9000 series has become the defacto standard for Quality Management Systems worldwide. They provide an excellent framework for managing work processes, regardless of the type of organization.

While Quality Management Systems are a necessary condition for effective improvement, they are not a sufficient condition. If an organization has an excellent Quality Management System in place, for example, but does not have effective team problem solving, it will not achieve the improvement levels and rates that are necessary to stay competitive in today's markets. The Quality Management System must be an integral part of Total Improvement Management, complementary to the other elements described in this book.

Now, let's discuss some of the ways different organizations look at Quality Management Systems.

Approach to Quality Management Systems

- *Losers:* React to customer and governmental requirements. They do just enough to get by.

- *Survivors:* Set up different individuals or groups to establish criteria for safety, environment, security, and quality. They then prepare the procedures and provide the interface to the outside.

- *Winners:* Have integrated safety, environment, security, and quality into one organization that is responsible for the total process.

Audits of the Quality Management System

- *Losers:* Management does self-audits. Little documentation used.

- *Survivors:* Many different internal audits conducted by different groups.

- *Winners:* A combined total internal audit is conducted at least two times per year. Management does regularly scheduled self-audits using documented procedures.

Upper Management Involvement

- *Losers:* Only when things are highlighted by outside audits.

- *Survivors:* Review the results of internal and external audits.

- *Winners:* Often participate as members of the audit team.

Quality Management Systems Measurements

- *Losers:* No measurements.

- *Survivors:* Continuous improvement measurements.

- *Winners:* Benchmarking used to set targets and develop action plans.

Impact of Failure on Management

- *Losers:* No impact.

- *Survivors:* Line managers performance evaluation is downgraded because they let the problem occur.

- *Winners:* Failure is accepted as a top and middle management problem. One hundred percent successful audits are a condition of employment.

Documentation

- *Losers:* No documentation.

- *Survivors:* Very detailed procedures.

- *Winners:* Concept and direction is well documented and flow charted. Training requirements are defined and followed. The details of how to ensure compliance is left up to the individual manager.

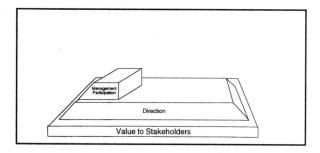

6

Management Participation: Management Must Set the Example

Too many managers expect their employees to correct the problems that are created by management. They cannot. Management must solve 80 percent of the problems that face most organizations.

DR. H. JAMES HARRINGTON

■ Introduction

It has been said, "We did not inherit this country from our ancestors. We borrowed it from our children and our grandchildren." As managers and employees, we have an obligation to pay back this debt with interest, making this country a better place to live in, providing more opportunities for growth and prosperity to our future generations. Management has the major responsibility to pay back this debt. Our employees cannot do it. Sure, they can help, but it is primarily management's responsibility. The problem is not the employee; it's management. Management must be willing to change their behavior and lead by setting the example before they can hope that their employees' behavior will change. Employees listen to our words and smile politely, but they watch what we do and follow.

Now is the time to pay back the interest on our debts. Managers need to answer these questions honestly to determine if they have done their part.

- Have I created more jobs?

- Have I improved the environment that I live in?

- Have I improved the reward system so that one of the two parents can afford to stay home and take care of their children under five years of age?

- Have I improved the safety of the environment I live in (not just at work, but also within the community)?

- Have the capability and competency of the people who work for me improved?

- Has the value-added content per employee continuously improved faster than inflation?

On a whole, government and business have performed very poorly over the past 20 years. Debt is at an all-time high. Most organizations are cutting back and laying off employees. Four out of five organizations recently surveyed stated that they would be cutting back over the next 5 years. The middle class has all but disappeared. It's time that management steps up to its leadership responsibility, thereby ensuring that *all* the stakeholders benefit from the organization's perform- ance, not just management and the investors. Management has stepped aside from its responsibilities for the improvement effort by saying, "The person closest to the job knows the job best, so they should be responsible for the improvement process." This is just a sophisticated dodge to get management off the hook. Eighty to ninety percent of all the problems were caused by management and can only be solved by them. Why would any organization put the burden for management's mistakes on its workers? They are only pawns that have been misused by management over the years. Most of them are really trying to do their very best. Asking employees to take part in the improvement process too quickly only leads to frustration when they see that the really important issues have not been addressed by management.

Management's role must change. They must remove the major roadblocks that they have put in the way of the employee before the employee's zest can be unleashed. A good rule of thumb is that the management team should solve 50 percent of its problems before the employees attack the 10 to 20 percent that they can control. Ron Hutchinson of Harley Davidson said, "If we really wanted to communicate a change in direction and a change in approach, what we needed to do as senior managers was demonstrate that we were going to live by a new set of rules and play by these rules."

I was conducting a focus group meeting at a Canadian cookie manufacturing company, discussing what needed to be done to improve the quality within the organization. A gray-haired woman in her early fifties, standing about five feet tall and weighing about 180 pounds, spoke up saying,

> The white hats (management wore white hard hats when they walked out into the manufacturing floor) don't really care about this company. They are all college graduates and they can go out and get a job any place within a matter of months if this factory closes down. But I have worked here all my life. The only thing I know is how to make cookies. If this place closed down, I'd never be able to get a job. We care about what we do. We want to do a good job. Management doesn't care.

Frank Squires, one of America's leading management consultants, stated, "Management is not against quality. Quantity just has higher priority. Management's order of importance has been quantity, cost, and quality." In the 1980s, management loudly broadcasted and communicated to everyone that quality was top priority. But their own priorities never changed. Management empowered employees and then cut back on the work force as productivity improved. As a result, management lost their employees' trust and loyalty. You can buy a person's time, you can buy their physical presence, you can even buy their mental effort, but you cannot buy their loyalty, their trust, and their enthusiasm. Those key behaviors for any improvement process have to be practiced by management first. It's hard for management to:

- Admit they make mistakes
- Apologize
- Shoulder the blame
- Take time to explain
- Be honest
- Admit they do not know it all
- Share information
- Take advice
- Not change the rules when it is to their advantage
- Alter management styles to meet different employees' needs

Yes, it is hard to do all of this, but it pays big dividends and it is what managers must do if they are going to succeed in the 1990s. Our professors and management consultants make this all sound easy. They talk about empowerment, employee involvement, participation and motivation, along with a neat little package of how all of this fits together in a simple three-step formula. But I assure you, there is nothing like a little experience to tear apart the theoretical. Although the ideas are good, the application strategy has to be uniquely molded to your personality and the environment that you exist in. As former President Reagan stated, "Quality demands efficient management, productive use of human resources, and responsiveness to consumer needs and preferences."

■ What Do We Call Them?

The debate rages on. What do we call the individuals that we used to call "bosses"? Some of the options are

- Associates
- Bosses
- Coaches
- Consultants
- Counselors

- Dictators
- Facilitators
- Leaders
- Managers
- Nonmanagers
- Organizers
- Sponsors
- Supervisors
- Suppliers
- Teachers
- Unleaders

In Brian Dumaine's *Fortune* Magazine article entitled "The Manager—Nonmanager," he wrote, "Call them sponsors, facilitators—anything, but not the 'M' word." In Walter Kiechel III's *Fortune* Magazine article entitled "The Boss as Coach," he advocates calling the boss "coach."

I don't know what a psychiatrist would call a man who runs up and down the room yelling at the top of his lungs, and in the next minute is sitting down crying, but I call him a basketball coach. Surely a coach is not the type of person we need in our organizations. A coach tells the team what they must do. A coach lays out the plays, makes up the playbook, and takes away the individual initiative. And if the players do not perform just as the coach directs them to, they will be pulled out of the game. Coaches play to win. A tie, or what I would call a win-win situation, is a loss for most coaches. "Coach" is certainly the wrong word to use for the individuals that will be working with our employees.

How about the word "leader"? Leaders are out in front. People follow them because they believe in them or are afraid of what would happen if they did not follow them. True leaders have earned the respect of their followers by having the right answers in the past. Certainly, "leader" is a better description of what we need, but it still does not fill the total requirements, because it does not cover building the capacity of the followers. In fact, leadership that is unearned often gets very close to dictatorship. The military is a good example.

We have even heard of people wanting to call their managers "suppliers," because they supply direction, resources, and feedback to their employees. But to me, that is going from the ridiculous to the sublime. When I was 30 and just made manager, I took a correspondence class on basic management skills. This 40-year old book defines a manager as "a person that accomplishes tasks through others." That definition still holds true today, and with that as a definition, a manager utilizes all of the activities previously listed as part of their job.

After carefully considering all the options, I see no reason to call our managers anything other than managers. People that are trying to pull us away from using this term are usually consultants that are trying to establish their own personal niche. When it comes right down to it, good management is getting the average person to produce excellent output.

■ Why Start with Management First?

I have worked in the business environment for over 47 years and I have yet to find one manager that performs at an equivalent quality level to the average production worker. A good production worker consistently performs at the parts-per-million error level. Most managers perform at the errors-per-hundred level. Missed schedules. Starting meetings late. Not following up on commitments. Lack of prompt feedback. The list of management errors goes on and on. Why do we live with these gross management errors? We believe it is because no one takes the time to prepare a management activity inspection plan and conduct audits of management's performance similar to those that are conducted in the manufacturing operations. We set low expectations for our managers. We just do not expect management to perform at the parts-per-million level. We have grown to accept mediocrity as superior management performance. The biggest single opportunity for improvement in business today is management performance.

If you look at what the major roadblocks are to improvement within most organizations, they are:

- Lack of employee trust

- Lack of management credibility

- Lack of training

- Poor communications

- Fear of risk-taking

- Lack of delegation

- Untimely decision making

- Misdirected measurement systems

- Lack of employee loyalty

- Lack of continuity

Each of these roadblocks can only be broken down by management. Without the removal of these roadblocks, the organization cannot make major progress. For example, the single area that is most often rated as the one that needs to be improved first by the employee in surveys we have conducted is trust in management.

You can buy employees' time, effort, and skills, but you cannot buy their enthusiasm, loyalty, and trust. These are things that must be earned by management. Salary increases and bonuses will not do it. Only a close, open, honest, personal relationship between management and the employees will do it. It is a two-way street that must be started by management respecting and trusting their employees. Too often, management gets quickly disappointed when they cast their bread on the water and don't see it immediately returned. This is a process that management has to prime many times to overcome the negative feelings we have built up in our employees in the past. It requires that management develop a true and sincere interest in their employees, not just as providers of services, but as individuals with problems, personal needs, families, and concerns. Too many managers today believe that loyalty is dead, that employees only care about "what's in it for me."

Management action over the past decade has certainly dampened employees' spirits and increased their distrust of management's loyalty to them. You can't pick up a newspaper without seeing articles about employee layoffs, government cutbacks, and organizations requesting that their employees put forth additional effort with less and less resources. In reality, loyalty and trust are down because management has driven them down. It reflects the frustration that employees have with management who haven't bothered to truly take an interest in developing a close personal relationship with and personal responsibility for their employees. This distant style of management is the result of an inner sense that if they keep the employees at arm's length, when and if something negative happens within the organization, management won't feel so bad about laying them off. Management needs to truly demonstrate that they are concerned about their employees, not just their own interests. Most employees want to trust management and be loyal to the organization, but they have been given little reason to do so.

As managers, we have to deserve their trust and loyalty by our actions. We earn their trust and loyalty by being sincerely and genuinely interested in the employees' present and future goals, by appreciating their viewpoints and helping them reach their career goals, and by dealing with them as adults and providing them with information so that they can make intelligent decisions. Building trust and loyalty is difficult in today's environment, but it can be done and is being done by managers that are truly interested in their people. Those managers that take the time and make the effort, find it is well worth the price they pay.

Management has destroyed their own credibility often without knowing or understanding why. Typical mistakes that management makes that destroy credibility are:

- Hiding bad news from the employees
- Half truths and outright lying
- Mission, values, and visions that are not lived up to
- Not taking action on poor performers
- Decision dodging

Credibility builds trust. Trust builds loyalty. Loyalty breeds success not only for the individual, but for the organization as a whole. This important cycle has to start with management and is a key function before we can call upon the employee to start to improve.

Caught in the Middle

Let's be honest with ourselves. Middle and first-line managers have been caught up in a pressure cooker as the country presses for flatter organizations, and programs like self-managed work teams become a way of life. In the 1980s, managers made up 10 percent of the U.S. industrial work force. In Japan, they constituted only 4.4 percent. Organizations like AT&T had more than 100 layers of management. Certainly, they were prime candidates for job elimination. Management guru Peter F. Drucker stated, "The cynicism out there is frightening. Middle managers have become insecure, and they feel unbelievably hurt. They feel like slaves on an auction block." Organizations like IBM, General Motors, Westinghouse, General Electric, Mobil, Ford, and DuPont have slashed their management ranks. Management jobs have been combined, creating what we originally thought would be more meaningful and challenging work. Instead we have created an atmosphere where managers feel over-burdened and under-appreciated.

In a recent study of 112 middle managers conducted by the National Institute of Business Management, they discovered that more than one-third of them felt that they would be happier some place else. As a General Motors middle manager stated in an article in *Business Week*, "When they started the re-organization, one of their objectives was to drive decisions down to the lowest level. But the reality is that decisions have continued to go higher and higher." A middle manager from General Electric stated, "Quite honestly, I feel overworked. I work hard and sometimes I don't enjoy it anymore." A Honeywell middle manager stated, "To survive you have to follow a narrow path. Those who diverge in style or thinking take a risk by stepping out." An IBM middle manager told me, "Being a wild duck used to provide you with growth opportunities. Upper management welcomed diverse ideas. Today, you can't afford to take the chance. The less visibility you have, the better off you are."

Yes, middle and first-level managers are very unsure of their status in most organizations today. The probability of being let go is much higher than the probability of promotion. Their futures have been put on hold and their life savings are in jeopardy. It is absolutely imperative that we do not skip over them and go directly to our employees, or we will lose their support. The early retirement programs that most organizations have implemented in the late eighties and early nineties have allowed the most talented, most knowledgeable managers to escape the organization. We cannot further alienate the managers we have.

■ Managers Are Ultimately Held Accountable

When all is said and done, the management team are the ones held responsible for the organization's performance. How well the organization performs is directly reflected in their promotions, salaries, and longevity with the organization. In this country, management's exposures are limited to reduced salary or loss of position. In China, poor performance is dealt with in a much more hostile manner. For example, in the mid-1980s a Chinese newspaper reported that "Eighteen factory managers were executed for poor quality at Chien Bien Refrigerator Factory on the outskirts of Beijing." The managers—12 men and 6 women—were taken to a rice paddy outside the factory and unceremoniously shot to death while 500 plant workers looked on. Minister of Economic Reform spokesman, Xi Ten Huan, said,

"It is understandable our citizens would express shock and outrage when managers are careless in their attitudes toward the welfare of others."

China is not the only country that holds its managers accountable. Russia feels the same way about its managers. *Pravda* newspaper in 1985 reported that three female factory managers were sentenced to two years in a labor camp and fined $14,000 for producing poor quality clothes at a government factory. In addition, they were fined 20 percent of all future salaries.

We are not recommending these types of stern actions on the U.S. government's part, but it is time that management steps up to their responsibility to improve the quality and productivity of our organizations.

If our management team is going to be held accountable for the improvement, then they must be involved in the implementation of any improvement process. This involvement must extend far beyond knowledge of its existence. They must become the leaders of the movement. They must be the shakers, the movers, and the teachers. If our employees are to excel, then our management must excel. As someone once said, "We are what we repeatedly do. Excellence, then, is not an act. It is a habit."

Why Managers First?

Why start with managers? The answers are simple if you just look at their areas of responsibility. Managers are responsible for:

- Allocating resources
- Establishing the organization's structure
- Selecting the leaders
- Developing the processes
- Setting performance standards
- Making job assignments
- Preparing the job description
- Providing the measurement and reward systems
- Setting priorities
- Selecting and training employees

Considering management's responsibilities, it is obvious that first we must execute these responsibilities in a superior manner if we hope to succeed in the improvement process. Only when management executes their responsibilities can we hope to release the enthusiasm and pent-up creativity that exists in our employees.

Lack of Management Trust in Their Managers

When managers were employees, they got the word through the grapevine. Then when they were promoted into management, they thought they were being promoted into the inner sanctum. They thought that they would be provided with all the inside data, but in most cases, that is not true. Top management considers middle and first-line managers as the employees' representatives, so they are not allowed to share in the organization's critical information because upper management fears that it will be disseminated to the employees. On the other end of the spectrum, they are quickly dropped from the rumor mill because they represent management and, of course, they already know. The result is that an information void is created. One manager shared his concerns with me, stating, "I find out more about what's going on in the organization by reading *The Wall Street Journal* at lunch than I do at all of our staff meetings."

Even Japanese firms do not involve middle management in their decisions when they are operating outside of Japan. In a survey conducted by the Japanese Machinery Export Association on Japanese companies doing business in Western Europe, 67 percent of the 94 companies participating in the survey stated that they are unwilling to let locally hired staff take part in decisions on raising funds for long-term purposes. Seventy-six percent stated that locally hired staff should not be permitted to have a voice in making decisions on new plants. Forty-two percent said that they are unwilling to allow local staff to take part in research and development. The president of Honda Corporation, Nobuhiko Kawamoto, has stopped using Japanese-style consensus management in his U.S. operations in preference to an American-type organization chart. As a result of this action, communication and decision making have become much faster within the modern Honda Corporation.

■ Why Is Management the Problem?

Dr. Joseph M. Juran has long stated that 80 to 85 percent of all problems are caused by management. Donald Stratton, Manager of Quality at AT&T Network Systems, reported the following findings in a *Quality Progress* article.

- 82 percent of the problems analyzed were classified as common cause. These are process problems owned by management.

- 18 percent of the problems analyzed were special cause. These are problems that were caused by people, machinery, or tools. Only a small portion of these problems can be solved by employee teams.

- Of the 82 percent management controlled:
    — 60 percent of the corrections could be implemented by first- and second-level management.
    — 20 percent could be implemented by middle management.
    — 20 percent could only be implemented by top management.

It is easy to see that the major problems within organizations throughout the world are the processes that management are responsible for modifying and controlling. Unfortunately, all the talk in the world and the desire to do something good does not get it done. The employees cannot correct the problems that management has created. Only management's personal involvement in the improvement process will bring about the required changes.

◼ Why Managers Fear the Improvement Process

Many managers feel threatened and uncomfortable with the improvement efforts that are underway throughout the world. In a survey of first-line managers conducted by M. S. Janice Klein, an associate professor at Harvard Business School, she found that "nearly three-quarters (72 percent) of the first-line managers surveyed viewed employee involvement as being good for the company. More than half (60 percent) felt it was good for the employee. But less than one-third (31 percent) felt it was good for themselves." How can we expect our first-line managers to implement programs that they feel are detrimental to their position in the organization? But they are intelligent individuals. They will not say they won't do it, but without them sincerely trying, the process is doomed.

First-level and middle-level managers are very concerned about how the improvement process will impact them. These are proud individuals who have worked hard to get where they are. They are accepted in their community as being successful, contributing individuals. And now they are suddenly faced with a number of uncertainties. Among them are their fear of:

- Loss of job security

- Loss of authority

- Increased workload

- Loss of responsibility and measurement

- Erosion of the one-on-one relationship

Yes, the fear and concern about loss of power, prestige, and control is real, and in most cases, very justifiable. It is something that must be dealt with early in the process in a very serious manner. All levels of management must be prepared, and most managers will need help in adjusting to the new management style. Management who have been in their jobs for more than ten years have spent most of their adult lives developing and refining the techniques required to survive and prosper in a hierarchical organization. It is unrealistic to expect them to discard these successful experiences to try some new "fad" without someone devoting a great deal of time to help them develop a new set of management skills.

To accomplish this, a comprehensive change management process must be initiated, directed at all levels of management. In addition to change management, most organizations will have to rewrite management job descriptions. In most cases, they sound like they were written for dictators. An organization will then need to modify its reward and penalty systems to reflect the new management job descriptions.

◼ Management's New Role

We often hear that quality should be first among equals (quality, cost, and schedule). But today's golden triangle of quality, cost, and schedule requires that management ensure that all three are met at the same time. It is easy to get one at the sacrifice of the other two. For example, you can get schedule with poor quality and high cost, or you can get high quality at high cost and long schedules. But today's customers demand all three at the same time. Managers that meet these demands will grow and prosper, and those that do not will be out of work.

Management's role is changing, and the survival of a manager rests in the ability to keep pace with this changing environment and to be a role model for the employees. There are two types of managers that are working today. Which type are you?

Old Management Style	New Management Style
Gives orders	Gets agreement on objectives
Holds back data	Openly exchanges information
Expects employees to work long hours	Requires results
Stresses individual performance	Stresses team performance
Gains approval decisions from above	Makes decisions after discussions with the affected employees
Primary job is to get the assignment completed	Primary job is to enable the employee to complete the assignment
Takes credit for employees' work	Gives credit to the employees
Tells how to do it	Explains why it needs to be done
Works within the organization's structure	Changes the organization's structure to meet the activities' needs
Chief reward is self-promotion	Chief reward is growing employee capabilities
Thinks of him/herself as boss	hinks of him/herself as a manager of human development
Follows the chain of command	Works with anyone necessary to get the task completed
Thinks of him/herself as a manager of a discipline	Thinks of him/herself as a manager of processes
Sets schedules	Stresses urgency of the job—approves schedules set by employees
Dodges unpleasant tasks	Takes immediate action on unpleasant tasks
Delegates unimportant, uninteresting jobs	Makes job assignments based upon individual capabilities and skills
Gives the best worker more assignments	Keeps a balance of work expectations between both good and bad performers
Pay is based upon time on job	Pay is based upon knowledge and contribution
Stays aloof from the employees	Employee and management share outside activities
Feels minorities and women have to be treated special	Treats everybody special

Worries about employees that could replace them	Develops a backup for themselves
Manages all employees the same way	Adjusts management style to meet the employee's personality and task assignment
Checks to be sure employee never fails	Allows employees to learn from failure, as long as the impact is not too detrimental on the organization

■ Building Trust and Understanding

One of management's top priorities in the improvement process is to build a competent, close-working team that performs both efficiently and effectively without stifling the creativity of the individual. This requires that every member on the team trust and understand each other. Management needs to trust their employees and share the power that information about the organization provides to everyone. In the past, management spoon-fed employees with only enough data to do their job, holding back most of the key operational information. This builds a false sense of power for management and fosters a feeling of distrust in the employee. Providing everyone with as much data as possible is always best, because it short circuits the rumor mill that acts something like this:

- Unknown (first person)—"I don't know if there will be a layoff or not."

- Rumor (second person)—"We could have as many as 1000 employees laid off."

- Fact (third person)—"There will be 1000 employees laid off next month."

As the Old Testament states, "People perish from the lack of knowledge."

To earn trust and understanding from their employees, management must provide them with a secure environment. Management must realize that any improvement process will cause the employees to ask themselves the following questions:

- What is in it for me if I make the organization more productive?

- Will productivity improvements cost me my job or reduce my standard of living?

- Am I willing to change jobs or relocate to stay with the organization?

- What will my future be with the organization?

These are known as "silent questions" that management must help the employee answer to provide a secure environment. This can best be accomplished by management's positive actions, not words.

Tell Them Why

Everyone wants to have self-respect and to be respected by others. This is a universal need that we all have, and it's a need management needs to fulfill more today than ever before. For without self-respect, it is impossible to build trust and loyalty. Showing that you respect a person is a sincere form of flattery. Think of how you handle people. The higher level a person is, the more you respect him or her, the more time you take to explain why you are doing something. When you don't feel a person is important, you have the tendency to tell them what to do. You tell your children to take out the garbage or to do the dishes. The more you respect an individual, the more time you take to explain why they should do something. Management often falls into this trap of telling their employees what to do without explaining why it is worth their time and effort to do it. It is always better to tell employees why they need to do something in preference to how to do it.

A boss tells an employee how to do something. The modern manager tells the employee why it needs to be done, taking the manager out of the role of boss and putting him in the role of a modern leader and associate. No longer are you ordering employees to do something. You are providing them with an understanding of the results that need to be accomplished, the impact the activity has on the organization, and a sense of urgency. Telling an employee how to do a job may get it done, but explaining why the job needs to be done gets it done with enthusiasm. People that understand why, develop their own approach for accomplishing the task, make fewer errors, and complete the assignment faster because they have *a sense of ownership*. They will also feel free to change their approach as the situation changes. If employees do not understand why they are doing the task, they charge ahead, implementing management's direction until they are stopped.

Do It with a Smile

A smile goes a mile, while a frown drives you down. A smile unlocks the door of acceptance. It denotes friendship, caring, and a willingness to listen to both sides of the story. Managers that have a smile on their face, a twinkle in their eye, and sincerity in their voice, breed an environment of friendship and cooperation into the entire workplace. The energy level of the total department surges. People like to work for managers that are friendly, likable, and have a positive attitude. Too often under the pressures of the job, managers forget that everyone looks to them to set the attitude of the organization. Other managers think that they won't be taken seriously if they don't appear to be a little aloof, firm, and stern. They rely on this harsh personality to give them stature and respect. But that's not true. Sure, you can get short-term results by threatening people, but you build long-term team relationships and performance by creating an enjoyable work environment.

People just perform better when they are happy and satisfied with their job, when they are not threatened by their job and their manager. What are the key management characteristics that make a work environment enjoyable?

- Sincere interest in the employee
- Easy to talk with
- Treats everyone as equally important
- Comfortable to be around
- Friendly and pleasant personality
- Not above doing any job
- Realizes that other people are busy too
- Does not bring personal problems to work
- Has a consistent personality that can be depended upon
- Remembers commitments
- Shares success and shoulders blame

President Dwight D. Eisenhower typified this type of person. There was never any doubt that he was in command and meant what he said, but he always had a smile for everyone. Although a smile or a frown can make management convincing, it's meaning what you say and following through to be sure it's accomplished.

Listening

As a manager, we need to be good listeners. God gave us two ears and only one mouth. I believe he was trying to tell us to listen twice as much as we talk. Our employees cannot tell us what their problems are or what they are thinking when we are talking. Make effective use of silence to encourage your employees to talk. Take the time to develop good listening habits. Some useful guidelines are:

- Gather as many ideas as possible before making a decision.
- Look directly at the person who is talking to you.
- Use words of encouragement like: "Yes, I understand, tell me more."
- Put your phone on hold when someone comes in to talk to you.
- Ask probing questions and don't jump to conclusions.
- Take time to chat with coworkers.
- Ration the time you talk.
- Understand what is behind their words.
- Listen with your eyes and ears.

Urgency and Persistence

Today's work environment is very fast-moving, and management must create a sense of urgency in every employee's mind. People that make things happen radiate an energy and a sense of urgency. A good manager is a person who doesn't put off until tomorrow anything that someone can be made to do today. The world is full of good intentions. Many well-meaning, brilliant people turn out to be unsuccessful because they have made it a habit of putting things off until tomorrow—people that don't start working on a project until it is almost due, and then something interrupts them so that they miss the schedule. Success comes to the manager who makes things happen on schedule without sacrificing quality or cost. The quality of that last-minute job is usually compromised. Good managers know how to communicate a sense of urgency without being obnoxious and overbearing. They do it by showing interest, by reviewing plans, by checking progress, by being there to help break down the roadblocks that get in the way of their employees.

The combination of urgency and persistence make a winning formula for management and employees alike. As important as a sense of urgency is, it takes persistence to get the job done. Calvin Coolidge put it this way

> Nothing in the world can take the place of persistence. Talent will not: Nothing is more common than unsuccessful people with talent. Genius will not: Unrewarded genius is almost a proverb. Education will not: The world is full of educated derelicts. Persistence and determination alone are omnipotent. The slogan "press on" has solved and always will solve the problems of the human race.

■ Recognizing Good and Bad Performance— The Feedback Process

Most people want to do a good job and feel that they are contributing to the success of the organization. Today's manager must provide ongoing, continuous feedback on both the negative and positive aspects of performance. If the employee's manager does not pay close attention to the employee's output, it is perceived as not being of value. So the employee reasons: What difference does it make how good or how bad the job is done? Employees that rarely receive feedback feel that their job is unimportant and that no one cares. On the other hand, if all they receive is negative feedback, they feel that they are inadequate and that management is down on them.

Use positive feedback to reinforce desired behavior patterns. Things like when the employee goes out of his or her way to help other employees, comes up with unique ideas, beats schedules, handles difficult situations, catches an error, puts out additional effort, or sacrifices self-interest for the sake of the assignment. The best positive feedback occurs in public, at a meeting, on a bulletin board, with a group of associates, etc. At times, it is best to send a personal note to provide private feedback. But don't wait. It always is best to provide positive and negative feedback as close to the time that the activity occurs as possible.

Most managers find it difficult to give negative feedback, but feel it is the primary reason that they have their job. As a result, they charge into it like a bull in a china closet, trying to get it over with as soon as possible. But it is important to realize that people are a lot like magnets: They are drawn by positives and repelled by negatives. The challenge for management today is to provide negative feedback in a constructive way. Management must talk about results, not about the individual's personal attitudes or actions. They must probe deeply into the employee's side of the situation so they thoroughly understand the circumstances surrounding the incident. They should couch the criticism in a way that allows the employee to save face. The manager is not there to win the battle. Their job is to try to help the employees change bad attitudes or approaches to their assignments. Remember, you can effectively make your point without nailing the employee to the cross.

■ Basic Beliefs

The sophisticated management methods rely on some basic beliefs that must be mastered before these methods can be applied. Managers that have not mastered these basic beliefs, stand a major exposure to being the individuals that are left out in the cold during the next restructuring cycle. These basic beliefs are:

- *Delegation.* Management must be able to accomplish assignments by delegating work to their direct reports. Management must be able to free themselves to do planning, break down barriers, teach, measure, and network.

- *Appraisal.* Management must be able to develop individual performance goals in cooperation with the employees, and provide honest, continuous feedback on performance compared to these goals.

- *Disagreement.* Disagreements between management and employees can be healthy. Management needs to understand both sides of the situation to make the very best decision. "Yes" men or women are not helpful.

- *Be decisive.* Management cannot be reluctant to make a decision. Often, "gut feeling" is an extremely important part of managing the organization.

- *Positive attitude.* If the manager conveys a feeling of failure, the department is doomed to defeat.

- *Five-way communication system.* Management must establish excellent upward, downward, sideways, supplier and customer communication systems. They must be willing to share information with their employees. Information is power. Every year, Robert Crandall, CEO of American Airlines, conducts 20 to 30 president's conferences in his 165-city route system, ensuring he maintains open communication with all of his employees.

- *Invest.* Management should invest heavily in their employees, provide them with training, and help them grow and mature. This is one of the best investments an organization can make. Art Wegner, CEO of Pratt Whitney Turbo Manufacturing, sent his design engineers into the plant to spend six months on assignment as general foremen. Certainly, this is a major investment in their future, but it has paid off in improved manufacturability of new products designed by these engineers.

■ Tomorrow's Managers

The term "management" should be treated as a set of concepts, not as a group of people. Traditional management, as we know it today, has evolved as follows:

- Individualistic Management
- Professional Management
- Scientific Management
- Human Relations Management
- Participative Management

In the nineteenth century, the management style could have been called, "Individualistic Management." The entrepreneur was responsible for creating most of the large organizations that we have today. Families like the Fords, Rockefellers, Carnegies, Durants, Mellons, Sloans, and Watsons were all creative individualists that built and managed great organizations. This approach, driven by the economics of business, gave way to the Professional Management era, where management was measured by the short-term bottom line. The professional manager's goal was to produce the maximum output with the minimum expenditure.

This led naturally into Frederick Winslow Taylor's "Scientific Management" approach to running an organization. Scientific management was based upon four principles:

1. Scientifically designing the job.
2. Scientifically selecting workers to match the job requirements.
3. Scientifically training the workers to perform the job as designed.
4. Work must be done in a spirit of cooperation.

This worked well with workers who had a low skill level and low intelligence. This approach divided the work process into small assignments that required little training. As management's style became more and more autocratic and the employee became better educated, the system began to break down. The employees began to resent management who, from the employee's standpoint, were taking advantage of them. As a result, "Human Relations Management" became the preferred management system. Human relations management is based upon the belief that if management treats the employee with respect and dignity, the employee's performance will be maximized. A simple idea for a simple situation. Unfortunately, today's working environment is anything but simple. As a result, the management style of the future must be a "Participative Management" style.

Tomorrow's managers will have to be much more effective than they are today. They will have a much bigger span of control or, as I like to think of it, a larger span of support and new, more demanding challenges. They will become more and more impacted with the soft side of management, since most technical decisions will be made by intelligent computers. As employees become empowered to be responsible and accountable for their jobs, management's role must change. Figure 6.1 shows how the management environment at all levels is and will be changing.

Activity	Yesterday	Today	Tomorrow
Management style	Dictating	Coaching	Assisting
Providing direction	Orders	Consensus	Define results
Goal setting	Management's goals	Common goals	Employees' goals
Evaluation	Criticize	Appraisal	Two-way evaluations
Decision making	Management decisions	Team decisions	Individual decisions
Compensation	Pay for years worked	Pay for performance	Pay for knowledge
Way to correct problems	Focus on the individual	Focus on the activity	Focus on the process

Figure 6.1. The way management is changing.

Management Style

In this environment, management style must have many facets. Today, we expect managers to adjust their management style to meet the personality traits of the employee. In the future, we will have to adjust our management style to the individual's personality and to their job assignments. People's working personalities can be divided into four categories.

1. *Planners.* People that excel in taking an idea and laying out a systematic approach to its implementation. Planners tend to be introverts.
2. *Networkers.* People that establish excellent communication systems between groups. They are excellent negotiators and politicians. Networkers tend to be extroverts.
3. *Doers.* People that take a plan and implement it. They like to be assigned a problem and get it corrected. They make things happen.
4. *Leaders.* People that through their charisma, appearance, or example, attract others to them. People follow them because it is unpopular to do otherwise.

Each of these personality traits impose very different needs on management. These needs can be classified into two types:

- *Social needs.* These needs are satisfied by management contact, public recognition, demonstrated interest in the individual and his or her career and personal life.
- *Technical needs.* The skills required to perform a given task.

Two factors drive the degree and frequency to which both needs have to be fulfilled. They are:

1. How well the individual is performing the assigned task.

2. The type of personality that composes the individual's make-up.

Figure 6.2 shows how management styles need to change versus the employee's performance level. It is easy to see that based upon how the individual is performing, management styles need to be changed to meet the employees' needs.

- *Employees that do not meet requirements need a "coach"*—Someone that will tell them what to do. Show them the correct way when they cannot accomplish a task. Minimize their chance of making an error. They need someone that will help them feel good about themselves, even when they are not doing well.

- *Employees that meet minimum requirements need a "teacher"*—Someone that can help them understand the concepts. Someone that will measure their performance and show them when they make an error. They need someone that recognizes their success and helps them to succeed.

- *Employees that meet requirements need a "boss"*—Someone that gives them assignments and follows through to be sure they are accomplished. Someone that helps them develop and improve the quality of their output and their productivity.

- *Employees that exceed requirements at times need a "leader"*—Someone that knows what needs to be done and has empowered the employees to take on responsibility and accountability for their jobs. The leader works with the employee to ensure that barriers to completing the job are eliminated. The leader focuses his or her effort on coordinating the employee's interfaces and providing feedback. The leader sets the example for the employee—for his or her technical and personal style while at work.

- *Employees that always exceed requirements need a "friend"*—At this level of performance, management can delegate the responsibilities for the assignment to the employee and hold them accountable for its outcome. Management should de-

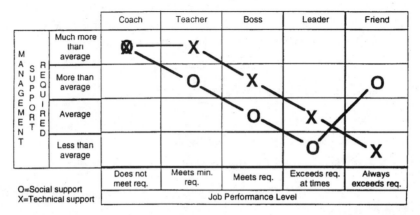

Figure 6.2. Management support required vs. job performance level.

velop an open, two-way personal relationship with the employee, sharing experiences and family concerns. Technical interest and understanding are developed by providing a ready ear to discuss project operations and exchange ideas, but the technical decisions are made by the employee. The employee is empowered to make decisions and take actions on all tasks assigned to him or her, without management direction.

It is easy to see that a management style must change from coaching all the way to friendship, based upon varying degrees of performance. But performance is as much the responsibility of the manager as it is the employee (see Fig. 6.3). If a person has a networking personality and is assigned to do networking (coordinating between areas), his or her chances of meeting requirements are extremely high. But if a networker is assigned to a planning activity, it will be hard for him or her to meet requirements. Unfortunately, in today's and tomorrow's complex environment, employees will be moving back and forth through many types of job assignments. As a result, management style for an individual employee will have to vary based upon the task that the individual is performing. Management cannot hold an employee responsible for poor performance if management misassigns the individual.

■ Understanding the Customer

All management must keep in close contact with the external customer and understand their desires, as well as the type of products and services that the competition is offering. One very effective approach to understanding customers is to assign each vice president and middle manager to a major customer, making him or her responsible for understanding the customer and reacting to any complaint or

		Type of Assignment				
		Planner	Networker	Doer	Leader	
PERSONALITY TRAITS	Planner	Outstanding	Very poor	Good	Poor	**PERFORMANCE LEVEL**
	Networker	Very poor	Outstanding	Poor	Good	
	Doer	Good	Poor	Outstanding	Very poor	
	Leader	Very poor	Good	Poor	Outstanding	

Figure 6.3. Expected performance based upon personality traits and types of assignments.

suggestion that the customer has. This is a good way to keep the management team close to the most important part of the business, the customer, and at the same time to provide major customers with the feeling that their business is valuable. It also ensures that any problem that key customers have is given priority treatment within the organization. This approach provides each major customer with their own ombudsman.

Direct feedback on customer complaints and praise needs to be provided to every level of management on a regular basis. In addition, customer survey data that monitors customer satisfaction and after-sales performance data are required so that trends are detected and understood before they become problems. Special surveys should be directed at the many individual processes that interface with the customers to provide specific improvement measures data (example: billing, shipping, order handling, etc.). Management must understand that customer expectations and desires are continuously changing. Products and services that were outstanding yesterday, just meet requirements today, and probably will be inadequate tomorrow. There is no saturation point to improvement.

■ Participation/Employee Involvement

Participation, empowerment, employee involvement—all are current buzz words today, but they are words that bring about a cold chill to the heart of the bravest middle and first-level manager. Theoretically, and even outwardly, these programs are usually supported by middle management, because upper management wants them done now. But down inside, middle managers interpret them as ways to eliminate their jobs. In the massive layoffs that have occurred, middle and first-level managers have taken the biggest percentage of the burden. In this environment of career uncertainty, how do you think managers feel about employee involvement?

With middle and first-level managers' deep concerns about participative management and employee involvement, how can you expect the processes to work? Can top management edict it into the organization? Yes. Can they edict it into the organization and will it work? No. Our middle and first-level managers are too wise to outwardly buck top management when they tell them that they want to install a participative management process. The middle managers think that they are not involved, that it's the first-level manager that has to do it. The first-level manager gets the word that top management wants them to give up some of their responsibilities to their employees. The word spreads like wildfire that this is the start of self-managed work teams—an environment where the first-level managers' jobs are eliminated. The first-level managers cooperate because the top of the organization tells them they have to. But both middle and first-level management hope with every team meeting that this program blows up in the face of the top management team. Top management dictates that the first-level managers must be more participative, but they do not change themselves, nor do they require middle management to change. Isn't there something wrong with this process? Yes, the process is flawed, and it can lead to spending a lot of money on training and nonproductive meetings.

When participative management is implemented correctly, it starts with top management. But, instead of giving orders, they give up some of their responsibility to middle management. Typically, they give up the month-to-month firefighting activities that take so much of their time. This creates a heavy workload for middle management, who already are overworked, causing them to delegate more of their work to the first-level manager. As first-level managers become more empowered, and have fewer checks and balances that they have to report up the ladder, they will see that upper management believes that participative management is the way of the future and is essential for their future growth. Because most people want to emulate the people that they report to, participative management flows willingly throughout the organization. In this environment, first-level managers willingly give up some of their responsibilities to their employees and welcome the employees' contributions to the planning and decision-making processes that relate to their work assignments.

If participative management is implemented correctly, upper management has more time to do the things they should have been doing all along, but have never had the time because they were too busy fighting fires and worrying about the quarterly bottom line. What should upper management do with the freed-up time that participative management gives them? They should use this newly found time as follows:

- Working with the employees to understand the real problems.

- Talking to the external customer to understand their present and future needs.

- Providing direction through the strategic planning process.

Marvin Runyon, CEO of Nissan Motor Manufacturing Corporation, spends time with each of his 3200 employees at least once a year. Does your CEO have time to do the same? If not, maybe he or she needs to be more participative.

Remember that participative management does not mean democratic management. The law of one person - one vote, does not apply to participative management because the managers are still held accountable for the actions of the people that report to them. Management should encourage the employees to freely contribute their ideas and empower them to implement them, but management must have the courage to reject ideas that are not the best solution for the organization. When ideas are rejected, it is very important for management to take the time to explain why the suggested solution was not the correct answer for the organization. If this feedback process is not handled very well, the idea stream will soon dry up.

The Push-Pull of Management

General Dwight Eisenhower used to show the difference between leadership styles with a piece of string. Pull it and your employees will follow you anywhere. Push it and the team goes nowhere. Management who prod and threaten to get the job done are pushing on the end of the string. People who push on the end of the string rarely get the best performance from their employees. The employees work against management, not with them. Their only desire is to get the boss off their back. Managers that pull the other end of the string have the employees working with them. Managers help the employees to be their very best. They break down roadblocks and concentrate on making it easy for the employee to perform.

Don't Turn the Organization Upside-Down

The participative environment has changed the way many organizations look on the organization chart. In the old-style organization chart, top management was at the top of the pyramid, indicating that everyone below them provided services to top management. The popular notion is that with participative management, the pyramid has been turned upside-down, with everyone servicing the employee, who in turn services the customer. But what could be more unstable than a pyramid resting on its pinnacle? It is obvious that the slightest vibration would cause the upside-down pyramid to topple.

We like to think of the organization structure more as a square (see Fig. 6.4). This way of looking at the organization structure indicates that all activities are important, that five-way communication exists, and that everyone in the organization has an obligation to make the best use of the organization's resources in their efforts to serve their customers. It also lends itself to the concept that organizations use processes that flow across boundaries to conduct the organization's business. It has the advantage of showing that everyone in the organization has an obligation to service the external customers as well as the internal customers.

How to Get Employees to Work

Management's role is to get work done through others. We must realize that the work ethic is something that has to be learned. It is not an inborn trait. Many

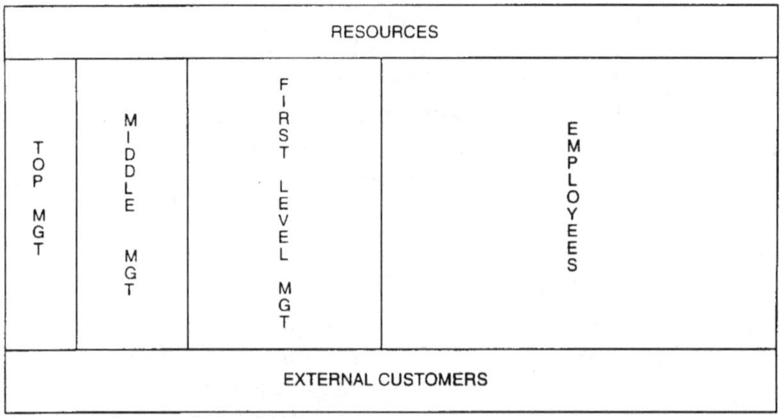

Figure 6.4. Preferred organizational model.

experiments with animals and people have proven beyond a doubt that the work ethic is a learned trait. In these experiments, when individuals are provided with all their desires without doing anything to earn them, they replace their work with less productive ways of occupying their time. Unfortunately, work ethics are developed during the formative part of a person's life, between the ages of two and fifteen. By the time people enter into the adult world, their work ethics have already been formed.

As America has developed and prospered, work ethics have degraded. At the start of the nineteenth century, people lived to work. The harder you worked, the better person you were. Honest labor was a valued trait that would lead to success. As survival became a given, and government stepped in to provide a minimum living standard for everyone, work ethics began to slip.

As America became more and more successful, parents expected less and less from their children. Honest labor was replaced with time in front of the TV. Dishwashers eliminated the need for children to work with their parents when doing dishes. Product reliability improvements and increased financial wealth have reduced the need for the parent to work with the child in repairing the car, the washing machine, the plumbing, etc. Store-bought frozen foods, cakes, bread, etc., have reduced the time that both the parent and the child spend in the kitchen working together. Fast food has for many families all but eliminated the need for a kitchen. This reduction in time and effort required to run the home has freed-up the wife to become a very important part of today's work force, which has led to the emancipation of women. This added freedom didn't come too soon because as male earnings dropped off, women were forced to step in and shoulder part of the financial burden. As a result, families with both partners working are the rule rather than the exception, in order to maintain their desired standard of living.

Coupling this new financial security for women with the decrease in family work projects has had a very negative impact upon family values, leading to a period where a one-parent family is not unusual. The result is an ongoing decline in work values. As a result of these factors, there has been a drastic reduction in the honest work time that children are involved in. We estimate that an average 12-year old today only works 25 percent of the time that a 12-year old worked in 1940.

As society became more affluent, people began to balance their lives between the various areas that occupy their time and account for the major interest in their lives. Figure 6.5A presents the theoretical arena of life. In the past, "work" was the overriding quadrant in the arena of life, and the other three quadrants were allocated to the leftovers (Fig. 6.5B). Today, work is used to make the other three quadrants possible (Fig. 6.5C). The quadrant of "self" that was almost ignored in the 1950s has become more important, with ever-growing portions of the individual's time being devoted to it, as they spend more time staying in good physical condition and indulging in self-pleasures like watching television, attending sports activities, and listening to high-quality music.

In the forties and fifties, employees used to line up to work overtime. Everyone wanted to work Saturday for time-and-a-half. Managers needed to keep very close records on the number of overtime hours each employee worked so they did not show preferences. Today, just the opposite condition exists. It is difficult to find

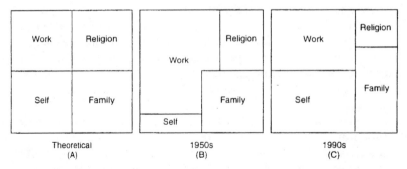

Figure 6.5. The arena of life (in percentage of hours awake).

anyone who will work overtime. Everyone expects two days' advance notice if they do have to work overtime. The work quadrant today has been neatly boxed in, and the other three quadrants are jealously guarded to ensure that work does not interfere with them.

In this environment, management needs to look closely at their management style to ensure they are in tune with the current personality of their employees. The question that management must face is, "How can I obtain the best results from the portion of time the individual is devoting to the work quadrant?"

Management Threats. The fear of losing one's job still motivates people. The boss with a whip still exists and produces results, particularly in an environment where jobs are hard to find. But this type of management style requires a great deal of ongoing effort on the part of the manager and does not, in the long run, produce superior performance.

Self-motivation. This is the best of all worlds. If all employees were self-motivated, management's job would be extremely easy. Figure 6.6 is a picture of how some managers view a really self-motivated employee. But in the real world, employees look around and find individuals that are not performing or working as hard as they are and still receiving similar benefits. As a result, in this environment, employees tend to be influenced by the poor performers and drop their level of performance down to the level of the poorest performing individual in the group. Self-motivation works with less than 10 percent of employees.

Employee Involvement. Employee involvement is, by far, the best way to enhance employees' performance. Involving the individuals in discussions about things that impact their jobs produces improved commitment and increases work ethics. Individuals respond and perform better when they are involved in designing their work process, and understand the reason why something is being done and the value of their efforts.

Once an organization realizes that its success rests in the hands and minds of its employees, and when the employees experience the stimulation and satisfaction

Figure 6.6. Self-motivated employee.

of helping the organization improve, participative management will become a way of life and its achievements will continue to grow.

5-Way Communications

For a long time, organizations have been trying desperately to establish effective 2-way communications (up and down) and just as we have started to make real progress, we find that this is not adequate. Many organizations are now trying to establish 5-way communications (up, down, horizontally, customer, and supplier). This 5-way or star communication process is a key part of a participative team environment that is based upon a strong supplier-customer relationship. Today, the complex organizational environment makes this 5-way communication system an essential part of an effective operation. In our information-rich society, management can no longer ration out just the information they believe that the employee needs to know. Likewise, every employee has an obligation to maintain active communication with all of his or her interfaces. The day of management being responsible for keeping the employee informed disappeared when we began to empower the employees. Management's responsibility is to provide the communication process and media. It is everyone's responsibility to use it. Everyone has a personal responsibility to manage the communication network within their own star. At a very minimum, every employee is involved in four of the five points of the star. The actions that need to be taken to effectively make the 5-way communication process work are:

- Management needs to trust the employees with information previously restricted to management.
- Management must establish the process.

- Everyone needs to be trained on how to use it.
- A reward system needs to be put in place to encourage the use of the process.

Today, technology and personal initiative can easily take care of the communication problem if they are both used correctly. Computer networks, voice mail, centralized databases, telecommunication, video conferencing, town hall meetings, bowling leagues, the list goes on and on and on. The 5-way communication process in today's environment is not only desirable, it is mandatory.

One of the weaknesses in most communication systems is the inability and/or reluctance of the employees to use it. In most organizations, communication training has been limited to managers. This was a very realistic approach in a one-way, downward communication process. Why would you want to train employees in communication skills? If they knew how to communicate, they would want to be listened to, and who had time to do that? But today, with everyone having shared responsibilities in making the communication process work, everyone needs to be trained in how to use it and have access to the required medias. The biggest problem we face in establishing a 5-way communication process is getting the data into the process. A well designed data system that is user-friendly groups information into relative packages that minimize the time and effort required to obtain needed information. The biggest roadblock to making this happen is upper management who, for years, have believed that information is power and were unwilling to share this power with their employees. Add to this a feeling that most employees cannot be trusted and/or are not intelligent enough to understand the business and it is easy for management to justify not putting most of the information they control into the communication process. The fact of the matter is that most employees can be trusted and the one that cannot be trusted has already found ways to get the information they want. When management starts to openly share their information, the communication process is legitimized and becomes effective.

■ Organized Labor's Involvement in Participative Management

As Donald Ephlin, the Vice President of The United Auto Workers, put it when he was asked about the union's position on participative management, "Participative management should not be seen as the union participating with management, rather it should be seen as management participating with the workers."

Although organized labor should have been involved in developing the organization's visions and three-year improvement plan, it must now become even more involved as management prepares to roll out the participative management process. Traditionally, management have been the actors and the unions are the reactors. Management defines the rules and the union demands that they are changed. This adversarial relationship exists because the organized labor movement was born in an era where some organizations took advantage of their employees, but here again, things have changed. Today's competitive environment calls for a different relationship between organized labor and management if the organization is going to survive.

The Quality of Work Life concepts included in the 1973 General Motors/UAW contract was an historical breakthrough in this new era of organized labor/management cooperation. Since that time, many labor unions have followed suit. (For example, the United Steel Workers of America, and the Communication Workers of America.) These cooperative endeavors exist under many names; i.e., Quality of Work Life, Employee Involvement, Relationship by Objectives, Labor-Management Participation Teams, etc. These programs exist based upon the belief that if organized labor is part of the rule-making and problem-solving process, they are much less apt to take exception to the resulting activities. To accomplish this new cooperative environment requires a major change in attitude, both in the organization's managers and the union leaders. It is very important that this change process be well-managed and coordinated. Often, under the pressures to recoup from a negative position, these relationships are pushed too fast, with disastrous results. The key principle here is that organized labor's participation must be voluntary or it will not work. This requires management and the union leaders to jointly develop a plan for the implementation of participative management before it is rolled out to the union members. To accomplish this, the local union leaders and local management team work together to understand the process and define how it will be implemented and how it will be presented to the rank and file union members.

Probably one of the first things that needs to be addressed is how a collaborative relationship between management and organized labor can coexist with the collective bargaining process. The next most important issue is how the participative management process will impact the union's grievance process. The other critical issue is what subjects the teams can work on and still meet the rules as defined in the National Labor Relations Act. The National Labor Relations Board rulings on the Electromation and the Dupont cases have broad implications for business and unions alike. If you want to run a full-fledged participative management effort in a unionized environment, labor and management must enter into a clear contract regarding the governing, administration, and scope of that effort. This agreement should include, but not be limited to:

- A joint steering committee with decision-making authority for the entire participative effort
- Joint structure for all changes related to work redesign technologies, work flow, work processes, etc.
- Joint training
- Joint communication
- An agreement on the decision-making process (majority, consensus, etc.)

If you have not negotiated the full breadth of the participative relationship, you will be operating illegally. For the most recent status on these labor relations activities, contact the Association for Quality and Participation in Cincinnati, Ohio.

Union leaders will have a number of concerns during this initial phase that they and management must find answers for. Among them are:

- What is in it for the union and its members?
- Is this just another way to speed up work?
- How will it impact the contract and the collective bargaining process?
- Will the rank and file view the union leaders as leaning too heavily toward the organization's management?
- How much work will be involved?
- Will it change the relationship for the better?
- How will it impact union politics?

Often, to help get these answers and to help develop the implementation plan, an experienced third party is used. This neutral third party helps facilitate the development of the relationship between the two parties. Frequently, the third party will conduct seminars for the union leaders explaining the advantages and disadvantages of participative management. Often, the union leaders are encouraged to tour other facilities that have successfully implemented a participative management system within an organized labor environment to discuss with these union leaders their negative and positive experiences. These visits to other locations that have successfully implemented a participative management system provide many of the answers that the union leaders are looking for. In addition, there are many articles and videotapes that provide relevant information for the union leaders. Frequently, senior management people will be invited to the union hall or to a neutral location to discuss their commitment and role in a participative management system. Slowly, in this open environment, the process will become defined and ground rules will be developed.

Once these questions and concerns have been answered, a document of operation and an implementation plan can be jointly developed by the union leaders and management. This is frequently done during an offsite meeting. The first part of this meeting will be devoted to creating guidelines for the cooperative effort and a statement of what will be accomplished as a result. The latter part of the meeting will be devoted to developing an implementation plan, including how the participative management process will be communicated to the employees.

As Glenn Watts of Communication Workers of America (CWA) points out, cooperation with management may seem a high-risk activity for unions, but it is essential if both parties are to survive in the long run. He said, "The only real risk is if the union does not participate." How do the union members feel about participation? They are for it. Labor leaders who support the program find themselves in a stronger political position than ever before. Based on comments made by United Auto Workers (UAW) leaders, officers who campaign with a quality-of-worklife theme are almost sure of winning election.

■ Overcontrolled and Underled

Management talks participation and empowerment, but in today's environment with organizational flattening, more and more instead of less and less decisions are being made by higher and higher levels of management. These negative trends need to stop, and the best way to stop them is for each level of management to question each decision they make to be sure they should be making it. If the decision could be made at a lower level, the decision should be directed to the proper individual. We realize that at first this may slow down the process even further, but people soon get the word and only the proper level of decisions will be escalated up the management ladder. In addition, management should not second-guess decisions that are made.

Management's Role

There are specific roles that management at all levels must play in the improvement process. Figure 6.7 is Compaq Computer's definition of the roles that the three levels of management must play in supporting the improvement process.

These roles clearly demonstrate a very participative environment down to the point that the employee teams are granted the right of developing their own work processes. In this approach to Total Improvement Management, managers truly provide the guidance, not the control of the process.

The following is a list of requirements for employee involvement and participative management to thrive within the work environment.

1. Management will have to share their power and responsibility.

2. Management must provide the employees with much more information than they have in the past.

3. Participative management/employee involvement needs to be practiced at all levels of management.

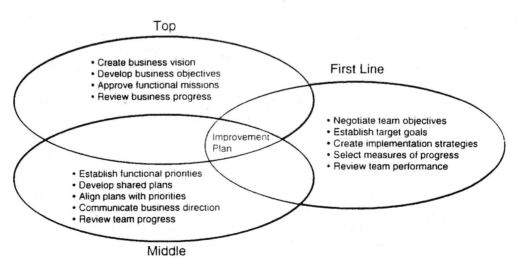

Figure 6.7. Management's role in supporting the improvement process.

4. Management needs to trust their employees in order to earn the trust of their employees.

5. Management needs to feel comfortable with decentralized decision making.

6. Management needs to stop performing hindsight appraisals.

7. Management must encourage an environment where failures are accepted as being part of the learning cycle and risk-taking is a dominant trait of the people that get ahead.

8. Time must be made available to train the employees in how to make decisions, learn new jobs, and perform business analysis.

9. Management must have the courage to reject poor solutions, but take the time to explain why the suggestion is being rejected.

For an organization to be successful, management must provide employees with:

1. Detailed job descriptions.

2. Relevant measures that can be used to evaluate progress.

3. Tools so that the employees can perform error-free work.

4. Job-related training so that employees understand how to use the tools and the processes that management provides.

5. An understanding of the importance of the job and why it is being performed.

6. Ongoing, continuous customer feedback so that employees can evaluate their performance.

7. Time to do the job correctly.

Town Meetings

The days of the pilgrims and wild west are on their way back as American business and Ross Perot embrace town meetings as a way of developing open communications between all levels of management and employees. The new town meetings are not used to manage the organization. They are used to provide upper management and the employee with a way of exchanging ideas and obtaining a better understanding of each other. Some organizations schedule town meetings that are open for anyone to attend, while other organizations invite specific individuals to attend. Typically, the agenda for these town meetings is a short presentation by upper management followed by an open question and answer session on any subject. Someone should be assigned to take notes at all of these meetings so that upper management can follow through on any problems that are defined by the audience and on any commitments that were made. After the minutes from these meetings are prepared, they should be posted on a bulletin board without the names of the people in the audience that asked the questions. This allows everyone to share the outcome from these meetings. Posting the minutes from the town meeting on the bulletin board has proven to be good practice because the questions asked are frequently questions that are on many people's minds but they have not had the opportunity to ask them or are too shy to ask them.

■ The New Middle Manager

Top management is the key to getting the improvement process started, but middle management are the ones that keep it going. If top management truly accept their role as planners and direction-setters, they become distant to the day-to-day problems that face the business. This puts the middle manager in the role of running the organization and ensuring that it continues to improve. The traditional middle-manager role of "kicking tail and taking names" has to change drastically. The old micro-management attitude needs to give way to a macro-management that has a wide viewpoint and understands the interfunctional relationships. The key traits of a new generation middle manager are:

- Develop close working relationships and understanding of their customers.
- Capability to focus on the "big picture" and manage it.
- Provide education, guidance, and mentoring to the first-line manager.
- Focus on the process rather than the activities.
- Help employees learn from failure rather than punish them for it.
- Concentrate on why problems occur rather than who caused them.
- Recognize continuous improvement as well as meeting targets.
- Embrace change and act as a change agent.
- Reject requests to make decisions that should have been made at lower levels.
- Place high priorities on networking with other functions.
- Provide a role model for the first-level managers and employees.
- Always be honest and aboveboard.
- Have the ability to sacrifice departmental performance to improve total organizational performance.
- Proactively stimulate upward communication.
- Openly share data with all levels.
- Proactively search out employee ideas and actively support the good ones.
- Always explain why an employee's ideas are rejected.
- Practice consensus decision making whenever possible.
- Empower the employee who is in contact with the customer to resolve customer problems.
- Encourage first-line managers to empower their employees.

- Communicate priorities and hold to them.
- Establish networks that identify potential negative trends before they become problems.
- Place high priority on prevention of problems.
- Recognize and reward employees who prevent and solve problems.
- Vary the reward process to meet the needs of the awardee and the contribution of the activity to the organization.
- Treat everyone as being equally important.
- Demonstrate the importance of meeting schedules and the urgency of getting the job done without compromising quality.
- Handle more negative situations with a smile than with a frown.
- Place high priority on expanding employee capabilities and responsibilities.
- Always have time to listen to an employee's problem now.
- Help those that request it if possible.
- Above all, management must be good listeners. They remember that the exact same letters spell the word "listen" that spell the word "silent."

The importance of the middle manager to the improvement process cannot be over-emphasized. Middle managers should provide the parental figure for the first-level managers and the employees. Middle managers are the ones that shape the management style of the organization's future leaders. They are close enough to the employees that they should be on a first-name basis with all the employees within their organization. They should have knowledge about the individual's performance, strengths, and weaknesses, as well as their career aspirations. Often, as we develop potential future organizational leaders, they frequently move from managing one department to another. But the middle manager is the technology and managerial expert that the new managers, experienced managers, and employees all look up to. In the improvement process, they are the teachers, the coaches, the friends, and the mentors. Whereas top management are the beacons of the improvement process, middle managers are the rudders. They truly make the difference between excellence and mediocrity.

■ Management's Change Process

Managers must change before they can expect the employees to change. If management keeps on doing the same old thing, the same old way, they will get the same old results. Improvement must start at the top of the management ladder and flow down layer by layer, washing away the undesirable traits, skills, and behaviors, before it flows on to the next level. We call this the "waterfall effect" of total improvement management. Only after a manager has accepted and embraced the change, can they expect their employees to embrace the change. The change should occur first in top-level management, then in middle-level management, and next, at the first level and supervisor level. The employees should not be expected to change until management has led the way. As the old saying goes, "The employees listen to the tongue in management's mouth, but they follow the tongues in management's shoes." And if both your feet and words do not travel in the same direction, the employee can only assume that management speaks with forked tongue.

To set the stage for this change, the organization needs to invest in training and preparing the management team for their new roles. Remember, "Training isn't expensive; it's ignorance that is expensive." This training should help the management team progress through the transformation cycle that the total improvement management process will create. This cycle can be compared to reshaping a block of ice. You can chip away at the ice with an ice pick to reshape it. But in this case, you lose a lot of potentially good material. We believe this is the wrong way to go. The correct cycle is to first thaw out the old personality, then reshape it into a more desirable new personality, and freeze the new personality into the daily working habits, practices, and beliefs of the organization. With this approach, about 95 percent of your management team can be salvaged and progress through the transformation cycle.

■ Developing the Desire to Change

Management, more than anyone else, is going to be personally impacted by TIM (Total Improvement Management). Many of their personal traits that were responsible for their success will have to be changed. Whereas the employees are subjected to very positive change experiences as they are empowered to take control of their own destiny, many managers will undergo the reverse experience. Most managers got where they are because they outperformed the other people in the group. Along with the management assignment came the ability to get more things done that they wanted done because suddenly they had many more arms and legs to work with.

In years past, managers felt that they knew what needed to be done (or at least what they thought needed to be done) and told other people to do it. When the job was complete, the manager personally and publicly was credited with getting the job done. Now, suddenly, we are asking management to step back and let the people they outperformed do things their own way, to help these employees develop without ordering them to do things, to give them the credit for ideas that were originally the manager's, to take the spotlight off themselves and put it on the employee. This is understandably a very difficult adjustment for most managers to make, particularly in a technical environment.

Before we can hope to bring about an attitude adjustment, we must first create a desire to change. This is where organizational change management can be a very effective tool. Management needs to be provided with enough data so they can understand that the pain associated with staying with the present process is greater than the pain they will be subjected to in the new process. To accomplish this, the first part of the training process must be focused on awareness training so that the pain associated with staying with the status quo is thoroughly understood.

Once management understands the pain related to their current situation, then they are ready to be introduced to the concepts that will be used to transform the organization from its current state into its future vision. Introducing the management team to these key concepts allows them to understand the pain that they will be subjected to during the change process, and the pain level that will be associated with the future state. This understanding is necessary for the management team as a whole, and each manager as an individual, so that each manager can decide if he or she will embrace the improvement effort, thereby allowing their attitude to change. If the balance of pain is not heavier on the status quo side, the desired commitment will not be obtained.

■ Management Education

Management education should always start at the top, as defined in the "Waterfall Concept." Remember you cannot sweep the stairs from the bottom up. Top management should attend all classes before they ask their employees (lower-level managers) to attend. The general rule should be that no employee should take a class that their manager has not already attended unless it is a very technical subject that is specifically job-related. Although all levels of management are subjected to the same material and concepts, the classes at each level of management will vary in depth significantly. Top management classes are designed to teach concepts, how to ask the right questions, and how to evaluate results. Middle management, on the other hand, will receive the most thorough training, with lots of case studies and practical assignments. In most cases, middle management has to be the group that best understands the improvement methodology and tools because after the initial introduction of the approach, they will be the ones responsible for training the new first-level managers in the use of the tools and facilitating the implementation of the concepts. First-level managers and employees on the other hand, receive very practical application-oriented training that is provided in conjunction with the opportunity for applying it immediately in the business environment. To make this approach practical, middle management must develop a plan to roll out the training within their area and be actively involved in the delivery and application of the training. Management training can be divided into two basic categories: Technical Capabilities, and Social Skills.

Education is the maintenance program for management. For the experienced manager, about five percent of their time needs to be devoted to training on social management skills, and another five percent to the technical skills related to their assigned area. This means that between four and five weeks a year of training and education need to be provided to the average experienced manager to keep them at peak effectiveness.

The big mistake most organizations make is that they do not provide their employees with an opportunity to study the social side of management until they are placed in a management role. This is like asking an employee who has no engineering background to design a bridge and expecting him or her to start doing an excellent job while still learning basic math and mechanics. What organizations need to do is to provide opportunities for aspiring and/or potential managerial candidates to study social management-related techniques before they become managers. In fact, an even better approach would be to require potential managerial candidates to successfully complete a series of classes before they are considered for a management assignment. More and more organizations are adopting this technique to minimize the negative impact new managers have upon the organization.

The following is a partial list of typical subjects that most management should master:

- Process Controls
- Data Collection and Analysis
- Experimental Design
- Process Engineering
- Participative Management
- Business Process Improvement
- Leadership Skills
- Effective Communication
    — Nonverbal Communication
    — Effective Listening
    — Interpersonal Communication
    — Interviewing Techniques
    — Negotiating
- Assets Management
- Statistical Decision Making
- Strategic Planning
- Program Planning
- Trust Building
- Use of Rewards and Recognition
- Variables Management Style
- Personality Analysis
- Financial Controls
- Safety Management
- Meeting Management
- Audit Methods
- Career Planning and Development
- Performance Planning and Appraisal

- Business Accounting
- Goal Setting
- Time Management
- Creativity Improvement
- Project Management
- Risk Management
- Conflict Management
- Motivation Essentials
- Equal Opportunity
- Etc.

As you can see, there are a lot of new skills that most new managers have to develop at a time when they usually are the busiest.

In addition to these basic skills, middle management needs training in the following areas so that they can perform their jobs as educators and facilitators.

- Facilitation skills
- Designing effective training programs
- How to present training with impact
- Networking
- Adult education
- Presentation delivery

■ Job Descriptions

The job description is the way most organizations define and communicate the form, fit, and function of each job and define its value to the organization. Each assignment is guided based upon the job description. In many organizations, job descriptions have proliferated to the point that they are dragging the organization down. Often in these organizations and in other organizations, job descriptions are out of date and as a result, not usable. The basic objective behind the job description approach to defining and evaluating jobs is a sound one, but now we need to make it a value-added tool.

Management Job Descriptions

Frequently, management job descriptions are outdated and certainly not in keeping with the new improvement environment. The typical manager's job description is written for dictators, for people that kick tail and make things happen. Frequently, the whole social side of management is completely ignored. Little or no credit is given in the job description for the manager that develops their employees so they can accept increased levels of responsibility. One of the very first steps in implementing the improvement process has to be restructuring the management job description so that it is in line with expected management performance. These management job descriptions should be divided into three sections:

1. Technical
2. Social
3. Control of Resources

Employee Job Descriptions

Equally important as the management job description is the employee job description. In an environment where we are empowering the employees to make all appropriate decisions related to their jobs, their job descriptions soon become obsolete. As a result, management needs to go back and rewrite each job description so that it reflects the increased responsibilities placed upon the individual. Then the job should be reevaluated to determine if the employee is getting a fair day's pay for a fair day's work. It is important to minimize the number of job description levels and associated pay ranges. This allows and encourages the employee to move from one assignment to another, increasing their knowledge and experience base while increasing their value to the organization. The use of job rotation programs helps keep morale high when there are few promotional opportunities available in the organization.

■ New Performance Standards—Error-Free Output

Management needs to set a new performance standard for themselves and their employees. The performance standard that we like to use is error-free performance. We pay our people to do the job correctly, not to make errors and create more work for themselves and others. To err is human, but to be paid for it is divine. The business world has accepted errors as a way of life. We live with them, we plan for them, and we make excuses for them. They have become part of the personality of our businesses. Our employees quickly realize our standards and create errors so that we will not be disappointed. I had an employee tell me, "If I did my job right all the time, I would put my friend Jane out of work."

We selected the word "error" in place of "defects" because most people view defects as occurring only in the manufacturing operation, whereas errors are something that everyone makes. Now that we have admitted that everyone makes errors, is it logical to have an error-free performance standard? The answer is yes, because we don't know how many errors are acceptable. Is it acceptable to drive through your garage and crash into the back wall one time out of ten? One time out of one hundred? One time out of ten thousand? Hopefully, your standard would be to always stop before you hit the back wall of the garage. The same is true of our business activities.

In businesses during the 1970s, error rates of 1 or 2 per 100 were acceptable. In the 1980s, error rates of 1 or 2 per 10,000 became the standard in manufacturing. In the 1990s, error rates of 1 or 2 per 100,000 are the starting goal. Motorola has started a program that they call Six Sigma. This program asks every employee to measure their performance and not be satisfied until their error rates are less than 3.4 defects per million. Is that good enough? The answer is no, not if you're talking about catastrophic failures in automobile brakes or elevators.

Our ultimate goal has to be error-free performance. But how do you set a seemingly impossible goal and have your employees accept the goal? The truth of the matter is that all of your employees are error-free performers. Everyone performs error free for some period of time—maybe it's five minutes, maybe it's five hours, maybe it's five days. Hopefully, it's five weeks and even better still, five months or five years. You see, we all are error-free performers and when we start talking about an error-free performance standard, we are talking about increasing the time interval between errors. This provides every individual with a way of measuring their error-free performance rates and setting goals for themselves for continuous improvement. Stop and ask yourself what your mean time is between errors. Does it meet Motorola's Six Sigma standard? I doubt if any executive performs to this very aggressive standard all the time. We estimate that executives make bad decisions only about 5 percent of the time but they only make the very best decision about 10 percent of the time. The other 85 percent of the time, they make an acceptable decision but not the best decision. This area presents a major opportunity to improve the organization's competitive advantage.

■ Measurement Systems and Performance Plans

In support of the new performance standard and the new job description, management has to revamp the individual's and group's measurement system and performance plans. The new performance plans need to place emphasis on teamwork, adhering to schedule, continuous improvement, and creativity. Management needs to be honest with the employees and evaluate each employee's performance, remembering that 50 percent of their people are below average. This does not mean that 50 percent of the people do or do not meet the requirements of the job. It means that the word "average" defines a center line where 50 percent of the employees are above the line and 50 percent are below.

Too many managers take the easy way out and evaluate most of their employees as being above average. This is unfair in two ways. First, it is unfair to the poor performers because they are not given the correct information that will help them understand how they are performing in comparison to the rest of the group and provide them with the sense of urgency that is needed to improve. Second, it is unfair to the high performers because they are not given full credit for the extra effort they put forth.

The other approach to performance evaluations is not to compare people to each other. In this approach, the manager compares the employee's job performance to the job description only. This type of approach leads to evaluating people in the following categories:

- Does not meet the requirements of the job.
- Just meets the requirements of the job.
- Meets the requirements of the job.
- Sometimes exceeds the requirements of the job.
- Usually exceeds the requirements of the job.
- Always exceeds the requirements of the job.

With this type of approach, 80 percent of the people in better organizations fall into the sometimes exceeds or usually exceeds the requirements of the job. The 5 percent of the people that always exceed the requirements of the job are the superstars that should be promoted to a more challenging assignment. The 5 percent of the people that do not meet or just meet the requirements of the job are people that must improve or they will be reassigned or let go. That leaves only 10 percent of the employees that meet requirements of the job. In organizations that are not among the better performers, this number can be as high as 50 percent. This is the preferable approach to performance evaluation, but it requires excellent job descriptions and detailed performance plans developed jointly by the manager and the employee.

■ Upward Appraisals

Many managers do not really understand how their employees perceive them. Dictators throughout history, from cult leader Jim Jones to Adolph Hitler believed their followers loved them, when in truth, they ruled out of fear, not devotion. This is true for many managers. To take advantage of the employee's viewpoint, many organizations have established an upward appraisal system. Successful upward appraisal systems provide upward feedback on specific management behaviors and stay away from personality issues. These upward appraisals not only provide valuable input to the managers on their strengths and weaknesses, but also information about the employees' needs and expectations. A good upward assessment procedure provides the employee with a way to define what their manager needs to do to further the employee's development and their performance in their present assignment. Effectively used, upward appraisals provide management with information that alerts them to development needs before they are noticed by the manager's boss and/or have a negative impact upon the manager's performance.

Kathy Colbourn, Organization-Development Administrator for the Federal Reserve Bank in San Francisco, points out that managers are often surprised when they find out what their employees really think of them. The higher up in the organization the manager is, the more important upward appraisals are. Higher-level management get less and less useful feedback. Janina Latack, Associate Professor of Management and Human Resources at Ohio State University's College of Business stated when discussing upward appraisals, "The higher management goes, the less likely that they'll get usable feedback. The higher they go, the more people assume they know what they're doing—and the more people are afraid to give a candid opinion."

At Dow Chemical, the upward appraisal process is followed by a structured coaching session with individual employees. This option is used in about 95 percent of the cases.

360 Degree Appraisals

In the most advanced organizations, the appraisal evaluation process is taken on a 360 degree input focused at the supervisor and management levels. In these cases, they do not rely solely upon information provided by their immediate manager and the employees that report to them. They also receive input from their peers and their customers. In some cases, this input is provided directly to the manager for self-evaluation and improvement without being processed by a formal organizational representative that is designated to evaluate the individual's performance. The customers', employees', and peers' evaluators are usually given the option of identifying themselves or keeping their identity secret. This provides the individual manager with a total review of his or her performance that is a very effective way of helping individuals who really want to improve. The downward appraisal is replaced with a personal evaluation of performance to stated business goals that is reviewed with the individual's manager as part of developing personal and group goals for the next evaluation period. This ensures that these goals are in line with the organization's critical success factors and short-range goals. The manager's salary and bonus are based upon a combination of his or her group's ability to meet these goals and the total organization's performance.

■ Management Improvement Teams (MIT)

Between 70 and 85 percent of all errors can only be corrected by management. How can we expect an operator to make good parts when the equipment that he or she is using is maintained poorly? How can we expect our salespeople to excel when they haven't been trained and don't know what is expected of them? The education process should have provided management a new awareness of their role and the need for immediate improvement in the organization's performance. Now is the time for them to get involved and become active participants in the improvement process. Each manager should become an active member of a Management Improvement Team (MIT). Often, middle management and top level management will lead MITs that consist of the managers that report to them, and become a member of the MIT that is lead by their manager. These MITs will meet on a regular basis—weekly at first. After the improvement process becomes part of the normal management practices, MIT meetings will be cut back to once a month. Minutes should be published for each MIT meeting. The purpose of the MITs are to:

1. Define appropriate, management-agreed-to, department missions.

2. Develop ways to measure the improvement within the organization. Share these measurements and trends with employees.

3. Define the improvement education needs for the organization.

4. Solve problems that cannot be solved at lower levels.

5. Share successful improvement methods.

6. Develop short- and long-term improvement strategies and tactics.

7. Develop ways to move from a reactive to a preventive management style.

8. Identify individuals who should be rewarded for their improvement activities and/or their abilities to prevent errors from occurring.

9. Provide five-way communications—up, down, sideways, suppliers, and customers—throughout the organization.

10. Prioritize improvement activities for maximum return on investment.

Of all the activities listed, the one that seems to be the most difficult for nonproduction areas is to measure the output from the area. Most managers will claim that there is no way to measure the efficiency, effectiveness, and adaptability of their area, but if that is true, how do they know who to promote or who to fire? The performance of every job, every individual, and every department is measurable. It may mean that new systems need to be developed to provide the required data, but all activities can and should be measured to ensure they have a positive impact on the total organization. When you realize that 85 percent of the errors that occur within the organization can only be corrected by management, it is easy to see that these management improvement teams will have a very full agenda.

When President Bill Clinton was governor of Arkansas, he formed a state government Quality Management Board that he chaired. He also attended a two-day training class on TQM and held regular meetings with his Management Department Improvement Team, even during his campaign for president.

Management needs to get off their soft chairs and go to the employees so they can get close to the problems. The biggest change I have ever seen in a management team occurred when the employees went on strike and management took over running the production operations. Following the end of the strike, the biggest capital equipment budget the organization ever had was approved. When management starts to work on a problem, they should get firsthand information by doing the following:

- Go to the work area where the problem is occurring.

- Put their hands on the problem, do the activity that caused it.

- Take part in the data collection process.
- Ask questions and request suggestions from the employees.
- Find the root cause.
- Personally try out the corrective action.
- Follow up to be sure the correction remains in place.

■ The Down Side of Improvement

It all sounds too good to be true. All you have to do is use the appropriate improvement tools in the correct sequence, and quality and productivity will soar. Everyone will earn more money and it is a win-win situation for all stakeholders. Theoretically, that is true under ideal conditions, but unfortunately in many organizations, that is not the reality of life. It probably is true if you can grow your market and market share at a rate that is equal to or faster than your productivity increase. However, if the market is constant and productivity improves, someone, somewhere, has to be put out of a job. Maybe that someone is not in your organization if you are growing your market share in a stagnant market at a rate that offsets the productivity improvement within the organization, but in some other competitive organization there will be a negative impact.

In the ideal situation, increased productivity allows prices to be reduced, resulting in a larger market for everyone to feed upon. The reality of today is that most organizations improve to stay in business. As a result, improvements in productivity and quality often allow the organization to operate with fewer and fewer people because the market has not expanded fast enough to consume the additional output. The reality of not being able to maintain full employment has been realized by even the best organizations around the world.

At this very early stage in total improvement management, managers need to step up to the reality that as performance improves, jobs probably will be eliminated, and they have to decide what will be done with these surplus employees. If management has been running a tight ship, they probably can adopt a "no-layoff policy." Typically, these policies read something like this:

No-Layoff Policy

No one will be laid off because of a quality or productivity improvement. People whose jobs are eliminated will be retrained for equivalent or more responsible jobs. This does not mean that it may not be necessary to lay employees off due to a business downturn.

When President Clinton was governor of Arkansas, he pushed the 1991 Legislative Session to pass quality management legislation. The law included the following provisions:

- Assured that no state employee will lose employment because of quality management efficiencies

- Provided ways to reallocate funds to support the quality movement within agencies
- Created a Quality Management Board

Many organizations that are involved in TIM and issue a no-layoff policy accomplish it by putting a freeze on hiring. In one organization that we worked with, during their first year, 253 people were identified as surplus as a result of the improvement efforts. A freeze on hiring allowed 204 of these surplus people to be placed in meaningful assignments. The remaining 49 were still actively engaged in organizational activities that really didn't need to be done. The creative solution to this problem that this organization arrived at was to have an organization-wide competition to select 49 individuals that would be sent to a local university. The organization paid the employees' regular salary while they were going to school, in addition to their tuition and books. The employees were told that when job opportunities arose within the organization, they would be brought back and put to work. Two semesters later, all 49 people were back doing productive work within the organization, and they had a much greater value added content than they originally had. In addition, 90 percent of these individuals went on to get their degrees on their own. The impact on the other employees of this approach to effectively use surplus employees was exciting. Suddenly, everyone was running around looking for ways to eliminate their jobs so they could go to school.

So what happens if you cannot afford to issue a no-layoff policy and your market does not grow sufficiently to keep your employees fully utilized? Is your only option a layoff? No, there are at least 17 other options that should be considered first.[2]

Improvement's Negative Impact on Management

The improvement efforts going on in western countries have hit management extremely hard. Most organizations are trying to flatten their organizational structure by eliminating middle managers. In a survey of 836 organizations conducted by the American Management Association last year, middle management made up 5 percent of the work force, but accounted for 22 percent of the past year's layoffs.[1] Lower-level managers are beginning to feel the impact of self-managed work teams. Unfortunately, when we lose our managers, we are losing the cream of our employees. In most cases we selected them because they were the most productive, most technically competent individuals in the group. We have invested vast sums of time, effort, and money in further developing their skills. They should be our very best employees. If they are not, it is not their fault, for at some point in time they were the best.

If we accept the reality that most organizations have a smaller span of control than they need, it is inevitable that some managers must go. Management needs to find a way to decrease the number of managers without losing this valuable resource and without embarrassing the individual. The answer to this dilemma is to establish a dual growth ladder combined with a management technical vitality program.

Many organizations talk about a dual growth ladder. One side is for management skills, and the other is for technical expertise. This approach has the advantage that superior technical people need not transfer into management and become poor managers to have a growth path within the organization. It also allows managers to move back and forth freely into technical roles without being unduly and/or unfairly impacted. The dual ladder also provides an excellent solution to the organization's problem of management technical vitality. Today a manager that moves out of management is looked at by the total organization as a failure. As a result, managers dogmatically use all their skills and contacts to hide their obsolete technical competence to stay in management. They work excessive hours to look good in front of other managers and misuse the time of the employees that report to them keeping them informed and explaining what they are doing.

A technical vitality program for middle management will solve most of these problems. Many organizations have started to rotate their middle managers into technical assignments for three years, after being in management for six years. No manager is exempt from this rotation. Every middle manager knows that every six years they will be placed in a technical vitality assignment, so there is no shame associated with rotating out of management. The organization benefits as soon as the rotation project starts because they automatically can reduce the number of middle managers by one-third. The organization also benefits when the middle manager returns from the technical vitality assignment, because they have a much better understanding of the technical side of the business. This truly is a win-win activity for middle management and the organization.

■ Management and Employee Opinion Surveys

Upward, downward, and sideways communication is a major problem in most organizations. Management needs to understand how the employees feel about the organization's performance and the environmental conditions within the organization that impact them. Often, there is a major disconnect between management and employees.

All organizations should conduct an employee opinion survey at the very beginning of the improvement process to provide a balance between the organization's social and technical obligations. This survey should be repeated every 18 to 24 months.

■ Management Self-Assessments

Most organizations have staff that audit individual functions and departments to ensure management is complying to procedures. Large corporations have staff perform preannounced audits that are preceded by a flurry of cleaning up and changing to be prepared for the audit. Still the audit team leaves an embarrassingly long list of problems. I call this the "seagull" approach to control management. The audit team flies in, does their job, and flies off. Just think how long their list would have been if the audit was conducted as a surprise audit, or to put it another way, if we really knew how each area was actually performing.

Audits and audit checklists are a way of life in most organizations today, but isn't it strange that the managers in charge of an area do not periodically perform a detailed audit of their own area? Certainly no one is better positioned to know if the job is being done correctly than the employees that are doing the job. The person that is next best prepared to evaluate the area's output is not an external audit team, but the department manager. We normally think of the department manager as working so closely with the employees that they always know what is going on, and that without any research, they can provide an accurate assessment of the department's performance level. The truth of the matter is that the manager often becomes part of the problem. What the manager needs is a systematic and objective way of assessing the department's activities, allowing undesirable situations to be corrected before they become problems. One way of accomplishing this is for all levels of management—from the president of the organization, to the first-line manager—to conduct a quarterly self-assessment of the area they are responsible for.

Written procedures are prepared to be followed, but many things that are not put in writing are important, and should be included in the checklist. Things like housekeeping, security, and safety are critical, but are not always documented in formal procedures. Often good business practices are critical to setting the required attitudes. When a Sumitomo manager was taking a group of Toyota engineers on a plant inspection, one of the visiting engineers noted a crack in a bathroom window and asked if it was supposed to be cracked. The Sumitomo manager said no. Immediately the engineer cancelled the tour, and a very lucrative contract was lost. Why? How can you hope that a supplier can provide excellent products when a simple rule like not having broken windows cannot be followed?

■ Summary

The improvement process must be embraced by management before the employees are exposed to it. Managers are the only ones that can correct 80 percent of the problems most organizations face today. With so much of the improvement opportunities resting in management's hands, there's no need to involve the employees until many of these major problems have been attacked and solved. The management team must demonstrate that they are willing and able to change before they ask the employees to change. Management has a beautiful bald eagle boxed into a 4x4 cage. Encouraging that eagle to soar produces little results and soon becomes discouraging to everyone unless management breaks down the bars that imprison the eagle.

Management must stop talking and start performing. They need to show their leadership by changing the way they manage. Management needs to change from a single management style to a variable style that meets the needs of the individual employee in the many different circumstances they are subjected to. This frequently means giving up much of the control and power management previously exercised over the employee, and allowing employees to participate in the decisions that are relevant to their work.

The following shows how managers in different types of organizations perform:

Problem Solving:

Losers—View most problems as being caused by their employees. Major problems are attacked by engineering and management.

Survivors—View management and the employees as sources of organizational problems. Employees are trained and are responsible for solving problems.

Winners—View managers as the major source of problems, feeling that 80 percent of the problems can only be solved by management action. Management and employees are trained to solve problems. Management takes a leading role in correcting the major problems.

Training:

Losers—Most of the training provided is on the job and job-related.

Survivors—Both job and improvement training are provided to the employees and line managers. Everyone is trained whether or not they can apply it.

Winners—All levels of management and employees are trained in the improvement methods and job-related subjects. Training is provided in conjunction with the opportunity to apply it.

Management Style:

Losers—A coaching style is used by management most of the time.

Survivors—Participative management is widely practiced.

Winners—Management style varies to meet the situation, ranging from dictatorial to friendly.

Management Attitude:

Losers—The employees cannot be trusted. They need to be watched and pushed to get the job done.

Survivors—Employees will work if management motivates them and controls their progress.

Winners—Employees are trusted, valued assets. They will perform well if they understand the goals and objectives.

Management Process Audits:

Losers—Do not audit.

Survivors—Have a department that does random audits.

Winners—Perform self-audits, with random checks conducted by independent third parties.

The key to successful management is simple: Get back to basics.

- Treating employees as you would like to be treated.
- Setting a positive work ethic example.

- Encouraging those who fail and praising those who succeed.
- Providing honest evaluation of an individual's efforts.
- Stepping up to the unpleasant situations.
- Being friendly and having a smile on your face.
- Freely giving credit to the people who do the job.
- Portraying a sense of urgency and importance regarding the work that is being performed.

As the famous Notre Dame head football coach, Knute Rockne, put it, "The trouble in American life today, in business as well as in sports, is that too many people are afraid of competition. The result is that in some circles people have come to sneer at success if it costs hard work and training, and sacrifice." No manager can be successful without a great deal of hard work, training and sacrifice. If you are not willing to give your all to the job, then management is not the right career path for you.

■ References

1. *Fortune* Magazine, February 22, 1993, p. 80.
2. Ernst & Young Technical Report TR H.J.H., "The Down Side to Improvement."

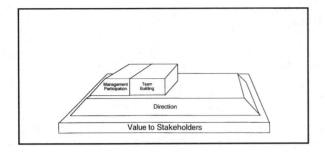

7

Team Building: Bringing Synergy to the Organization

Kenneth C. Lomax
Senior Manager, Ernst & Young
National Practices Group

Being part of a team or a group that provides security,
acceptance and a sense of belonging is a basic need for
most human beings. DR. H. JAMES HARRINGTON

■ Introduction

In the seventies and eighties, we learned how important teams were to the overall success of the improvement process. In the ensuing years, hundreds of organizations throughout the world have validated this. In the eighties, we realized that teams had much more to offer than just their abilities to solve problems. We gave teams the authority to make decisions and manage their own process. By doing this we increase our ROA (Return on Assets) and increase the overall organization morale. It's a win-win for everyone.

Japan's Dr. Kaoru Ishikawa, the "father" of the Japanese quality approach and Quality Circles, is no longer with us but the legacy he left is one of a more humanistic approach to quality. We no longer need to ask the question, "Are teams right for our organization?" Of course they are. Do teams, or should teams, be expected to solve all the problems of the organization? Of course not. A recent survey, conducted as a joint effort between the consulting firm of Ernst & Young and the American Quality Foundation (ref. "The International Quality Study) found that, for organizations just starting their quality effort, building the human resource infrastructure and organizing teams into effective work units is one of the basic strategies. Typically, these organizations have less than five percent of the work force participating on teams. As we move up the ladder to organizations that are beyond the starting point in their continuous improvement effort (medium performance), we find continuing the focus on teams is important and of benefit to the organization.[1]

Most "medium performance" organizations say that approximately 25 percent of their employees participate in department-level teams. Problem solving and department teams play a key role in these organization improvement efforts.[1]

Higher performing organizations also see advantages from numerous forms of employee involvement. This is less in the form of employee department teams and more in higher "macro-level" teams such as Process Improvement Teams.[1]

While there is certainly a time when organizations can back off of wide-spread department team involvement, the organizations surveyed all felt problem-solving teams played an important role in the continuous improvement effort.

The following shows the percentage of organizations in several different countries, including the United States, indicating that 25 percent or less of their employees are involved in quality-related teams.[2]

Country	Present %	Future %
Canada	59	23
Germany	81	58
Japan	64	62
United States	51	30

The following shows the percentage of organizations with more than half the work force participating on natural work teams (Department Improvement Teams).[2]

	Manufacturing %	Service %
Past	22	3
Present	30	8
Future	53	33

A detailed analysis of this data can be found in the Ernst & Young Technical Report by Dr. H. James Harrington entitled, *"Different Strokes for Different Folks—The International Quality Study."*[2]

Canada, Germany, and the United States plan to greatly increase the level of team participation during the next three years. If Germany lives up to its plan, it will have a higher percentage of employees involved in quality-related teams than

Japan. At the present time, North America has a higher percentage of its employees on teams than Japan does and if its projection holds true, North America will have approximately 50 percent more team involvement than Japan will be utilizing. This lack of team involvement flies in the face of what we have been told about Japan, but verifies Dr. Kaoru Ishikawa's statement that Quality Control Circles were developed to train Japanese employees, not to solve problems.

It looks like North America is over-correcting again. The most extensively used type of team is the natural work team (Department Improvement Teams/Quality Control Circles). Manufacturing has been using these types of teams since the 1960s. They have been growing in usage for the last 30 years, but during the next three years their usage will almost double. The service industry has been slow to adopt this tool but will expand its use by over 300 percent during the next three years, starting to close the gap between itself and the manufacturing industry.

You see, it's really not how many teams you have, or what types of teams you have. It's having the right number of teams trained with the skills to solve organizational issues and problems.

■ Elements of a Team

In Dr. Harrington's *The Improvement Process* he stated:[3]

> Team participation should never occur until the management team is totally participating in the improvement process, if you don't want the employees to believe they are being manipulated. Management must provide visible evidence of the company's thorough commitment to a policy of preventing problems rather than reacting to them.

We find there are four key elements in the team environment. The element that must be in place prior to teams being formed is the Executive Improvement Team's (EIT) commitment to the process. The second is the team members themselves, the third is the team leader, and last, but certainly not least, is the facilitator. Let's look at them individually.

The Executive Improvement Team

The executive team has overall responsibility for the entire improvement effort. If the effort succeeds much of the credit should go to them, and, if it fails, all of the credit should go to them. Why?

When the team environment is first being established in an organization, its early success will depend on the amount of support and encouragement given teams and team members by management. Early team planning by the EIT will eliminate many of the problems that plague organizations with a poorly organized team structure. However, there is one ingredient that comes before establishing the team, and that's training.

Basic Team Effectiveness Training. The smart executive team will train as many employees as possible in the basics. By basics we mean:

- Understanding the Organization's Goals and Objectives
- Understanding the Improvement Process
- Team Dynamics
- Team Effectiveness
- Effective Meeting Skills
- Basic Problem Solving

Establishing the Team Process. Establishing the team process should take place prior to forming the first team. The EIT should know what type of team is being established (see Fig. 7.1: "Types of Teams and Their Characteristics"). The EIT should also decide how the team process will be managed. For example: Will the

Characteristics	Task force	Task team	Process improvement team	Department improvement team	Quality circle	Autonomous work team
Membership	Selected members based on experience	Selected members based on experience	Members involved in the process	Department members	Department members	Department members
Participation	Mandatory	Mandatory	Mandatory	Mandatory	Voluntary	Mandatory
Management direction	High	Moderate	Moderate	Moderate	Low	Low
Task selection	By management	By management	By management	By team	By team	By team
Urgency	High	Moderate	Moderate	Moderate	Low	Moderate
Scope of activity	Organization wide	Organization wide	Process wide	Department wide	Department wide	Department wide
Activity time	Long meetings short period no other assignment	Short meetings long period	Short meetings intermediate period	Short meetings intermediate period	Short meetings ongoing	Short meetings ongoing
Process facilitator	Optional	Optional	Recommended	Recommended	Recommended	Recommended
Team leadership	Appointed	Appointed	Process owner or designee	Supervisor	Supervisor or designee	Shared or rotated
Implementation	By others	By team or others	By team or others	By team	By team	By team

Figure 7.1. Types of teams and their characteristics.

EIT itself be responsible for team training and day-to-day management? Probably not, but it is important to decide who is! Listed below are several areas an EIT should consider in establishing the team process.

1. Who is responsible for overall management of the teams?
2. Who is responsible for setting the team mission?
3. How empowered is the team?
4. What type of team reporting will be required?

Each team, however should be required to submit for approval their Team Charter and Project Plan. These are discussed in more detail later. In addition, minutes of all team meetings should be prepared and distributed to members as soon as possible.

Setting and/or Approving Team Structure

For organizations just starting their improvement efforts, we recommend the EIT set up and/or approve each individual problem-solving team. Yes, this causes more work for an already overburdened executive team, but we believe in the long term it will pay off in huge dividends.

The EIT identifies and/or approves the issue the team is to focus on, picks a team leader (typically someone with knowledge of the issue) and, with the leader's assistance, identifies the team members.

This approach tends to conserve the organization's limited resources while allowing the overall team process to "settle down." After the organization has been involved in teams for six months to a year, the EIT can (and should) start empowering lower level management to establish teams.

Establishing the Team Mission

As we mentioned in the preceding section, the EIT should either pick or approve the task the team is to focus on. Once this is decided the EIT should determine the mission of the team. The team mission should simply give the team a clear perspective of why it (the team) exists.

One of the most frustrating experiences a team can have is to be assigned an unclear task or problem by the EIT. An example of this is forming a team to look at the "communication" issue. This is referred to as a *divergent* problem or, a problem that tends to grow in size, and complexity as the team moves forward in their efforts to solve or control it. In other words, it keeps getting bigger and bigger. It's up to the EIT to ensure the team is working on problems that are *convergent* or those that tend to become more clearly defined as the team moves toward solution and implementation. In the case of "communication," a more convergent problem would be, "We have a problem communicating between management and the employees." The EIT would then give the team a mission that might say:

"The mission of the communication task team is to identify ways to enhance communications and understanding between management and employees."

The mission sets the stage for how you want the problem or issue to change. It should be very brief, no more than two or three sentences.

Once the mission is set, the team can establish the rest of its own Team Charter. The charter consists of three key elements:

- *The mission*—Why the team exists
- *Team goals*—What the team hopes to accomplish
- *Team guidelines* (or Code of Conduct)—How the team will manage and measure itself

As you can see, the clearer the team mission the easier it will be for the team to complete the task. Once the team charter is established it is typically signed-off by each team member and the team's EIT sponsor. This sign-off shows "ownership" by both the team and their sponsor and helps the team in identifying team process issues that may inhibit their performance.

Another element that supports the team charter is the *project plan*. This plan gives the team specific direction in the following areas:

- Team Meeting Schedule
- Resources Required
- Schedule of Activities
- Completion Time Frame
- Measures of Success

Both the completed Team Charter and the Team Project Plan should be reviewed and approved by the EIT. This review and approval authorizes the use of the organization's resources for a specific period of time to complete a specific task.

Providing Resources

Chances are, no matter what problem or issue the team is working on, some resources will be expended by the team. Not providing adequate resources to the team will send a very clear and negative message as to how much the EIT supports their efforts.

It is the EIT's responsibility to understand enough about the team's problem or issue to adequately budget the money, manpower, time, etc. to complete the task. It is not a good idea to expect the team to accomplish corporate miracles on their own time.

If the EIT is concerned or unsure about the amount of resource it will take to complete the task, the team should be asked to complete a cost analysis. This analysis should include, but not be limited to:

- Estimated time away from the job (for each team member)
- Estimated cost of outside analysis (if any)
- Estimated cost of team supplies
- Estimated cost of outside consulting (if required)
- Potential savings

The cost analysis should not include any implementation estimates at this time. At this stage the team would not have progressed far enough in the problem-solving process to make any implementation estimates.

After the team has completed its task and it is ready for implementation, a new cost analysis should be conducted and presented to the EIT. This analysis will assist the EIT in identifying and planning for the resources needed for final implementation.

Approving Team Projects

After an organization's improvement efforts have been underway for awhile, teams will be formed to solve organizational problems. Approval of these projects may take place at several levels, but first, let's look at the three basic types of problems.

- *Type I—Team controls.* The team has information and knowledge about the problem as well as expertise, resources, and authority to solve the problem.

- *Type II—Team can influence.* The team does not have full control of the problem or issue, but can influence the outcome, with some outside assistance.

- *Type III—Team neither controls nor influences.* The team neither controls nor influences this problem or issue, and should not take on a task of this nature.[4]

If the team and the project is at the department or work unit level, the manager or supervisor may be empowered to approve the task. This would be a Type I problem. If outside resources are required the task would be a Type II problem and the project may need to be approved at the senior management level. The team should not undertake any Type III problem since this problem type is outside their scope. The Type III problem should be turned over to management to form a new team to correct the problem if it is justifiable.

Any project requiring cross-functional team membership should be approved by the EIT. In cases where the organization is multi-divisional, any project requiring cross-divisional teams may have to be approved at the corporate level.

In Dr. Harrington's report, *"The Collapse of Prevailing Wisdom,"* he provides us with insight into the controls placed upon the formation of teams in different parts of the world. He reported:

Percentage of Companies in the Automotive Industry Where Management "Always or Almost Always" Approves the Formation of Teams

Country	Past %	Present %	Future %
Canada	34	18	10
Germany	41	31	28
Japan	68	68	64
United States	45	30	21

Percentage of Companies in the Computer Industry Where Management "Always or Almost Always" Approves the Formation of Teams

Country	Past %	Present %	Future %
Canada	38	24	11
Germany	24	24	15
Japan	68	68	72
United States	24	14	5

Using this data, it is easy to see that Japanese management maintains strict control over the formation of new teams and plans to continue the practice. In contrast, the other countries are reducing their controls.

EIT's Support to Teams

The EIT should be willing to run interference for the team and guide them on policy issues and potential barriers. As the team progresses, the EIT should hold periodic reviews on progress to eliminate any surprises during the solution and implementation stages. Team recommendations should be reviewed and approved by the EIT prior to implementation. This should not make the team feel less empowered. Remember, the EIT still has final responsibility for utilizing the organization's resources wisely.

Last, but certainly not least, the EIT helps drive the implementation effort. This becomes much easier, even for a team totally empowered, when management has been a part of the process from start to finish.

Team Leader

The individual selected to lead the team may be elected by the team, or more often, appointed by the EIT. In the case of natural work group teams it may be the supervisor or manager of that organization or department. It is usually someone with a deeper knowledge of the problem area, someone having more experience in the problem-solving process or someone that has an excellent understanding of the team process.

For those of you with a more autocratic management style, learning to be an effective team leader can be a challenge. While one of the harder jobs in the team process, we feel it is one of the most rewarding. Two of the most important traits of an effective team leader are being able to guide the team without dominating it and acting as an effective role model to the team.[4]

Some of the duties of the team leader are to:

- Coordinate team meetings and activities
- Teach members
- Promote and sustain the team synergy
- Encourage individual member participation without coercion
- Follow up on meeting action items
- Assist the team in monitoring and measuring its progress
- Ensure the team process is being followed

Team Members

The team member is certainly the "heart" of the team. The idea of "Participative Management" is based on allowing employees to help management make better decisions. The whole concept of "synergy" is based on two heads being better than one. If the team leader is there to guide, the team members should assume responsibility for successfully completing the task. Some specific team member responsibilities are:[2]

- Willingness to express opinions or feelings
- Active participation
- Listen attentively
- Think creatively
- Avoid disruptive communication
- Be willing to call a time-out when necessary
- Be protective of the rights of other members
- Be responsible for meeting the goals and objectives of the team

Team Facilitator

Probably the most difficult role in the team environment is that of the facilitator. There are several different thoughts as to what the role of the facilitator should be. Some organizations use the facilitator as a full member of the team.[4] Other organizations believe the facilitator should be an expert or have a lot of knowledge in the team issue. There are essentially three types of facilitators. They are:

1. Integrator/Coordinator
2. Group Process Specialist
3. Session Leader

The Integrator/Coordinator can, and in many cases does, fill the role of assistant team leader. Duties for this individual are typically in the administration of the team process and in communicating with others, outside the team process. It has been said that this role is that of the Team Secretary or Team Assistant.

The Group Process Specialist is the more traditional facilitator role and is primarily that of facilitating the team meeting with a focus on process. This is the facilitator role that was developed as part of the Quality Circle and group problem-solving process developed in Japan. This facilitator is not a full-time member of the team. His or her role can be described as that of teaching, coaching, and supporting the team.

The Session Leader Facilitator is simply one who leads a session, often in the form of a training session or workshop where a more traditional facilitator is not required. He or she is usually a subject matter expert on the problem or activity being presented.

We believe the more traditional view of the team facilitator, the Group Process Specialist, is best when an organization is developing a team culture. First of all, the team facilitator should *not* be a part of the team. The key responsibility of the team facilitator is to assist the team in focusing on *process*, not *content*. The most effective facilitators are those with the ability to tune-out most of the team's content discussion and focus on the overall effectiveness of the team. In other words, is the team reaching its goals and objectives? Do they stay on track? Do they start and end on time? Are they using the proper tools and techniques at the proper time?

We have included a chart that shows the team roles and responsibilities (see Fig. 7.2). The key is to remember that the team leader's and members' major concern is *what* decisions are made (content) and that the facilitator's concern is *how* decisions are made (process).

■ The Problem-Solving Process

The team environment really exists for one reason only—to improve the organization's performance. Let's face it, we probably wouldn't have a participative management process like teams if it only gave the organization a "warm and fuzzy, good feeling" and didn't produce tangible results that translate into dollars. In other words, *"if it doesn't make the organization money, don't do it!"* There are as many approaches to problem solving as there are to structuring teams. The one we've chosen is probably the most detailed approach.

	Team facilitator	Team leader	Team member
Purpose	To promote effective group dynamics	To guide teams to achieve successful outcomes	To share knowledge and expertise
Major concern	*How* decisions are made	*What* decisions are made	*What* decisions are made
Principal responsibilities	• Ensure equal participation by team members • Mediate and resolve conflict • Provide feedback and support team leaders • Suggest problem-solving tools and techniques • Provide TQM training	• Conduct team meetings • Provide direction and focus to team activities • Ensure productive use of team members' time • Represent team to management and EIT • Document team activities and outcomes • Assist team in developing and maintaining measurements	• Offer perspective and ideas • Participate actively in team meetings • Adhere to meeting ground rules • Perform assignments on time • Support implementation or recommendations • Maintain measurements
Position type	Organization-wide	Team-specific	Team-specific
Selection criteria	Personal characteristics	Job title and/or description	Ownership of/in the process

Figure 7.2. Team roles and responsibilities.

The Opportunity Cycle

Let's change the way we look at problems. Let's think about each problem we face each day as an opportunity to contribute to making the organization more successful. As these opportunities arise, we need to have a systematic way of addressing them so that they are not just put to bed, but buried. If you put a problem to bed, it can and will get up some time in the future to cause the organization more disruptions. It may be next week or next month or next year, or perhaps in five years, but it will come back unless the process that allowed the problem to occur initially is error-proofed. When you have error-proofed the process that allowed the problem to occur, then and only then have you buried the problem so that it will not come back. That's what the "Opportunity Cycle" is all about (see Fig. 7.3).

When you investigate each problem, go through the six distinct phases indicated in Figure 7.3. Each phase contains a number of individual activities. The total cycle consists of 25 different activities.

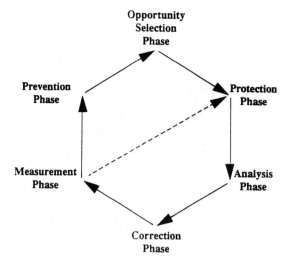

Figure 7.3. The opportunity cycle.

Phase 1: Opportunity Selection Phase

Activity 1: Listing the problems

Activity 2: Collecting data

Activity 3: Verifying the problems

Activity 4: Prioritizing the problems

Activity 5: Selecting the problems

Activity 6: Defining the problems

Phase 2: Protection Phase

Activity 7: Taking action to protect the customer

Activity 8: Verifying the effectiveness of the action taken

Phase 3: Analysis Phase

Activity 9: Collecting problem symptoms

Activity 10: Validating the problem

Activity 11: Separating cause and effect

Activity 12: Defining the root cause

Phase 4: Correction Phase

Activity 13: Developing alternative solutions

Activity 14: Selecting the best possible solution

Activity 15: Developing an implementation plan

Activity 16: Conducting a pilot run

Activity 17: Presenting the solution for approval

Phase 5: Measurement Phase

Activity 18: Implementing the approved plan

Activity 19: Measuring cost and impact

Activity 20: Removing the protective action (installed in Phase 2)

Phase 6: Prevention Phase

Activity 21: Applying action taken to similar activities

Activity 22: Defining and correcting the basic process problem

Activity 23: Changing the process documentation to prevent recurrence

Activity 24: Providing proper training

Activity 25: Returning to Phase 1, Activity 1

When teams follow these six phases (or a similar process), their life becomes much easier. Unfortunately, the more experienced the team becomes, the more likely they are to take short cuts. Process short cuts have probably led to the demise of more teams than can be counted. When a team elects to circumvent the correct problem-solving process they automatically reduce their ability to function in a continuous improvement environment. The team may ultimately be successful but it will be by accident, not by design.

■ Types of Teams

We have identified what we believe are the six types of teams most often used in businesses today. They are:

- Department Improvement Teams (DIT)
- Process Improvement Teams (PIT)
- Quality Circles (QC)
- Task Teams (TT)
- Self-Managed Work Teams (SMWT)
- Task Force (TF)

Department Improvement Team (DIT)

One of the most valuable teams in the entire process, the "Department Improvement Team" is made up of the employees in a particular department reporting to the same manager. They are also called Natural Work teams.

Typically, these types of teams start by performing an Area Activity Analysis to develop a mission, their customer-set, and their measurements. If they have problems meeting their measurements, then they are trained to solve them.

This team tends to focus on Type I problems only. These are problems the team has knowledge about, resources to use, and is empowered to implement their solutions with little or no outside approvals. This team is normally led by the department manager or supervisor. In cases where the department has more than 10 employees, membership in the actual team that meets periodically may rotate every 90 to 180 days. This gives everyone a chance to participate.

The team normally meets for about one hour, once a week for an indefinite period. Departmental problems are identified and prioritized. Management has the final veto in case the team selects a problem that is outside their scope or will not meet the ROI requirements.

Since this team is looking at issues that affect their own efficiency and effectiveness, there are huge opportunities for saving organizational resources.[6]

Process Improvement Team (PIT)

Another very valuable team to any organization is the "Process Improvement Team." While other teams tend to have more of a "task" oriented mission, the Process Improvement Team is allowed to focus on a specific process. These teams are also called Cross Functional Teams.

Membership in this team is directed by management or individuals intimately involved in the particular process. In some cases, short meetings are held over long periods of time (typically one to two hours per week for six months or more). These types of PITs very often will identify process issues that can be corrected through the use of a task team. While the process team remains together, the task team would only meet until the particular process issue is resolved.

Often organizations will prioritize their critical business processes and assign PITs to redesign or reengineer one to three processes at a time. In these cases, the PIT members usually are assigned to the PIT for between 50 to 100 percent of their time for three to six months.

Like the Department Improvement Team, the Process Improvement Team has great opportunities to reduce internal costs by increasing the efficiency, effectiveness, and adaptability of the process.[3]

Quality Control Circles

This is the team concept that allowed Japan to excel in the 1970s and 1980s. It also started the participative management movement in North America as we know it today. Unfortunately, Quality Control Circles (QCC) got a bad reputation in the late seventies and early eighties because most North American organizations used the concept wrong. Management did not provide the required skill training, direction, and support for successful implementation. In addition, management expected the Quality Circle Teams to solve problems that management had been unable to solve for years. As a result, management became discouraged with circles because they did not rack up big dollar savings. Dr. Yoshio Kondo, one of Japan's leading quality professors and consultants, stated, "Quality Control Circles are to motivate employees, not to reduce cost." Japan has been very effective in using Quality Control Circles to train their employees on how to solve problems. In North America, the Quality Circle movement failed not because of the employees, but because of lack of management's understanding of the process, and, as a result, misused it. As North American management matures, they will understand why these types of teams are so important to their overall improvement success. The self-managed team concept is based upon the Quality Circles concept.

This type of team is mostly made up of volunteers who hold short meetings over a definite period of time and work on either departmental or organizational issues. Management direction tends to be low on this type of team. This is probably what got the QCC teams in trouble in the first place. We have found the more management shows interest in the process, the more likely the team is to succeed. Most U.S. organizations have moved away from calling a team a Quality Circle and, even if they are the exact same thing, call them something else.[3] If and when U.S. organizations start to empower their employees, these are the types of teams that will most often be used.

Task Teams

A task team is put together to resolve an issue, then disbanded. Management identifies the team members who are selected, based on experience in the issue.

The issue and/or problem usually are not urgent. The length of team assignment may be short meetings over long periods (i.e., one hour per week for 30 days or more) or if the issue is of a more immediate nature, longer meetings over a short period.

Self-Managed Work Teams

The Self-Managed Work Team seems to be the "brass ring" many organizations are grabbing for. We believe there is a big future for self-managed work teams in the United States but we don't believe it's here quite yet.

We have heard this type of team called everything from an Autonomous Work Team to Self-Directed Work Teams. In the truest sense, the Self-Managed Work Team is one that manages its own business without outside interference from upper management. The team is responsible for setting its own departmental budget, managing its own resources, and even hiring and firing its employees. Most organizations in the United States that use this type of team allow them to function with low direction from management. And they function very effectively in most cases.

A word of caution: While Self-Managed Work Teams can be a real money maker for certain organizations, they don't work for everyone. Before the organization implements this type of structure, they should be very far along with their quality improvement initiative, and there needs to be a high degree of trust between management and the employees. This type of organizational structure is not for neophytes.[6]

The Task Force

The task force is a team most often designed to work on a very important issue or problem. They meet for long periods, sometimes as much as twelve hours a day, seven days a week, for a short duration (typically 30 days or less). This team is usually called on to solve "survival" issues or issues that must be corrected or resolved as soon as possible. Usually, this team's activity takes precedence over all other activities going on in the organization. Typically, task force type teams are formed when a manufacturing process is closed down for what could be a long period of time due to problems or a customer safety problem.

The task force is usually formed by management with membership being mandatory and the leader and members being selected based on experience in the issue at hand. Direction from management is as high as the urgency of resolving the issue.

The use of task forces indicates that management has major problems with the organization's processes. Organizations that have good business processes should never need to use task forces to manage and correct the organization's problem, because problems will be eliminated or recognized before they become critical.

We have included some information that may assist you in identifying the team that is right for you and your situation (see Fig. 7.1: *Types of Teams and their Characteristics*).[6] This data is designed to guide you in understanding basic team characteristics.

■ Training Teams

In every organization where we've conducted improvement assessments, the topic that always makes the employees' "Top 5" list is "training," or the lack thereof. It most often is their number one issue. At Xerox, effective Quality Improvement Team participation grows, in part, from effective quality training. At a minimum, every Xerox employee has received 28 hours of quality training. Xerox's initial training investment is estimated at more than four million employee hours and $125 million. The training of teams is just as important as training individuals to do their job! There are really only two ways to train teams. Let's look at both.

Formal Classroom Training

Teams can be trained as a group, where all members of the team go through training together; or as individual participants, where several (normally 10 to 20) employees are trained together and assigned a team at a later date. Either way works well; however, in both cases the training should not be conducted until the participants are within 30 days of using what they have learned. Don't waste valuable organizational resources by training and waiting 3 to 6 months to put the training to use.

On-the-Job Training

Actually, this should be called "on-the-team" training—more skilled or experienced team members training new members. While not as effective as formal classroom training, the team member can catch up with others on the team.

Here the untrained team member is taught the skills he or she needs by other, more skilled or experienced team members. Some organizations formalize their on-the-job training process by assigning a training curriculum for teams to follow. This ensures each new team member receives the same amount of training as the others.

Obviously, the down-side to this approach is having one or more team members that may not be as effective as the members with more formal training, which will slow down the rest of the team.

An additional word of caution: Timid, or less aggressive team members may never become completely competent through this approach. We recommend team and problem-solving training be a part of the New Employee Orientation process thereby providing everyone in the organization with a common language and approach.

■ Using the Team Approach to Organize and Run Meetings

One of the quickest paybacks to the team process is better meeting management. A study conducted a few years ago identified ineffective meetings as the number-one time waster of management in American businesses. Most of us can quickly relate to this since we all spend an inordinate amount of time in meetings.

If you are a senior level manager you probably spend anywhere from 60 to 85 percent of your time in meetings. Mid-management spends from 50 to 70 percent of their time in meetings. When asked, most managers will tell you that maybe 10 percent of the meetings they attend are beneficial and those would be more valuable if they were conducted with some structure.

We feel there are three pieces to the meeting "pie." They are:

1. *Logistics*. This includes the *where*; i.e., things such as getting the meeting notice out on time, preparing the meeting room, etc.

2. *Process*. This is the *how*. Here we're looking at how the meeting is conducted, what type of meeting style is in use. What tools and techniques will be required and how do we use them?

3. *Content*. Here is the heart of the meeting, or the *what*. What do we hope to accomplish, what action do we need to take, etc.

Five Elements of a Successful Meeting

Equally important are what we call the Five Elements of a Successful Meeting:[6]

1. Before the Meeting:

A. Plan the meeting.

B. Develop a complete agenda.

C. Prepare for the meeting.

2. At the Beginning of the Meeting

A.   Start the meeting on time.

B.   Utilize a scribe or recorder.

C.   Reach agreement on the agenda and the key objective of the meeting.

D.   Define each participant's role.

E.   Define the meeting process style. There are basically four types of meeting styles. They are: *informational*, *discussion*, *problem solving*, and *decision making*.

3. During the Meeting

A.   Be as positive as possible.

B.   Stick to the agenda and meeting time limits.

C.   Keep the discussions on track.

D.   When necessary, conduct a process check or "time-out."

E.   If new items come up, post them on an "Issues List" (parking lot) for discussion at a later date.

4. At the End of the Meeting

A.   Reach agreement on the results.

B.   Establish action items.

C.   Review the process. Take time to evaluate the meeting.

D.   If required, set a time and place for the next meeting.

E.   Prepare the agenda for the next meeting.

F.   End the meeting on time.

5. After the Meeting

A.   Follow-up on any action items.

B.   Prepare and distribute the minutes of the meeting.

■ Evaluating Team Meetings

Earlier in this chapter we talked about the advantage of having a "Team Charter" consisting of a mission statement, guidelines or code of conduct, and goals or objectives.[6] For a team to function properly, it must first decide how the members will react and respond to each other. The guidelines, sometime called "The Code-of-Conduct," set the stage for member interaction. Figure 7.4 is an actual code-of-conduct developed by a team during their first formal team meeting.

At first glance this looks like a nice "wish" list, but it's much more. Once the team has developed its own guidelines, it has a way to measure efficiency and effectiveness. The team takes the guidelines and turns them into an evaluation form.

Evaluate the meeting if you are really interested in holding better meetings.

Each member participates	Critique ideas, not people
Practice respectful listening	Team ownership of ideas
Be respectful of each other	Try to understand others
Think win-win	Be a proactive team
Make and keep commitments	Be willing to speak-up
Keep the "mission" in mind	Call for "process" checks
Start on-time, end on-time	Be proactive
If possible, consensus decisions	Maintain confidentiality
Keep the spirit	Have fun

Figure 7.4. Meeting team guidelines/code of conduct.

■ Basic Problem-Solving Tools

No work on teams would be complete without mentioning the tools used by the team in solving problems or resolving issues. Below we identify the Basic Problem-Solving Tools and give a brief overview of each.

Brainstorming

Brainstorming is simply a group technique used to generate a large number of ideas on a given subject, problem or issue. It is probably the most used of all the tools.[7, 2]

Check Sheets

These documents are simple forms for collecting and organizing data. In some cases they can be used in analyzing the data, but more often are used in the preliminary steps of data gathering. There are three basic types of check sheets: *recording, checklist,* and *location.*

Graphs

Graphs are visual presentations of data collected by some means, probably by check sheets. The relationship between different sets of data can be easily identified with the aid of a few well-drawn graphs. The seven basic types of graphs are:

1. Line Graphs (the most simple type of graph)

2. Column and Bar Graphs (typically used to compare two or more measurements)

3. Area Graphs (used to show "totals amounts" or 100 percent of something)

4. Milestone and Planning Graphs (these graphs help organize and coordinate projects and activities)

5. Pictorial Graphs and Pictograms (uses pictures or symbols to represent data)

6. Histograms (special type of column graph showing the variable measurement of a given object or process)

7. Pareto Diagrams (another special type of column graph and line graph, used to prioritize data)

Nominal Group Technique

Nominal Group Technique is a structured method combining individual idea generation and idea selection through voting. It can be used in problem identification and selection or in problem resolution. Its main strength is that it generates a large number of ideas in a short time and prioritizes them. It is basically a five-step process that works best with groups of seven to ten individuals.[3]

Force Field Analysis

Force Field Analysis is a very simple and effective problem-solving tool. It can help you to better understand a problem, develop a problem statement, determine root causes of a problem, generate and evaluate solution ideas, and prioritize and plan the implementation tasks required for change.[3]

There are two basic influences, or forces that influence the problems, causes or solutions in question. These forces are identified below.

1. *Driving or facilitating forces*. Forces which promote the occurrence of the particular activity of concern.

2. *Restraining or inhibiting forces*. Forces which inhibit or oppose the occurrence of the same activity.

An activity level is the result of the simultaneous operation of both Facilitating and Inhibiting Forces.

The two force fields push in opposite directions and, while the stronger of the two will tend to characterize the problem situation, a point of balance is usually achieved which gives the appearance of habitual behavior or of a steady state condition.

In order to appreciate just what kinds of forces are operating in a given situation and which ones are susceptible to influence, a Force Field Analysis must be made. As a first step to a fuller understanding of the situation, the forces—both facilitating and inhibiting—should be identified as fully as possible. Identified forces should be listed and, as much as possible, their relative contributions or strengths should be noted.

Cause-and-Effect Diagrams

Cause-and-Effect Diagrams are graphical pictures showing the relationship between the effect (the problem) and its potential causes. These diagrams help analyze problems by organizing their causes so that they can be systematically investigated. These diagrams are also called "Ishikawa Diagrams" or "Fishbone Diagrams." The step most often left out of the problem-solving process is the Cause Analysis step. When presented with a problem, many of us find it very difficult to sort out all the possible causes of the problem.

This approach almost always results in correcting or resolving only a part of the problem. Most often the implemented solution is, at best, a "temporary fix." Remember, in problem solving you really aren't trying to *solve* the problem. You are trying to correct the things that *caused* the problem. Eliminate the key causes and the problem will go away.

We have now looked at what most consider the basic problem-solving tools. There are other simple tools that can be used, as well a many complex tools. In the next section let's look at some of the Macro or Big Picture Improvement Tools.

■ Macro Improvement Tools

The macro improvement tools identified here are designed to assist the organization in improving, not only quality, but also productivity. Given the amount of room available, only an overview of each macro tool will be presented.

Business Process Improvement (BPI)

Business Process Improvement is a long overdue improvement tool. Very little attention has been paid to the processes that maintain the business. It is estimated that only 5 to 15 percent of the organization's dollar is spent in the manufacturing processes. This makes sense because the business processes are where most of the organization's decisions are made and are the sources of the majority of the costs. (This subject is covered in detail in Chap. 11.)

Area Activity Analysis (AAA)

Area Activity Analysis is one of the few, if not the only, improvement tool that is specific to an individual natural work group and focuses on the internal and external customer and supplier. It's based on a model that has a supplier providing input into an area that adds value to produce an output that a customer needs. This tool works best with organizations that are well along with their overall improvement effort. It is used to understand how a natural work group fits into the overall picture and to establish its measurements. AAA is divided into eight individual activities. They are identified below:[9]

Activity 1: Developing the team's mission statement

Activity 2: Defining the team's major activities

Activity 3: Developing the customer-related measurements

Activity 4: Estimating real-value-added per activity

Activity 5: Developing the poor-quality cost per activity

Activity 6: Developing a real-value-added/poor-quality cost matrix

Activity 7: Developing efficiency measurements

Activity 8: Developing suppliers, input requirements, and measurements

AAA is known as "The People's Improvement Tool." Why? Because this is the one tool that helps individuals within a particular work area to identify what they do for the organization and assists in determining the area's improvement opportunities.

Poor-Quality Cost (PQC)

Poor-Quality Cost is a tool designed to help reduce the cost associated with poor quality. We would all probably agree it costs money to make mistakes or produce poor quality products or services. Poor-quality cost is defined as "… *all the cost incurred to help the employee do the job right every time and the cost of determining if the output is acceptable, plus any cost incurred by the organization and the customer because the output did not meet specifications and/or customer expectations.*"[10] (This subject will be covered in more detail in Chap. 14.)

Flowcharting

Flowcharts are graphic representations of processes which show the activities of both business and product/service processes and the relationship between them. They have value in almost any step of the problem-solving process. They may be used to identify problems, define measurements, generate ideas, provide a view of the desired future state, and select the proper solution.

As you can see, these macro tools can be used in any type of organization. They are, however, very dependent on two things. The first is understanding the need for using the tool and the second is training the individuals using the tool in its methodology and use. Additionally, these macro tools use many of the basic tools to support them. An understanding of all these tools will enhance any team effort.

■ The Seven New Management Tools

We've seen the basics and have been trained in their use. What's next? Are there other tools and techniques out there that may assist a team with its problem-solving process? Yes, and they're called the Seven New Management Tools. The following is a very brief description of each of the seven new management tools. For more detailed information about them, we suggest reading the book, *Management For Quality Improvement* (editor, Shigeru Mizuno, published by Productivity Press) or Bob King's book, *Hoshin Planning—The Development Approach* (published by GOAL/QPC, 1989).

Affinity Diagram

Also known as the KJH method, the *affinity diagram* is a team process tool that organizes ideas, generated through brainstorming, into natural groupings in a way that stimulates new creative ideas. Categories and new ideas are obtained by the team members working silently.

Interrelationship Diagram

This tool displays the cause-and-effect relationship between factors relating to a central issue. Factors that have a high number of relationships (arrows going into and emanating from) are usually the most fundamental or critical.

Tree Diagram

A *tree diagram* shows the complete range and sequence of subtasks required to achieve an objective. A derivative of this tool is fault-tree analysis, which depicts all of the ways that a product or service can go wrong so that preventive measures can be planned.

Matrix Diagram

The *matrix diagram* is an excellent way to show the relationships among various data. For example, Quality Function Deployment (QFD) is a process to understand the voice of the customer and to translate it into technical design parameters, subsystems, parts, components, processes, and process controls. It depicts a tree diagram showing the relationship between primary, secondary, and tertiary customer needs and the technical design parameters or substitute quality characteristics which, if met, would ensure that the customer's needs will be satisfied.

Prioritization Matrix

This tool uses a tree diagram of alternatives and a list of weighted criteria. Prioritization matrices are used to reduce the number of alternatives to those that are most significant in a structured, quantitative way.

Process Decision Program Chart (PDPC)

This chart is used to plan the implementation of new or revised tasks that are complex. The PDPC maps out all conceivable events that can go wrong and contingencies for these events.

Arrow Diagram

The *arrow diagram* is used to develop the best schedule and appropriate controls to accomplish an objective. It is very similar to the Program Evaluation Review Technique (PERT) and the Critical Path Method (CPM).

■ Reaching and Managing Decisions

One of the most difficult obstacles a team will face, early on, is how to reach decisions and, once reached, how to manage them. There are basically three different ways to make decisions, but before we discuss them let's go over a few basics that need to be in place first.

If a team is expected to make good decisions, we must assume the team understands its task and is empowered to some describable level. We must not only have a team that clearly knows why it exists but also one that has been trained in the basics; i.e., Team Dynamics, Group Effectiveness, and Problem Solving.

Now we have a team that is, hopefully, functioning like a well-oiled machine and can focus on reaching decisions. As we mentioned earlier, there are basically only three different ways decisions are made. They are: *autocratic*, *voting*, or *consensus*.

In the perfect "team" environment all decisions would be made based on group consensus. There are two types of consensus. The first is "hard" consensus. This is where a group of individuals or a team are all in absolute agreement about an issue. It's easy to move forward when "hard" consensus is reached. Not so easy is the second type of consensus. This is called "soft" consensus. It's when most of the group or team is in total agreement with one or more members liking a different decision better. These dissenting members, however, are willing to support the group decision as if it were their own.

As we all know, we don't live in a perfect "team" environment and therefore must be prepared to compromise when appropriate. This compromise may be in the form of putting the task or issue to vote with everyone agreeing that, no matter what, majority rules. It could also mean that we, as a group, may have reached the wrong decision and should study the issue more thoroughly.

Once we have reached a decision we must find a way to manage that decision in the most efficient and effective manner. The first step is to properly manage the agreement we just made in reaching the decision. Teams taking the time to manage agreement will spend much less time making mistakes.

An associate professor of management science at the George Washington University, Dr. Jerry B. Harvey, wrote a paper on this subject entitled "The Abilene Paradox: The Management of Agreement."[12] This paradox is described by Dr. Harvey as:

> ...Organizations frequently take actions in contradiction to what they really want to do and therefore defeat the very purposes they are trying to achieve. It also deals with a major corollary of the paradox, which is that the inability to manage agreement is a major source of organization dysfunction.

The essence of this paper is that often we make a decision in which everyone fully agrees and move forward in implementing an action only to discover that in truth nobody was really that thrilled with the decision in the first place.

There are many ways to avoid the "Abilene Paradox." I believe the most effective way is effective communication between the team members. Whenever possible, take the time to reach consensus. Then discuss the decision and try to look at it from all angles including how the key stakeholder(s) may feel about it. If anyone protests the decision or really feels he or she cannot support it as if it were their own, take a second look. Go back and look at all the supporting data. You may find that by taking more time you save yourself a "trip to Abilene."

■ How to Implement a Team Process

There are many ways to implement a team process within an organization. Some organizations train everyone and send them on their way. Other organizations do it without training anyone. More organizations do it by selecting a few people that will be trained and assigned to teams. The combination of task teams, process improvement teams, department improvement teams, and executive improvement teams provide many other options. There is no one right way to implement the team process. The team implementation plan is always unique to the personality of the organization. The best practices that we want to impress on you are always to provide formal training before you assign anyone to a team, and never train anyone until they will have the opportunity to put the training to work.

The following is the approach that we recommend using if it is applicable to the organization:

1. Form an Executive Improvement Team (EIT) and train them on basic team, meeting, and problem-solving skills.

2. Have the EIT define some quick-win problems that will have significant impact on the organization. They should then assign Task Teams (TT) to solve these problems. To get the TTs started, they should be provided with:
   — Basic team, meeting, and problem-solving skills training for the members of the Task Team.
   — Task Team leaders should also have team leadership skills training.
   — Facilitators should be trained and assigned to work with the task teams.

3. Conduct management level Area Activity Analysis (AAA). In this case, each manager that has managers reporting to him or her should prepare an AAA with the managers that report to him or her.

4. Develop Management Department Improvement Teams (DIT). Now that the deviation from requirements are defined, management DITs can be formed to work on these deviations. All people assigned to a team should have team and problem-solving training. A facilitator should be assigned to each DIT.

5. Form Process Improvement Teams (PIT). As a result of the measurements that were defined in step 3, and the business needs, key processes should be selected by the Executive Improvement Team and PITs assigned to streamline these processes. The PIT's members will be trained on basic team, meeting, and problem-solving skills. In addition, the PITs will be trained to use the Business Process Improvement Ten Fundamental Tools and selected Advanced Tools. A facilitator should be assigned to each PIT.

6. Train the Trainers. After the management team has experience with the team methodologies, interested managers and/or key staff should be selected to be trained as team methods instructors.

7. Establish Employee Department Improvement Teams (DITs)/Natural Work Teams. When management has confidence in the team process and their ability to exist in the team environment, the process will be expanded to all first-level areas (departments). It will start by each area doing an AAA. All employees will be trained in basic team skills and AAA methodology before the AAA activities are started. After the efficiency and effectiveness measurements and requirements have been established for each major activity, the employees on the DITs will be trained in problem-solving methodology and start the problem-solving phase of the team activities. A facilitator should be assigned to each DIT.

8. Form Quality Circles (if applicable). As the team and problem-solving cycles become integrated throughout the organization, groups of employees will be encouraged to identify problems that they would like to work on that are not being addressed by the DITs.

9. Develop Self-Managed Work Teams. With the proper management support, the employees that are members of the quality circles become more and more effective at correcting their own problems. As this occurs, they can take over more and more of the manager's responsibility. With proper financial and general operations training, the normal day-to-day activities for the work group can then be turned over to the employees, allowing them to evolve to self-managed work teams.

■ How to Measure Team Success

The measurement of team success depends largely on the strategic business objectives of the organization. What does the organization hope to accomplish through its team process? What are management's goals? Are they interested strictly in raising quality awareness or do they want teams that can assist in problem solving, reducing customer returns, reject rates and defects? Or perhaps eliminating communication barriers between employees and management. What about reducing employee absenteeism and raising productivity? As you can see, there are as many ways to measure teams as there are teams. It is pointless to try to devise a single way to measure team success. Some teams, such as a reject rate reduction team, may be easily measured and have tangible results where others, like a management support improvement team, may not.

No matter the type of team or the problem or issue they are working on, measurement systems must be developed and applied during the start-up of the team. Some simple measurements may be applied to almost any team, regardless of type. They are:

▪ Meeting team milestones

▪ Proper use of problem-solving tools

- Team member attendance
- Meetings start on time and end on time
- Effective use of time and other resources

Other, more complex measurements may be:

- Process cycle-time reduction
- Reject rate reduction
- Customer satisfaction increase
- Cost savings

In measuring a team, the important thing to show is that the team is adding value to the overall organizational improvement effort. The team should be able to prove that they have an important role in improving the performance of the organization, its work environment, product and/or service quality and, very important, the people.

There are many reasons for measuring teams and looking at some of them will provide a foundation for what and how to measure. First of all, one reason is that measuring is the expected, management thing to do. After all, if you don't measure it, you can't manage it.

The common denominator in all the possible reasons has to be "dollars." To be value-added to an organization, teams must show a positive Return on Investment (ROI). The difficult part is determining how much the team affected the ROI. Teams do not exist in a vacuum. Ultimate team success will be directly attributable to the way the process is managed and supported.

Most organizations with an existing team process have already determined the cost effectiveness of the effort. This cost effectiveness is a key measurement of the overall team effort. The cost effectiveness measured is the "Return on Investment" ratio of cost reduction and savings resulting from Team suggestions compared with the cost of the process.

Organizations report a "cost effectiveness" range of a low of 2 to 1 to a high of more than 40 to 1. The average seems to be around 6 to 1.

■ How to Deal with Problem Teams

Every organization that has ever had teams and every organization that will have teams will sooner or later face this question—"What do we do about our problem team?" Organizations that tell me they have never had a problem team or a major problem with a team—well, they either aren't paying close enough attention to their teams or they're not telling the truth! Anytime you involve individuals in a process the potential for conflict exists.

Realizing there are many dynamics affecting group process, from individual personality types to basic communication failures, let's look at 13 of the most common problems with teams. We've listed the problems in what we feel is a "priority" order. A team may have any one of the problems listed below, but some are the effect of other causes.

1. The team doesn't have a good charter (mission, goals/objectives, and operating guidelines).
2. The team or team members don't understand the mission.
3. The team hasn't learned or isn't using the tools and techniques.
4. The team has failed to set goals and measure results.
5. There are too many goals with unrealistic expectations.
6. There is a lack of team leadership and accountability.
7. The team runs out of new ideas or problems.
8. The team mission or task is causing difficulties.
9. The team isn't integrated with the organization's vision.
10. The team is isolated from other employees.
11. There is a lack of understanding or support from management.
12. There is a lack of team recognition and rewards.
13. The team becomes inactive or dormant.

Problem teams must be dealt with swiftly and efficiently. The sooner the problems are dealt with, the sooner the team can start earning its keep. Tackle the problems like any other organization issue. Identify the problem, the cause, and then develop a recovery plan for the team.

■ The Future of Teams

Do teams have a place in the future of an organization's improvement efforts? The Ernst & Young *"Best Practices Report"*[1] makes a point worth considering. It states *"Building the human resource infrastructure is essential ...."* According to the report, lower performing organizations show that less than five percent of the work force is participating on teams while higher performing organizations show over 25 percent.

The United States was introduced to participative management and the team concept almost three decades ago. It's changed the way we do business. Richard M. Davis, President of Martin Marietta's Manned Space Systems stated:[14]

Ten years ago, when our employee involvement effort started, we had varying ideas about the program and expected results. Few of us envisioned the atmosphere and attitude of participative problem solving and cooperation that we have at Manned Space Systems today.

Ford Motor Company almost cut the Mustang automobile from its line-up. A team known as the "Gang of Eight" researched innovative ideas and changes, presented the changes to top management and, after very tough questions from CEO Harold Poling, they got the go ahead. The team promised a 37-month turn-around. This was several months quicker than any previous turn-around on new car design. Working together the team slashed bureaucracy and delivered the new Mustang in just 35 months. Will Boddie, Ford's Mustang boss said, "We made decisions in minutes around the coffee pot that would normally take months."[15]

Richard DeVogelaere[16] and the folks at GM took a team approach to taking on water leaks in the Camaro and Firebird model automobiles. They called themselves the F-car SWAT team. Not only did they fix the water leaks but also that "screee" noise made when a window glass rubs wrong against the rubber. They also took care of some shakes and squeaks in T-top models by using under-body braces. DeVogelaere's comment:

> You say to yourself, "If we'd done this five years ago, how many more could we have sold? How many more thousands of owners would be out there saying what a great, exciting car this is?" We thought we were meeting the customer's expectations, but we weren't really listening, I guess.

The stories go on and on. Every organization involved in teams can give you success stories like the ones above. Do they also have failures? Of course. To succeed you have to try and to try is to sometimes fail. From our failures we learn to improve.

So, do teams have a future? We believe so. The payoff from teamwork is substantial and proven.

■ Summary

If your organization needs to move from a loser category to a survivor category, it needs to do very different things than it would to move from a survivor to a winner category. The winning organizations need to do very different things to stay a winner. For example, Department Improvement Teams (natural work teams) are very important and should be strongly encouraged if the organization wants to move from the loser category to the survivor category. For those organizations that are already in the survivor category, expanding the use of natural work teams does not provide a significant positive or negative result. For those organizations that are already winners, expanding the use of natural work teams often produces negative results, because it can decrease individual creativity.

The following shows how the three types of organizations approach teams:

Department Improvement Teams (DITs)/Natural Work Teams

Losers: Less than 10 percent of their people are involved.

Survivors: Between 40 and 100 percent of their people are actively involved in DITs.

Winners: All employees have been trained for and served on a DIT. The percentage of the employees active on DITs is decreasing as there are fewer problems that need to be addressed.

Process Improvement Teams (PITs)/Cross Functional Teams

Losers: Not often used.

Survivors: Used on occasion only.

Winners: Used very frequently, but with discretion.

Task Forces

Losers: Often used.

Survivors: Used now and then.

Winners: Seldom ever needed or used.

Team Training

Losers: Informal training on seven basic tools.

Survivors: Formal training for about 50 to 60 percent of the work force on seven basic tools. Often this training is not directly tied in to application.

Winners: Total work force trained on seven basic tools, plus other, more sophisticated problem-solving methods. Training is always tied in very close to application of the training.

Approach to Forming Teams

Losers: Management assigns team members and projects.

Survivors: No management approval required (hit or miss approach).

Winners: Management charters all teams.

Most Effective Problem-Solving Tools Used

Losers: Brainstorming is used to define corrective action.

Survivors: Cause and effect diagram along with brainstorming to define corrective actions.

Winners: Business process improvement is the most effective approach to improving performance, although many other tools are also understood and used.

General Electric offers workers in Bayamon, Puerto Rico, management through self-managed work teams and an incentive plan that rewards employee learning and performance. The plant manager states "I'm going to have the best work force in all GE."[11]

Reengineering processes have been going on at Kodak for several years. Their "Zebra" team comprised of 1500 employees who make black and white film work in, what the organization calls "the flow." Within the flow are "streams" or customers (Kodak business units). In the streams, most employees are part of self-directed work teams. A recent *Fortune* Magazine article[17] says these Zebras *"... have good reason to horse around. Since black and white film manufacturing set up a horizontal organization—called 'the flow'—productivity, profitability, and morale have galloped ahead."* Also, Hallmark Cards expects to halve new-product development time with cross-functional teams. At Xerox teamwork is an essential element. At any given time about 75 percent of all Xerox employees are engaged in team projects.

Teams and the team process work, and the "winners" can prove it.

■ References

1. Ernst & Young, *Best Practices Report.*

2. Dr. H. James Harrington, *Different Strokes for Different Folks—The International Quality Study.*

3. Dr. H. James Harrington, *The Improvement Process*, McGraw-Hill.

4. David T. Farrell, Kenneth C. Lomax, Ralph Ott, *Team Effectiveness Manual*, Ernst & Young.

5. Wayne S. Reiker, *Employee Involvement Teams—Team Member Manual*, Ernst & Young.

6. David T. Farrell, Kenneth C. Lomax, Norman Howery, Ralph Ott, *Team Effectiveness Manual*, Ernst & Young 1992.

7. Wayne S. Reiker, *Employee Involvement Teams—Team Study Guide*, Ernst & Young.

8. Dr. H. James Harrington, *Business Process Improvement*, McGraw-Hill.

9. Dr. H. James Harrington & Norman Howery, *Area Activity Analysis—The People's Improvement Tool*, Ernst & Young.

10. Dr. H. James Harrington, *Poor-Quality Cost*, ASQC—Quality Press.

11. Shigeru Mizuno, *Management For Quality*, Productivity Press.

12. Jerry B. Harvey, *The Abilene Paradox: The Management of Agreement.*

13. Wayne S. Reiker, *Employee Involvement Teams—Team Facilitator Manual*, Ernst & Young.

14. *10 Year Anniversary—SRT Yearbook '79-'89*, Martin Marietta Manned Space Systems.

15. *Will Boddie article*—Sacramento Bee, Oct. 8, 1993.

16. James R. Healey, *USA Today*, July 9, 1992, "Simple Solutions Mark Cars' Road To Recovery."

17. Rahul Jacob, "The Search for the Organization of Tomorrow," *Fortune*, May 18, 1992.

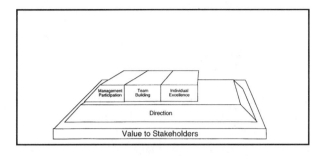

8

Individual Excellence: Going Beyond Teams

Dr. H. James Harrington
Principal, Ernst & Young

and

Norm Howery
Improvement Consultant

Teams make an organization good. Individuals make an organization great.   DR. H. JAMES HARRINGTON

■ Introduction

An organization can only excel when it taps the full potential of each individual within the organization, sparking their creative juices and providing them with a high degree of personal self-worth and pride. As Maslow so long ago pointed out, all people's first instinct is a self-survival instinct. Once that need is satisfied, the desire for camaraderie and friendship that a team involvement provides becomes their top priority. But the highest level of performance is self-actualization. This is the point at which the individual performs superbly; not because he or she is driven to perform by promises, threats, or praise, but because excelling in their chosen job provides personal satisfaction and fulfillment.

Don't fool yourself. The job you have today is your chosen job. Very few of us have not had options and career choices to make as we first enter the work force, and we still have options available to us today. You can choose to leave your present assignment and go to work in a fast-food restaurant, roam the streets of San Francisco looking for a handout, or maybe your option is to start your own business. Everyone has options. It is you that decides to continue doing what you are doing or not. It is your choice and you must accept your responsibility to excel at the job. Too many of us gave up many of the choices we could have had early in our lives by not doing well in school or putting fun ahead of work. Too many of us are unwilling to put forth the additional effort that is required to have the best options available to us. The only exception to this is the few people that are limited physically or mentally. The rest of us have no excuse. We are free people—we have freedom of choice. We need to weigh the consequences and live with our own decisions.

As Martin Luther King, Jr. put it, "If a man is called to be a street-sweeper, he should sweep streets even as Michaelangelo painted, Beethoven composed music, or Shakespeare wrote poetry. He should sweep streets so well that the host of heaven and earth will pause to say, 'Here lives a great street-sweeper who did his job well.'"

For employees to perform effectively, management must provide them with the three "T"s:

- Training

- Tools

- Time

The three "T"s get the employee to the starting gate. They are required for an employee to perform well. To excel, an individual has to build upon these basics using individual creativity, pride, and sacrifice as they reach for the star of self-fulfillment. The trick is to build personal challenge into your present job that will throw off the chains of boredom and mediocrity. What could be more b-o-r-i-n-g than hitting a walnut with a stick, then running after it and hitting it again for eight hours a day. Well, put eighteen holes in the ground and that b-o-r-i-n-g task becomes golf, a sport that millions of people wait anxiously to pay money to play.

We are not suggesting that any individual invest their entire life in their work. Everyone needs to spend time in each quadrant of the Arena of Life (Chap. 6).

In today's environment, the fastest-growing quadrant is the "self" quadrant, at the sacrifice of the other three. Individuals need to define the correct balance between the individual quadrants. Well-adjusted people cannot devote all of their efforts to any one quadrant if they are going to have a healthy, normal life. Employees that spend too much time at their workplace soon burn out and lose their individual creativity. Because of the overemphasis on the work quadrant in Japan, a firm called Japan Efficiency Headquarters rents out actors to visit aging parents and children of people who are too busy with their careers to do it personally. This Tokyo-based organization charges $385 for a five-hour visit by one person, $769 for a couple, and $1155 if the client also wants to rent a baby or child. For example, a 35-year old Tokyo computer salesman sent a couple to visit his 64-year old father who lived ten minutes away from him. Company President, Ms. Satuski Oiwa stated, "Our purpose is to fill a hole in the heart."

But things are changing, even in Japan. The people that are entering the Japanese work force today are looking at their jobs differently. Sachihiko Kataoka, a 24-year old college graduate, puts it this way, "When I was small, I got to see my father's face only once a week—on Sundays. It was lonely," he recalls. "I don't want to become like that. I'd like to spend time with my children if I have them. I plan to make a clear distinction between work and play. I have no intention of sacrificing everything for the company."

When it comes right down to it, everyone spends time in the "work" quadrant to provide funding for the other three quadrants. The average hours per day devoted to work from the time an individual leaves their home until they return are 13 hours in Japan, 11.3 hours in the United States, 10.5 hours in Germany, 10.4 hours in France, and 10.2 hours in Britain.

Selling half of your life to support your needs should be enough without working additional overtime. What we need to do as individuals is use our time effectively and creatively while we are at work so that we do not need to work overtime. I doubt that we have ever heard a person laying on his or her deathbed say, "I wish I had spent more time at work."

In discussing the "1994 Accreditation Manual for Hospitals," Paul M. Schyve, MD, Vice President for research and standards for the Joint Commission stated, "While the standards clearly emphasize systems and processes, rather than individuals in health care, you cannot ignore the role that an individual professional's knowledge and skills play in outcomes. In the interest of driving out fear from an organization, you can't choose to ignore issues of individual competence. The standards represent the need to balance those issues."

Yes, you have to have a good process and teamwork to get into the race, but when it comes right down to it, it is the individual's personal excellence that makes or breaks the organization.

Dr. Kaoru Ishikawa, the man who started the quality circle movement in Japan and contributed more to the Japanese quality movement than both Deming and Juran combined, openly recognized that individuals were more productive and effective at solving the Japanese quality problems than the team movement was. The need to excel is an idea that everyone can associate with on a fundamental, very personal level. Excellence applies to everyone's occupation. The need to excel, to be the very best we can be, is not something that can be imposed upon us. It comes from within us.

The key is not to sacrifice and take away another minute from the family, religion, or self quadrants, but to better use the time that you are presently devoting to the work quadrant. For example, Germans spend an average of 32 minutes a day socializing; Japanese, 1 minute a day. That is 31 minutes taken away from the other 3 quadrants.

Probably, the Japanese are not a good role model to follow. Most Japanese workers are unhappy with their jobs but due to the system, they are as much a slave to the organization as the American black slaves were on southern plantations before the Civil War. In Japan, people work long hours at great sacrifice to their family, with little or no personal sense of accomplishment. Jobs are boring and growth in the organization is slow for all but a few select candidates. As Masashi

Kojima, president of Nippon Telegraph and Telephone Corporation stated at a ceremony for new recruits, "You might feel bored during your first three years in this company because you will not be given jobs that require your brain."

If Japan is not the role model that we want to use, what is the answer? We all need to work to improve the way we perform our jobs so that it does not interfere with the activities that go on in the other three quadrants. At the same time, we have to increase our value added to the organizations so that we can improve our financial status and be able to accept more responsible assignments. To accomplish this, each person must:

- Be educated to perform their assigned task(s).
- Understand the organization's business plan and how it relates to their job.
- Understand how well they are performing.
- Not be afraid to take risks.
- Be willing to learn new assignments.
- Be uncomfortable with the status quo.
- Think creatively.
- Be willing to share credit.

Indira Gandhi said, "My grandfather once told me that there are two kinds of people: those who do the work and those who take the credit. He told me to try to be the first group; there's less competition there."

■ Training—Opening the Door to Individual Excellence

David Kearns, past CEO of Xerox and current Deputy Secretary of the U.S. Department of Education stated, "To be competitive as a nation, we must do two things; improve quality and improve education." Ex-IBM CEO John Akers stated that, "Market-driven quality begins with education and ends with education. It is everyone's job to teach it and coach it incessantly." One of the primary things that made IBM the leader in the computer field was its total dedication to education. As Thomas J. Watson, Sr. put it, "There is no saturation point to education."

Deplorable Public School Systems

The U.S. public schools are deplorable. Teachers perform more as baby sitters than teachers. In a recent United Nations test of 9- and 12-year old students from 16 developing nations, the United States came in number 13. The average U.S. high school graduate has an education level equivalent to a ninth-grader in France, Japan, Germany, and China. Individuals in the United States that do not go to college are not competitive with workers in the developed nations around the world. In 1980, 18 percent of U.S. citizens over 18 that did not go to college, earned less than the poverty level. By 1990, that had grown to 40 percent. In this 10-year period, the buying power of these workers dropped 19 percent. As fathers began to earn less, mothers were forced into the labor market. Today, to keep acceptable living standards, both parties have to work. This has led to an unprecedented high rate of divorce and family break-ups. The single-parent family is now a common and accepted way of life. This results in a higher and higher percentage of children that enter grade school who are not ready for the experience.

Our public school systems have led to a growing third-world type of poor people within the United States. The rich get richer; the middle class get rich; and the poor get poorer, all due to the education they have received. Students go to school a smaller percentage of the time than students in Europe and Asia do. To compete with the rest of the world, we would have to increase our students' time in school by at least 25 percent. We entertain our children, not educate them. Industry is provided with high school graduates that cannot read or divide (they have spent more time learning how to multiply sexually than they do mathematically). The school system has taught them that 3 right out of 4 is good enough, but as they enter the business world, they are told that they must be right 99.9 percent of the time, and they feel it is not fair. There is no doubt about it—the biggest problem that faces the United States today is its deplorable, sub-standard education system in its grade and intermediate schools.

Business Making Up for the School Systems

Management cannot wait for the public school system to correct its problems, because two-thirds of the labor work force in the year 2005 are already on the job. Until the school system can be corrected, U.S. organizations need to fill the void that today's education system has created. Motorola University is a good example of how one organization is helping to make up for the errors that are being made in our education system. Many organizations provide basic reading and arithmetic classes for their employees. This additional expense must be borne by the organizations that need to use the labor source until the school system can be corrected. It is putting a huge, unjustified financial burden on American organizations which in turn provides their foreign competition with a major competitive advantage. Is it any wonder that manufacturing is moving out of the United States to the better-educated and lower-paid human resource markets?

As Labor Secretary Robert Reich put it, "American companies have got to be urged to treat their workers as assets to be developed, rather than as costs to be cut." Education is not expensive—it's ignorance that is expensive. At one firm we visited in mainland China that produced some of the most expensive Reebok shoes, we found that a new employee gladly paid the equivalent of one month's salary to obtain a job with the organization. This advance payment was used to cover the cost of their initial training. If the employee stayed with the organization for three years or more, they had the advance payment returned to them. This practice was implemented because management realized that the cost involved in bringing on and preparing a new employee to become a productive member of the organization's team is a major expense.

Once the employee is on the job, training cannot stop or obsolescence sets in. Training and education are the key to keeping our people competitive. All employees, young or old, new or with 40 years of experience, need to be engaged in a very active technical vitality process. William Wiggenhorn, president of Motorola University stated, "When you buy a piece of equipment, you set aside a percentage for maintenance. Shouldn't you do the same for people?" Motorola estimates that for every $1 they spend on training, they receive $30 back within 3 years in productivity gains.[1]

Education and training can be divided into improvement-related training, job-related training, and career growth training.

Countries in the Americas have to do much more training than they have ever done before. An estimated 8 to 20 percent of the U.S. work force is functionally illiterate. Only one out of 14 U.S. workers have ever received formal training from an employer. The Massachusetts Institute of Technology Commission on Industrial Productivity reports that America's ineffective approach to on-the-job training does not meet the requirements of today's competitive environment. At the present time, U.S. firms spend about 2 percent of their payroll on training (about $30 billion per year). Japanese organizations spend about 6 percent. To make matters worse, only 0.5 percent of U.S. organizations account for 90 percent of the $30 billion spent last year on training. IBM, for example, has an education and training staff of over 2000 people.

President Clinton would like to require all firms with more than 100 employees to spend the equivalent of 1.5 percent of their payroll on training, or pay this sum into a training fund. This would cost U.S. businesses an estimated $17 billion and even then, 40 percent of the U.S. workers would not be covered. He would also like to increase training for people on welfare by $4 billion and drastically increase the number of Federal Manufacturing Assistance Centers. In addition, he would like to increase by $2 billion the grants to states for youth apprenticeship training programs.

Training also pays off in employee loyalty and reduced turnover rates. For example, when Volvo and General Motors built a truck plant close to a Will-burt location (note: Will-burt has an extensive training program) and offered workers three dollars an hour more, only 2 out of the 350 employees left Will-burt.

What's in it for the employees? Studies show that workers that are properly trained earn 20 to 30 percent more. But even more important, it can mean the difference between having a job or having it moved overseas. Why should any organization pay 100 to 200 percent or as much as 500 percent more for hourly workers that are not as well educated?

As the Chairman and CEO of Corning, James Houghton, put it, "The government contract is to send competent people to my door. That's what a quality supplier does. I'll take care of lifelong-learning after that." It is unfortunate but true that the government is not living up to their part of the contract. On the other side of the coin, as individuals we need to remember that the Lord provided us with two ends–one to sit on and one to think with. Our success depends on which one we use the most (from an Ann Landers' column).

Improvement-Related Training

Improvement-related training is directed at upgrading the employee's basic skills. It starts with the very basics to help offset the illiteracy problems facing many employees and continues through to help increase the individual's creativity.

Improvement-related training is directed at helping the individual perform better in any job that they are or could be assigned to within the organization. Improvement training should start the day a new employee is hired. The winning organizations have 3- to 5-day indoctrination programs that start the first day the new employee reports to work. Most of the improvement training activities should take place during the first 2 years with the organization, then continue at a slower pace as new concepts are introduced.

We cannot overemphasize the importance of creativity training and development. Surviving organizations train their people to understand and conform to the procedures. Winning organizations train their people to be creative, causing the procedures to be continuously challenged and upgraded.

Job-Related Training

Job-related training focuses on preparing the employee to meet the requirements of a new job and to understand process and product changes related to the present job. Often these process/product changes are driven by changes in the consumer's expectations. Refresher job-related training is usually required to ensure common approaches are used by all employees doing the same job and to help the employees excel in their jobs.

Job-related training is directed at all the special skills and knowledge required for an employee to perform his or her present job. Many organizations today have a section in the job description that defines the knowledge required to be assigned to a job and the additional training that should be provided before the employee is certified to do the job on their own. The advanced organizations have employee assignment certification programs that include observed performance analysis and written evaluations that document the effectiveness of the training process.

In Japan, new professional employees, or what could better be called recruits, start off in an almost military basic training program that introduces them to the organization. They dress alike, wear corporate pins in their lapels, eat together, and take long hikes together. Employees are trained in the Sanyo Corporate Etiquette Manual (as well as other things). This manual tells the recruit how to do everything–from how loud to say good morning, to how to reply to a superior's call, to how to bow. This is just the starting point for a socializing process intended to transform young Japanese students into dedicated corporate warriors or "kaisha senshi." It is difficult for most Americans to realize and understand the strict obedience that is taught during the first year of employment with a Japanese organization.

■ Career Growth Training

One of the best ways to develop long-term employee commitment and loyalty to the organization is to transform a job into a career for an employee. The best organizations realize that they have invested a great deal of money and time into each employee and do not want to lose this valuable resource. The best way to keep employees is to help them understand they have a future with the organization and get them to commit their personal time to preparing themselves for the future challenges that will face them in the organization.

To accomplish this is a two-way street. The organization needs to commit resources to prepare the employees so that they can effectively compete with outside resources when a career opportunity occurs. All organizations should compensate employees for some, if not all of the additional costs incurred when they take formal classes on their own time that help them to prepare for potential job opportunities within the organization. On the other hand, people who want careers within the organization need to be willing to invest their own time preparing themselves to be competitive for a desired career opportunity.

How Training Will Be Provided

With the increased need for more and more training, our approach to training needs to change. We need to find more effective ways of providing formal training than a group of people listening to a lecture because, by their nature, lectures are:

1. A poor means of communication.
2. Very costly.
3. Often not provided when needed.
4. Not adaptable to training a single individual.
5. Not consistent from one class to the next.

Because of these limitations, educational systems within organizations must drastically change in the years ahead, placing more emphasis on other media like closed-circuit television broadcasts, computer interactive programs, audio/video training tapes, audio tapes, and videodisk.

Federal Express uses videodisk systems to provide quality training to its entire U.S. work force of 75,000 people, at a cost 80 percent less than the same training would cost using classroom training methods. At Hewlett-Packard, 80 percent of the training provided to their work station and computer system sales force is accomplished through satellite television and video methods. Ameritech (a Baby Bell) uses an interactive video program to train their sales personnel on how to sell ad space in the Yellow Pages. More and more manufacturing processes are being videotaped and put into training programs so that each time a different employee is assigned to a job, he or she can study the way the job should be done without the assistance of an instructor. This reduces cost and assures that training is presented the same way each time.

■ Developing Individual Performance Plans

The keys to an effective individual performance process are:

- The right things are measured.
- Both the employee and the manager agree on the performance standards.
- There is an ongoing measurement and feedback system that provides information to the employees and management.

We strongly believe that any business plan should involve everyone from the board room to the boiler room. With this as a basic starting point, it is easy to see that the individual's performance plan should be based upon how the individual performance is going to support the organization's business plan. Concepts of this nature have become very popular in Japan under what they call Hoshin Kanri (Hoshin Planning or Policy Deployment). In our case, we are taking this approach one step further, bringing the individuals' goals and measurements in line with the organization's goals. After all, the best way for an individual to grow within an organization is to increase his or her perceived value added to the organization. The performance plan should also be focused on the organization's commitment to internal and external customer satisfaction and developing a strong, effective internal team.

The individual's performance plan should be prepared by the employee and the appropriate manager working together to understand what assignments the employee has and will be assigned to in the near future (next 12 months' maximum). Using the business plan and input from the individual's customers, a list of performance measurements should be prepared for each of the individual's major projects and/or activities. As major new projects are assigned, a new performance plan should be prepared for the new project. In many cases, this will require that several performance plans are prepared for one individual each year. The job description and input from the individual's customers should be used to define the "meets requirements" performance level for each measurement that will be established.

The process of aligning organizational objectives as they apply to the individual can be a long, challenging and difficult one, but one that management will find worthwhile for several reasons.

We do not like a preprinted performance plan. In fact, we believe that they are detrimental in the long run. A performance plan should be customized to the individual and the job. For this reason, we recommend that the performance plan form consist of only three columns, the Task Name, the Task Description, and the

Task Priority. The rest of the form should be left blank to be filled in by the manager and the employee.

Performance planning for managers usually consists of three major sections. They are management of business issues, technical management, and personnel management.

■ Performance Evaluations (Appraisals)

The ideal time to evaluate an individual's performance is as soon as a job is completed. To take advantage of this timing, the annual performance evaluation should be based upon a series of evaluations that occurred throughout the year. An evaluation should be completed each time a project is completed. Evaluations should not only occur at the end of a project, but also at key points during the project, thereby allowing the individual to correct errors and eliminate undesirable performance as early as possible. At a very minimum, performance evaluations should be scheduled every three months.

Who Should Do Performance Reviews?

The employee is in the best position to evaluate his or her overall performance. As a result, the employee should document his or her performance compared to the targets that they and management have agreed to. If the employee feels that he or she is exceeding requirements, the employee should explain what was accomplished over and above the required performance. The employee should also record any roadblocks that prevented him or her from performing as well as he or she could have and make suggestions of what action should be taken to improve future performance.

To supplement the individual's input, customers of the individual are often asked to evaluate the individual's performance from their viewpoint. Using these two types of inputs, the manager should complete a Performance Evaluation Form.

After this form is filled out, the employee and the manager will meet to review it and the other input. During this meeting, particular attention will be paid to any activity where the employee or the customer rates the employee's performance higher than the manager rated the employee. Any differences in perception and/or interpretation of data will be resolved during this meeting. The manager and the employee will also discuss the roadblocks that the employee faced and the suggestions the employee made to improve future performance. The results of these discussions will be recorded on the performance evaluation short form. As a result of this discussion, action plans will be developed to help improve the employee's future performance. Also during this meeting, a list of short-term performance objectives will be developed and the next performance evaluation date will be scheduled.

Turnabout is fair play. The appraisal process should provide the opportunity for the employee to make suggestions on how management can contribute to the employee's overall performance. Because the individual's performance is greatly impacted by the type of direction and support they receive from management, at each evaluation the employee should suggest at least one way that management can change or improve to help the employee perform better.

Annual Performance Reviews

Once a year, the manager will summarize all of the individual reviews to be sure that all of the objectives defined in the performance plan are met. The result of this summary should be reviewed with the employee. This review should run very smoothly because it is simply a summary of many individual reviews.

Organizations that require all annual appraisals to occur at the same time do it so that performance can be considered during the salary planning cycle. Organizations that do this create many problems for themselves. First, because of the heavy additional workload, the manager lets other tasks slip or does them poorly. Second, because of time limitations, the appraisals are poorly prepared and given.

Through the use of many ongoing evaluations, the time required to perform an annual review can be greatly decreased, and they do not need to all be done at the same time. A manager that has 12 employees reporting to him or her can do one a month. Some organizations have refined the process down to the point that a total review is conducted only when the employee is recommended for promotion or is being reassigned to a new manager.

The key to this performance evaluation approach is that the manager is never comparing the individual to other employees. The baseline used by the manager is the required performance level as defined by the job description and/or the employee's customer. A manager could have an entire department consisting of "far exceeds requirements" performers. There is no longer a need for a performance rating distribution that takes the shape of a normal curve, because the concept of an average performer is completely ignored.

■ The New Employee

Ann Landers wrote in one of her columns, "Anyone who believes that competitive spirit in America is dead has never been in a supermarket when a cashier opens another checkout line." Yes, people are competitive by nature, but too often we put away that competitive spirit when we enter the organization's front door. We become part of the pack. We are afraid to stand out as individuals. We don't want to be enthusiastic about our job because the other employees will think we are strange. But enthusiasm makes the ordinary person extraordinary. As individuals, we all have the same needs that must be fulfilled if we are going to excel at our job. They are:

- *Economic security.* We need to feel that we are getting a fair day's pay for a fair day's work.

- *Personal self-esteem.* We all want to be viewed as value-added to the organization. None of us wants to be average.

- *Personal self-worth.* We need to feel that we are contributing to a worthwhile goal.
- *Personal contribution.* We want to be listened to, to have our ideas heard. We can accept the fact that everything we suggest may not be implemented, but we need a fair hearing.
- *Personal recognition.* We all need feedback to show that good work is appreciated, that what we are doing is worthwhile.
- *Emotional security.* We all need to be able to trust the managers we work for and to feel they will be honest with us.

Only when these six basic needs are satisfied can an individual have a chance at excelling at his or her assigned task(s).

That's Not Fair

The new employee does not expect the world to be fair because it isn't. Those that dwell on the unfairness in life use it as an excuse for their lack of drive and success. No matter where you are in the world there will always be people who are above you who are not as deserving as you are (in your eyes), and people at your level or below you who do not do their fair share. Most of us believe that we have more than our fair share of problems. In truth, there are many people who have overcome more obstacles than we will ever face and have become more successful than we will ever be. They have used these obstacles to build stamina and the drive to succeed, to forge a will and personality that are unstoppable. On the other hand, there are many individuals who have had much lighter burdens to carry than we have faced who have failed miserably. No, the world is not fair, but the new employee accepts and understands this fact, making the best use of his or her talents and opportunities to provide themselves with a positive attitude and a personal dedication and commitment to success.

The Open-Minded Employee

In today's environment, growth is going to be very limited. Management and employees need to look for other ways to stimulate job satisfaction and recognition. Employees need to have a very open mind about what is going on around them and how they can contribute. Employees who do not find their job interesting are employees who have closed their minds to its possibilities. Employees and managers alike make excuses for their closed minds. Some of the more frequently used excuses are: we tried that before, let's hold it at bay, let's give it more thought, management would never do it, and you can't teach an old dog new tricks.

It is time to open your minds and stop using these phrases. Every time you utter or hear one of these popular phrases, it's time to challenge what's going on. Stop putting up roadblocks and detour signs to change and start knocking them down. Ask how you can make it work now if it didn't work before. Ask if it isn't time you tried something new if it is a first-time suggestion. Embrace the positive and cut the legs out from under the negative. You may not always win, but you will never win if you never try.

■ Career Building

Today you are building your career within the organization. The best way you can ensure the success of your long-range career is by doing a superb job today. Keys to a successful career within any organization are:

- Do an excellent job in every assignment you get.
- Make sure you and your manager understand where you want to go.
- Be willing to make the desired sacrifices.
- Ask for the opportunity to compete for the desired assignments.

Career Planning

Every individual needs to stop and reflect periodically on how things are going and where they want to go in the future. When it comes to how things are going, the organization's internal measurement system should provide the required information about the job that the individual is performing today. But that is a very short-term look at the employee's career. It provides no input as to where the individual is going. From the individual's standpoint, performance appraisals leave two of the most important questions unanswered. They are:

1. Am I progressing at the right speed?
2. Am I heading in the right direction?

It is for these reasons that everyone needs to develop a career plan that plots their course to retirement and, often, beyond. Too often, people get so bogged down doing the day-to-day activities that they never stop to determine if what they are doing today will help them to meet their career objectives. A career plan lays out the route that an individual needs to take to reach their personal career objectives. For some people, this career plan can be very simple.

Career planning is a very significant part of the Total Resource Management Methodology. The objectives of career planning are:

1. Help fulfill the individual's desire to develop their potential and grow in the organization.
2. Ensure a continuous supply of qualified people as a resource for the future and key leadership assignments.
3. Make the best use of the employee's ability now and in the future.
4. Enhance the employee's feeling of personal value.
5. Provide resources that allow for promotion from within.
6. Show that the organization has respect for the individual.

There is a distinct difference between career planning and performance planning. Performance planning addresses the immediate job and its responsibilities. Career planning deals with the individual's skills and preferences for the future as well as for today. Although there may be some overlap in the two activities, the primary intent is very different. Career planning is a shared responsibility. The basic responsibility rests with the individual. The manager's role is one of giving the employee encouragement, information, and support, as well as being a reality tester. The organization's role is to develop an environment for personal growth, provide educational support and promote from within whenever possible.

Career planning strengthens the employee/manager relationship by placing the manager in a guidance role. It is a useful tool to the employee and the employer in improving the utilization of the employees and developing the employees' full potential. Without an effective career planning process, there is a high probability that the organization will have an under-utilized, disenchanted work force that is very prone to making errors and job hopping.

■ Building a Bond with Your Manager

No one has more influence over your next career step than your present manager. The relationships you establish with your manager can make or break your career. To have a career-building relationship, you do not have to be a "yes person"; in fact these type of people are soon discarded as "no-value-added" type individuals by all but the extremely insecure managers. In Mark H. McCormack's book, *The 110 Percent Solution*, published by Villard Books, he gave the following advice to people who want to establish a career-building relationship with their manager.

- *Be loyal.* Disloyalty is a major character flaw that will not be accepted by any manager.

- *Keep the boss informed.* The boss should always know everything that is going on within your span of responsibility.

- *Embrace change, even if you do not understand it.* Managers are measured more and more on how effectively they implement change. Help them with this responsibility. Do not resist change.

- *Respect your manager's time.* Spend your manager's time like you would your own money.

- *Don't tread on his or her turf.* Honor the fact that your manager has divided up the available work into specific job assignments.

- *Follow up quickly.* When your manager gives you an assignment, get it done and out of the way.

These simple rules provide the key building blocks for developing a good relationship with your manager and will apply equally well whether you are an assembly worker or the vice president of a major corporation.

■ Reinforcing Desired Individual Behavior

We all work for the same things—to gain security and self-esteem. Few of us report to work at 8:00 AM, five days a week, because we can't find anything better to do. I worked for IBM for 40 years and really enjoyed my assignments. But the last day they paid me to come to work was the last day I got up at 5:30 AM, put on my dark blue suit and wing-tip shoes, and left home to go to IBM. Why? Because IBM and I agreed that it was no longer a desired behavior that I would be compensated for.

If we want individuals to excel, we need to recognize and reward them for the additional effort that they put forward. Japan recognizes the individual workers by crowning their best workers with the title of "Ginohshi." It is awarded by Japan's Ministry of Labor for passing a series of rigorous examinations related to industrial performance of the highest quality. Since 1971, IBM Japan's Yasu site has had over 200 employees honored with this coveted title. In addition, individuals are recognized by Japanese prefecture and regional governments for excelling at their assignments. For example, five IBM "Ginohshi" workers were chosen as outstanding skilled workers by the Governor of Shiga Prefecture in one year. Obviously, being singled out as outstanding skilled workers has an extremely positive effect on these individual's present and future performance. Likewise, not being recognized or taking away previously established recognition systems diminishes the importance of the act. For example, in IBM's zest to cut costs, G. H. Larnerd, site manager at IBM San Jose, issued a letter to all employees on March 17, 1992 that made the following changes to IBM's established recognition program:

- Quarter Century Club Annual Dinner and Family Day will be held biennially in alternate years. (They were both held annually in the past.)
- Watson Trophy Dinners for winners and guests will be replaced by Luncheons for winners only.
- Children's Christmas Parties will no longer give out gifts.

Although we believe that he did not mean to, what did he telegraph to the IBM employees?

1. Longevity with the organization and building an IBM family feeling was not as important as it used to be.
2. IBM's outside athletic program that helped build IBM's team spirit was less important than it used to be.
3. The individual's family was less important to IBM than it used to be.

Great care must be exercised not to establish individual recognition systems that may be cut back in the future. Any negative change in the recognition system is interpreted very personally by the employees.

Salary as a Reinforcer of Desired Behavior

Employees sell their lives for their salary. Today most people sell themselves into limited slavery just as the indentured servants used to in the seventeenth and eighteenth centuries. Many homeless people have rebelled against this form of slavery, feeling that their freedom cannot be bought. They are not going to let someone else control their lives for a few dollars. Most of us work to put food on the table and to live in the style that we have selected for ourselves. There are very few of us that have all the money we want and work for others only to get a feeling of self-worth. Management communicates the employee's worth hour by hour, by the amount they pay their employees. Everyone understands that the organization pays the people and the jobs it values the most, the most money. Every time you pay one person $10 a week more than another person, management is communicating that the first employee is more valuable.

What options does management have in determining where an individual should be paid within a salary bracket? The following are the most common options used today:

1. Time with the organization

2. Time in the assignment

3. Fixed pay per assignment

4. Age of the employee

5. Employee's knowledge of the organization

6. Pay for performance—quality of output, meeting schedule, cooperation, versatility, creativity, and productivity

7. Team performance

8. A combination of 5, 6, and 7

Combination.   More and more often, the individual's compensation is based upon a combined approach plus a bonus system that is tied into the organization's performance. Japan Incorporated makes very effective use of the bonus system to control expenses. Many organizations in Japan pay employees a very minimum wage, then supplement this wage with a bonus based upon the organization's performance. The annual bonus may exceed 100 percent of the employee's annual salary. Typically, these bonuses are paid twice a year, just before summer vacation and during the first part of December in time for the holiday-added expenses. The good thing about the bonus system from the organization's standpoint is that salary expenses can be cut 50 percent when the organization is not performing well and money is most needed.

In addition to the organization's "performance bonus" concept, we suggest a combination of the "pay for performance" and "pay for knowledge" concepts. This is accomplished by relating the employee's salary to their performance evaluation. Additional compensation is given to the individual based upon the number of work assignments he or she has mastered (see Fig. 8.1).

Assignments mastered	Added compensation per week
1-2	$ 0.00
3-4	10.00
5-6	20.00
7-8	30.00
Over 8	50.00

Figure 8.1. "Pay for knowledge" typical compensation plan.

The trend today is the combined approach to compensation. Even in Japan, which has traditionally based its compensation on employee's seniority with the organization or on age, things are changing. Today, things are changing very fast. For example, Honda Motors has announced that they will begin paying some of their managers on an annual basis rather than monthly. Individuals will negotiate each year with their supervisors either a raise or cut in pay based upon their performance.

■ Cross-Discipline Training

It is no longer enough for employees to perform one predefined activity and be oblivious to the activities going on around them. Employees today need to understand that they have customers that need to be pleased, and suppliers that they are dependent upon. To properly interface with their customers and suppliers, employees need to gain an understanding of what they do and what their problems are. Employees need to expand their view of the organization, getting away from the microscopic view of just their one assignment, to understand the macro-processes that the organization is built around. This broader view is necessary so that everyone is working in harmony to achieve the organization's goals.

To accomplish this, cross-discipline training is becoming an essential part of developing the employees within the organization. Cross-discipline training has a number of positive benefits to the individual and the organization. They are increased productivity, increased communication, reduced rivalry, and reduced bureaucracy.

For example, when a petroleum inspection firm sent their accounting staff out on sales calls with their salespeople, they quickly realized that the controls they had imposed on purchase orders were unrealistic and wasteful. As a result, the purchase order dollar amount that required accounting approval was raised significantly.

How does cross-discipline training help the individual?

- Revitalizes the individual
- Provides personal growth
- Expands personal contacts
- Increases visibility

- Opens new career opportunities
- Clarifies organization's key processes
- Provides increased personal flexibility

A major factor in succeeding as an individual or as an organization in the twenty-first century is the flexibility of the individual within the organization's processes. Cross-discipline training provides a key answer of how to break away from today's status quo while growing our most valuable resource, the individual employee.

An excellent way to start the cross-discipline training process is at the functional manager level. Many functional managers are empire builders. They are usually quick to complain about what other functions are contributing. Managers that reach this level in an organization should be professional managers, not technologists. These are individuals that have the potential of becoming the CEO and/or COO. It is important that the organization train these individuals in how the total organization functions. It is for these reasons that we recommend that the cross-discipline training program starts with them. We have seen organizations that rotate these managers regularly. Each functional manager tries to work with the other functions because if they complain about one too much, it probably will be the one that he or she will be assigned to next.

■ Turning Employees' Complaints into Profits

William Safire has said, "Language developed because of our deep inner need to complain." Yes, to complain is just human nature, but all too often, management looks at it as a negative habit. I have often heard managers say, "If only our employees spent as much time working as they do complaining." It is time for management to change the way they think about complaints. Complaints are a valuable first step in the improvement process. In reality, a complaint is an employee's way of telling someone else that they have recognized a problem.

The challenge that management faces is how to change a complaint into an improvement suggestion. Fortunately, that is not as hard as it sounds. For example, if the employee complains, "This process is so complicated that there is no way we can get good results," the manager has an opportunity to answer, "What would you suggest to simplify it? I am really open to any suggestions you might have."

Each time an employee comes to management with a complaint, they enter the office with a monkey on their shoulder. Frequently, management will reply to the complaint with, "I'll look into that," or will imply that the problem is not important with a comment like, "John, don't let that bother you. I'm sure you can meet your schedule." When this occurs, the monkey has just jumped off the employee's shoulder and now rests heavily on the manager's back. The trick is not to allow the employee to put the monkey on your back. If you make a consistent habit of not letting the employee put the monkey on your back, the employees will get the word that management expects them to solve their own problems, or, at a very minimum, that they come to management with a suggestion on how to correct the problem.

Ask your employees to try to solve their problems themselves. If they need approval to use resources outside of their control, tell them to come back and discuss it with you so you can help them. Let them know that if they cannot solve the problem, they should bring it to you and together you will attack the problem and try to find a solution.

Toyota has implemented what they call a "Thank-you Movement." When a problem is presented to management, they buy the employee a cup of coffee or tea as a way of thanking the employee for identifying the problem. Then the manager sits down with the employee to drink the beverage. During this meeting, the manager asks the employee, "Why do you think the problem occurred and what should be done to correct it?"

Many employees have already been trained on how to solve problems and present their recommendations as part of their team training process. Those who have not had the advantage of this training should now be introduced to these tools, thereby helping them make good, valid suggestions.

Improvement Effectiveness Program

Management needs to encourage and empower their employees to be creative related to their assignments. We hire engineers, accountants, MBAs, etc. to help improve the quality and productivity of the total organization. They are expected to do this as part of their assignment. Some of these employees do an outstandingly creative job. Others do a good job, while others do just enough to get by. For years, management has relied on the theory that "Cream floats to the top" to help them identify and promote the best candidates. This is an excellent theory if the milk is not homogenized and the management ranks are expanding to make room for these high potential employees. The problem is that this is not the case in most organizations today. As a result, we need to search for ways to motivate, allow our professionals to compete, and develop a database that ensures the very best candidates are identified for each promotional opportunity. The way to do this is through the implementation of an Improvement Effectiveness Program (IEP).

The improvement effectiveness program is available to all individuals and teams alike. It is a way to recognize employees for improving the things they are responsible for. In this program, the employee(s), after they have implemented a suggestion that was within their job scope, make an estimate of the first-year net savings that resulted from the suggestion (savings minus implementation costs). These estimates should be reasonably accurate (about ± 10 percent). The suggestor(s) should then fill out a form documenting the idea and the savings.

The department manager will review this document and sign it if he or she concurs with the estimate and can verify that the change was implemented. This form is then sent to Personnel where the information is added to the Total Improvement Management database and the individual's personnel record. The improvement effectiveness ideas that have general and/or multiple applications will be noted at this point. These multiple-use ideas are then documented in a quarterly report that is circulated to management. This report provides the stimuli for many spin-off improvement effectiveness ideas.

Suggestion Programs

To this point in time, we have been talking about suggestions and ideas that fall within the employee's job description. Now we want to discuss ideas that are outside of the employee's responsibility. For example: a secretary that suggests the use of a different printer because it would improve productivity; a test operator that suggests redesigning a test fixture so the parts cannot be put in backwards; or the repair technician that suggests using a different part because it will last longer in a specific application, etc. These are all good ideas that will save the organization money and/or improve its reputation. As a result, the organization should be willing to share the savings from these ideas with the employees who made the suggestions. Usually, the scope of the suggestion requires that someone other than the suggestor implements the suggestion or at least approves the suggestion before it can be implemented. As a result, these suggestions are submitted into a formal suggestion program.

All individuals and all ideas are eligible for the improvement effectiveness program. Many suggestions are not eligible for the suggestion program. The best approach is to evaluate an idea to see if it is eligible for the suggestion program and if it is not, then turn it in under the improvement effectiveness program after it has been implemented. The key elements of the suggestion program are:

1. The suggestion must not be part of the suggestor's responsibility.

2. The suggestion does not have to be implemented to be considered.

3. The suggestor shares in the savings resulting from the suggestion.

4. The suggestion cannot be pre-dated by activities or plans already underway.

How does the suggestion process work? National Cash Register Company developed the concept back in 1896. The value of the suggestion program is that it offers the person closest to the work activity the opportunity to suggest improvements. This results in more effective utilization of assets, increased productivity, waste reduction, lower product costs and improved quality. As Paul Petermann, then manager of Field Suggestions at IBM Corporation put it, "Ideas are the lifeblood of the company, and the suggestion plan is a way of getting these ideas marketed."

The formal suggestion process requires that employees document their ideas for improvement and submit them to a central suggestion department that is responsible for coordinating and evaluating the ideas and reporting back to the employee. The suggestion department reviews each suggestion and chooses an area within the organization that is best suited to evaluate the suggestion. The evaluation area studies the recommended changes to determine if they will provide overall improvement in quality, cost, or productivity. If the suggestion is accepted by the evaluation area, the evaluator will determine what tangible savings will result from implementing the idea.

In some cases, suggestions will be adopted even though the savings are intangible. These ideas benefit the organization but the savings cannot be measured or estimated in a precise dollar amount. If the idea is rejected, the investigator records the reason why the idea was rejected on the evaluation form. Both the accepted and rejected suggestions are then returned to the suggestion department, where the evaluations are reviewed for completeness and accuracy. A letter is then sent to the employee's manager describing the action that was taken on the suggestion. For an accepted suggestion, a check normally accompanies the letter. Each suggestion is then reviewed with the employee by the employee's manager. When major cash awards are received, the manager will usually call a department meeting to present the award to the employee, to publicly recognize the employee as well as to provide an incentive to get the other members of the department participating in the suggestion program.

Paul Revere Insurance Company employees submitted 20,000 suggestions during the first three years of their improvement process. The suggestions were a major contributor to the organization's improved performance.

- Income up 200 percent with no additional staff
- The organization moved from No. 2 to No. 1 in their field of insurance.

Frank K. Sonnenberg, in his article entitled, "It's A Great Idea, But ..." wrote, "A new idea, like a human being, has a life cycle. It is born. If properly nurtured, it grows. When it matures, it becomes a productive member of society." He points out that at 3M, some people claim that the company's "11th commandment" is, "Thou shalt not kill an idea."

Japan's Suggestion Process.   The following is a quotation from Toyota's Creative Suggestion System manual: "The system came to Toyota from the United States back in 1951 when Toyota was still a newcomer in the automobile industry. Two Toyota officials traveled to the United States to study modern management methods and at Ford Motor Company, they saw a suggestion system being used that inspired them to try a similar system in Toyota."

Starting with this very modest introduction, the Japanese, in a very methodical way, expanded this application and used it much the same way they did statistical process control and total quality control. Starting from a zero base line, they expanded the idea to the point that today it is the most effective employee involvement tool used in Japan, surpassing even the quality circle movement. In a study done by the Japanese Suggestion Association, they reported, "As viewed from the relationship with small group activities, which is the nucleus of suggestion activities, 50 times as many suggestions are made for every solution of one problem by one circle." Dr. Kaoru Ishikawa, the father of the Japanese quality process and the Quality Circle concept stated, "In Japan, only 10 percent of the quality improvements come from teams. The remaining 90 percent come from individual suggestions."

Now let's compare Japan's and the United States' suggestion programs.

Activity	Japan	U.S.
Suggestions per eligible employee	32.4%	.11%
Percent of workers participating	72.0%	9.0%
Percent of suggestions adopted	87.0%	32.0%
Average award value	$ 2.50	$ 492.00
Average net savings per suggestion	$ 129.00	$7,103.00
Yearly net savings per employee	$3,792.00	$ 276.00

Now this data may be interpreted that the United States goes after the big, important problems and the Japanese workers focus on the insignificant problems. But look at the bottom line. The average eligible Japanese worker saves their organization more than $3,500 per year over the savings generated per American worker.

Problems with U.S. Suggestion Programs.   The major reasons that suggestion programs are not as effective as they should be in the United States are:

1. Lack of management involvement

2. Long evaluation cycles

3. Lack of goal-setting

4. Lack of recognition

Too often, management uses the suggestion program as a way of putting off the employee. Instead of listening to the employee's ideas, they say "Write it up and turn it in as a suggestion." The manager's job in the idea generation process is to:

- Encourage employees to express their ideas.

- Help them clarify their thoughts.

- Determine if the idea is eligible for the suggestion program or if it should be used in the improvement effectiveness program.

- Support good ideas to help them get implemented quickly.

We have found that the quantity of employee suggestions is directly proportional to the manager's interest in the suggestion process. Japanese suggestion programs are so successful because everyone commits to them. Each department should set a target for the number of suggestions that the department will submit every three months. This helps make the suggestion program a challenge for the department and its members.

Idea Submittal Training

We have already talked about the importance of training people on how to solve problems, but management needs to provide them with training that addresses:

- How employees submit suggestions and improvement effectiveness ideas.
- How to evaluate what will be saved as the result of an idea.
- How to estimate the cost of implementing an idea.
- How ideas can be presented so that they are easily understood.

■ Getting Ideas Flowing

For the average organization, it is easy to embrace the concept of tapping the hidden powers of the employees' ideas. The problem is, how do you do it? How do you keep your credibility with the employees if they start turning in ideas and swamp the process? A good way to tap into this reservoir of ideas and not open the floodgates is to hold an "idea week." In this approach, management announces to the employees that a specific week will be set aside to see how many improvement ideas can be generated. For example, "The week of January 16 to 21 will be set aside to see how many ideas can be turned in that will improve safety and quality or reduce costs."

This is an excellent approach to getting the idea process flowing. It allows the organization to develop the idea processing system under a controlled environment. It will also help to define any problems that need to be corrected before the formal process is implemented. Many organizations will repeat this cycle two or three times before introducing the formal ongoing suggestion process.

Idea Sharing

An important part of developing a creative environment within an organization is the open sharing of the ideas that are generated. Many organizations accomplish this by maintaining a list of new and creative ideas that is made available to the entire organization. Often this data is stored in a computer database that can be sorted in many different ways, providing a valuable database to help solve future problems.

3M Corporation has made use of "Innovation Fairs" to exhibit new ideas. Employees from Product Engineering, Marketing, Production, and other departments attend these fairs to gain new ideas and to discuss the ideas that are being exhibited with their creators.

■ Problems without Known Solutions

No matter how good an organization is or how well employees are trained, there will always be a few problems that cannot be solved by the person who recognizes them, and questions that the employees would like to get answers for. In these cases, the first approach that an employee should use is to talk with his or her manager. Often, an employee's level of trust is low or they believe that their manager will just put them off. Other employees are just too meek to discuss the situation with their manager because they feel that they will bother him or her with little things or things that they should already know the answer to. To offset this situation, we need to provide all employees with other ways to get their problems solved and/or their questions answered.

Request for Corrective Action

Most managers think that they know all the problems that are plaguing the organization. We have seen managers who have told their employees, "Don't bring problems to me without your suggested solutions. I already know what the problems are. What I need is help in solving them." The real truth of the matter is that most managers do not know about most of the problems that are preventing their employees from doing an excellent job. In a study designed by Sidney Yoshida, a leading Japanese consultant, he reported that:[2]

- 4 percent of the organization's problems were known by top management.
- 9 percent of the organization's problems were known by middle management.
- 74 percent of the organization's problems were known by supervisors.
- 100 percent of the organization's problems were known by the employees.

Of course we all know that priorities are set by these same top management that know about only 4 percent of the total problems. The use of a Request for Corrective Action (RCA) process provides a way for the employee to inform management about the problems that the present process shields them from.

The RCA process is a very effective way of identifying the submerged problems before they tear the bottom out of the organization's ship. Any employee who is having a problem or knows of a problem can fill out an RCA form and send it in to the improvement control center. The writer has the option of signing or not signing the RCA, with the stipulation that he or she will remain anonymous unless the employee designates a desire to discuss the situation with the investigator.

Organizations that have implemented this type of program indicate that over 90 percent of the items submitted can be acted upon and brought to a successful conclusion.

Speak-Up Program

Another way to relieve pent-up emotions and provide employees with answers to questions is a process called, "Speak-Up Program." This program encourages employees to share with the organization the problems they are having or questions they may have about the organization and its activities so the situations can be corrected or explained. The speak-up program is a very confidential process that provides an ombudsman to represent the employee without divulging the employee's name. This is very effective at identifying and correcting personnel problems and items that should be discussed more openly with all employees. Many of the questions that are submitted are questions that many people have, but never take the time to ask, creating a sense of uncertainty.

■ Safety

CFO of HON Industries, John W. Axel, stated, "We are now putting safety on an equal footing with quality and productivity. Our total focus is on eliminating injuries, rather than reducing costs." HON was spending about $5 million per year on workers' compensation costs when they initiated their emphasis on the safety program. By implementing a comprehensive safety program, they were able to reduce this cost by $1.5 million in the first year. The number of accidents decreased by 50 percent.

We agree with Mr. Axel that safety is important, but we feel it is even more important than quality and productivity. There is no doubt about it. Safety must be management's first concern when it comes to protecting its most valuable resource: its people. But the documented results puts the United States in a very gloomy position compared with many of the developing nations. According to the U.S. Department of Labor's *Monthly Labor Review,* America has seven times as many private-sector injuries and illnesses as Japanese businesses have. Japanese work injury rates continue to decrease, while the U.S. rate continues to climb. This is true of most industries, not just the hazardous work environment industries. The following data reflects the computer and telecommunications industry's lost-time injury rate per year per 100 employees:[3]

Country	1987	1990	3-year trend
Canada	2.34	1.98	down
France	1.10	1.10	same
Germany	2.70	3.30	up
Italy	2.50	2.50	same
Japan	0.03	0.07	up
U.K.	1.70	1.30	down
U.S.	3.10	3.80	up

IBM, whose safety record is well below the average of all these countries for their industry, estimates it saves $50 million a year in workers' compensation as a result of these differences.

U.S. organizations need to step up to their responsibility for providing their workers with a safe environment. Management should not be content with anything less than 1000 percent improvement in the organization's safety record over the next three years. To accomplish this, organizations should

- Provide regular safety training for all employees.
- Provide a continuous safety focus campaign.

- Pay rewards to individuals who identify safety problems.
- Have an experienced staff investigate each accident and report what corrective action was taken. This report should be reviewed with upper management.
- Post accumulated accident-free hours.
- Have each manager conduct a safety review of their area at least once a month and turn in a report on their findings.
- Have middle management conduct an audit of their area and one other middle manager's area every three months and turn in a written report.
- Document a negative analysis of each area every 24 months.
- Have a third party conduct an audit of the entire organization every six months.
- Fire employees who will not correct unsafe working habits.

When it gets right down to it, the elimination of accidents is a very personal thing that requires each employee's attention. Management can help by error-proofing the business environment, but accidents will never be eliminated unless the employees use the tools and processes as they were designed to be used. Everyone's goal should be to eliminate not only lost-time accidents, but all accidents—right down to things as seemingly small as a paper cut.

■ Empowering the Individual Closest to the Customer

The individual who is closest to the internal and external customer must be empowered to meet the customer's expectations if the organization is to be viewed as a world-class leader. True service excellence is a very personal, individual thing. Procedures and good processes help, but the performance of the organization rests with each individual and how they relate to their customers.

There are very few processes that are so good, that sometime, some way, somewhere, under some condition, someone won't foul them up. It is for this reason that we must go beyond teaching our employees how to use the organization's processes. We need to also explain to them why the process exists and what the expected outcome from the process is. Then management needs to empower each employee to act on his or her own to ensure that the customer is satisfied with each individual's output. This personal empowerment does three things:

- It aligns accountability with responsibility.
- It builds pride in the employee.
- It maximizes customer satisfaction.

Empowerment occurs when management provides employees with all the required information, knowledge, and resources that are needed to perform their assignments and allows them to execute their assignments in any way necessary to achieve the desired results, as long as it is done in keeping with the organization's values. Usually, empowerment is limited to a set, specific boundary and/or accepted operating norms. (Example: The boundaries that could be set for a salesperson might be, "You can give full refunds as long as the customer has a receipt.") All employees today are empowered to some degree. The advanced organizations are continuing to relax the boundaries that each employee works within. (Example: The new boundaries placed upon the salesperson could be, "You can give full refunds as long as you believe it is the correct thing to do for the customer and the organization.") You will note that with the relaxed boundaries, the employee has a greater responsibility to exercise good judgment, while having increased authority to expend more of the organization's resources.

There is no doubt about it. If an organization wants to excel, each individual within the organization needs to be empowered and have the required knowledge to perform their job in a superior manner. The general can have the very best military strategy and plan, but if the soldiers do not hit the target, the war is lost.

■ The Start of Individual Excellence

The start of individual excellence is the hiring process. Most organizations have a base population that they want to develop and should work with them to help them excel. The degree to which this activity will be successful is greatly influenced by factors that are far outside the control of the organization. In an IBM technical report entitled, *Theory H*, it was pointed out that the highest individual performance level ("H" level) was a basic trait like honesty, diligence, religion, work ethics, etc., that is developed during a child's formative years. By the time an individual enters the work force, his or her "H" level has been established.

For the rest of the person's career, he or she will perform at a point someplace below his or her "H" level, based upon other outside forces. For example, how much pressure management applies on schedule or quality. The employee will only perform above the "H" level for short periods of time under extreme pressure. (Example: The employee is told that he or she will be fired if his or her quality of work does not improve, or if his or her life is in danger.) This "H" level will not change unless the individual suffers a serious emotional experience. It is for this reason that the organization must be very careful when it selects new employees.

There are a number of factors that need to be considered in the selection of a new employee. Some of them are:

1. Does the candidate have the personal traits that will have a positive impact upon the rest of the organization?
2. Does the candidate have the background to do the job?
3. Does the candidate have the physical traits required to do the job?
4. Is the candidate compatible with the organization's visions?
5. Can the candidate be moved to another function within the organization when this job is completed?

6. What type of long-term contribution can the individual make to the organization?

7. How well has the candidate performed other tasks that she or he has undertaken?

Japanese organizations realize the importance of selecting the right employees because they plan to make careers for them with the organization. Most Japanese export organizations rely heavily on recommendations from employees and associates when hiring new employees. They go to their schools and talk with a potential employee's teachers, not just at the college level, but also at the grade-school level. They talk to the employers of the candidate's siblings to determine how the candidate's relatives are performing. Strong consideration is given to the reputation of the family, as well as the individual.

In the United States, we used to be much more careful in selecting new employees. I remember when I joined IBM. They interviewed six of my neighbors, my minister, and three teachers before making a job offer. Strong consideration was given to me because my father had been an excellent employee with IBM for over 24 years, and my uncle also had worked for IBM for over 20 years.

Today, U.S. organizations frequently do not exercise enough control over the hiring process. They have a vision of a cooperative team environment and then hire a new employee who has not been involved with team activities within the school system. Management has a tendency to hire an individual to fill an immediate need without considering how other functions could use the employee later on. All new employees should be reviewed by at least two functions that feel the employee has potential for working in their area. We suggest that you select your new employees as carefully as you would a new son-in-law or daughter-in-law.

■ Creativity

The single biggest advantage that human beings have over the rest of the animal race is our ability to create new concepts based upon past experience. Constructive use of fire, the wheel, the telephone, the light bulb ... the list goes on and on of how individuals' creative minds have provided the fuel to move mankind ahead. But creativity does not just manifest itself in breakthrough concepts. It is around us all the time. Mary finds an easier way to print standard letters. Jim develops a new sales pitch that increases book sales. Fernanda discovers how to get waffles out of the waffle iron without burning herself. Yes, creativity and discovery go hand in hand. Creativity is the very personal things that every individual does to some degree on a regular basis. Creativity is not limited to a few geniuses. It is as natural as waking up in the morning. It's just that some do it better than others, because the more creative people have a tendency to think about things in a different way.

Realizing that creativity is a different way of thinking about everyday occurrences opens a whole new perspective to everyone. Why? Because everyone can be trained to use creative thought patterns that will greatly increase their creative abilities. With this realization, management around the world have altered their training programs to include creativity training.

■ Creativity for the Individual

Do you agree with this statement?

> Participating in creative activities is a lot of fun, but it is just a diversion from the "real" work to be done.

Well, part is true and part is false. It is true that creative activities *are* a lot of fun. It is false to think that they are not part of the "real" work. This book deals with how to improve the work that we do. This chapter deals with how to increase your capability to make improvements. Not by a little bit here and a little bit there, but with major steps forward. This occurs when being creative becomes not a diversion, but the way you think and do your job. Consider these facts:

In a *USA Today* survey of 100 executives on what is more important, being creative or being smart, 59 percent said creativity was more important, as compared to 28 percent responding that being smart was more important. That is better than a two-to-one ratio for being creative!

Fortune, April 19, 1993, had an article entitled "Japan's Struggle to be Creative," where the various organization programs for enhancing creativity are discussed. Those companies involved are Shiseido (cosmetics), Omron (electronics), Fuji (film), and Shimizu (construction), all leaders in their field. These types of programs are truly revolutionary for Japan.

Having said this, let's find out how we can become more creative by first defining some basic premises.

Many of the books and papers on creativity consider the role of the individual as being a member of a team or group involved in creative activities. No particular emphasis is made as to how being creative is best done—with a team or as an individual. Before we become involved in which way is best (if there is a "best" way), let's look at just what creativity is. The dictionary definition of *create* states "to bring into existence," and "to produce through imaginative skill." Not too much help; nothing we didn't really know before. Let me offer a definition that is helpful. It comes from Albert Szent-Gyorgyi von Nagyrapolt, an American biochemist (born in Hungary) who won the Nobel Prize in 1937 for physiology and medicine. He described creativity in terms of discovery in this way:

> Discovery consists of looking at the same thing as everyone else, and thinking something different.

I believe Szent-Gyorgyi has taken a vague definition and made it into a concept that we can understand and use. There are two parts to what he has said. The first part is quite easy, but the second part, well, that is the challenge we will discuss for the rest of this section. The first part, "looking at the same thing as everyone else," means that if I hold up a pencil or any object you choose, and ask "What is this?" most of you would say it is a pencil. Now if I asked you to "think something different" about the pencil, there just might be a bit of a mental struggle to do this. How to make this less of a struggle and to make "thinking differently" the rule and not the exception, is the whole reason why I wrote this and why you are reading it.

Thinking differently, obviously, means making changes, and through my experiences in teaching classes on creativity, I've come up with three basic mind-sets needed for making changes.

The Basis for Change

- *Attitude.* One must work toward developing the attitude that change can be a positive force in our lives, that whatever happens can be for the good. Remember the 1940s song that said "eliminate the negative, accentuate the positive"? Well, this is the way that change needs to be accepted.

- *Continuous improvement.* We must always be striving to improve whatever we do. My motto for many years has been that "there is always a better way." Nothing has been done that cannot be done better. The only way this happens is through the constant effort of finding that better way.

- *Reach out.* Our goals need to be high enough to cause us to "stretch" to reach them. We will often find that it is the journey toward the goal that is more rewarding than the achievement of the goal.

All three of these, what I call mind-sets, are needed for anyone that truly wants to think differently. Each person needs to internalize these thoughts, that is, keep the concept, but use your own words so it becomes your thinking, not just what I have written here. You may even want to add other concepts to these three regarding change, and if that helps in further developing the internalizing of your personal "basis for change," then that is great!

So now let's say that you have your own change mind-sets and move on. What we have just discussed is needed by each of us, as individuals, whether or not we are part of a team.

The opening paragraph asked the question: Is there a "best" way for being creative? Not all creative acts happen as the result of being on a team, but being on a team can, in some cases, speed up certain parts of the creative process. For example, developing a list of possible solutions to a problem will proceed at a much faster pace through the use of brainstorming by a group of people rather than one person trying to do so alone. This is especially so if the members of the team are relatively inexperienced in the brainstorming process. On the other hand, those individuals that have been involved in brainstorming activities on a regular basis can do almost as much as the team. What this means is that creativity, the act of thinking differently, always starts with the individual. It is the experience and the way creativity is approached that determines the "best" way. It is appropriate then, that we focus on the traits that you, as an individual, need to develop in order to facilitate your creative capabilities by thinking differently.

So what does it take to "think differently"? Almost anyone that has been involved in the creativity field will have their own set of ingredients for doing this. I don't think, though, that too many would disagree with the ones we are going to discuss now. There are (at least) five elements that each individual needs to have in order to think differently on a regular basis. The more we develop these elements, the easier it is for us to make creativity a normal part of our life and not a special effort that strains our brain (a bit of a serious pun!). Let's define these elements first, then look at how we can make them part of our life.

Elements of Thinking Differently

Curiosity. I think that all of us are curious to some extent, especially when something happens that really boggles our mind. For example, most normal sunrises are from the east and have the distinctive, soft orange-type color, and we don't give it another thought. What, though, would we think if the sun rose in the south and had a bright green color? Most likely, and almost without exception, each of us would say (leaving pure panic aside), "I wonder what is causing this (very!) strange event?" And we would probably speculate as to the cause of it, coming up with, maybe, two or three reasons for it. We need to have this level of curiosity, not for just the very unusual event, but for many of the usual, normal parts of our lives. We need to develop an intrinsic sense of curiosity as part of the way we think, the way we observe the things that go on around us. Becoming naturally curious is the first step in being able to think differently.

Risk Taking. Let me say right away, that I am not in any way talking about sky-diving or race car driving, or anything like this. I am talking about the risk we need to take in thinking about and doing the things we do during a normal day. Thinking differently (and in some cases acting differently) does involve a certain element of risk. Risk in the sense that we might be embarrassed, ridiculed, feel left out, be talked about, and all those other things that potentially happen when one stands out from the norm. The nature of the society that we live in today is strongly biased toward conformity, where everyone is expected to stay within certain bounds. Those that go beyond these bounds are considered different and are, in some way, "punished." But then thinking differently makes it necessary that we think beyond these bounds. We need to find ways to do this without damaging our mental selves and this we can do, as we will find out later on. Being able to and wanting to take these risks is the second step toward thinking differently.

Paradigm Shifting. A paradigm is a shared set of assumptions, the way we perceive the world. Having paradigms helps us explain the world around us and helps predict its behavior (from *Powers of the Mind* by Adam Smith). Another definition is that a paradigm is the basic way of perceiving, thinking, valuing, and doing associated with a particular vision of reality (from *An Incomplete Guide to the Future* by Willis Harmon). Though not the first to do so, but the most recent that popularized the concept and role of paradigms in our current times, Joel Barker in his book, *Discovering the Future—the Business of Paradigms*, has done an excellent job in explaining the need for and yet the problems of having paradigms. Having paradigms, and we all have many of them, is necessary, because they provide the stability we need in living somewhat normal lives. At the same time they create shackles on our mind and actions, which strongly constrain our ability to think and act differently. We need to identify our paradigms and develop the means to shift them (develop new ones) on an ongoing basis. Too many of us get stuck with the current paradigm and only change when society makes the change. Those who develop the ability to think differently are the ones that can, independently, make their own shift in paradigms. Being able to do this is the third step toward thinking differently.

Continuous Exercise.   You might wonder, just what does exercise have to do with thinking differently? Well, thinking differently, being creative, deals with the use of the mind. The brain as the physical entity of the mind, that 3 to 4 pound, grapefruit-size part of our body that sits inside of our skull, and it is as much a muscle as are the muscles in our arms and legs. It is well recognized that muscles improve with continuous exercise, whether it be lifting weights, running, swimming, or any of the many ways now used to strengthen, shape, and improve our bodies. The key, though, is continuous. The very same principle holds for our mind. The more we use it, the more it improves. As there are specific exercises for the other parts of our bodies, so there are specific exercises for our mind; specific ones for the left side (the verbal, analytical side) and for the right side (the visual, imaginative side). Later in this chapter, I will provide you with several mind exercises. Engaging in continuous exercise of our mind is the fourth step required for thinking differently.

Perseverance.   There is not a pill you can take, there is not a book you can read, and there is not a classroom course that you can attend that will, by themselves, make you a creative thinker, fluent in thinking differently. All of these can help, but acquiring and developing the ability to persevere is absolutely necessary. Persevere means to persist in your actions in spite of counter influences, opposition, or discouragement. Mastering the first four steps will be a major part in developing your perseverance, which is the fifth step toward thinking differently.

So where are we now? Each of us has developed our personalized "Basis for Change" factors and we are now aware of five "Thinking Differently" elements. Let's now look again at each of these and see what we have to do to make them become our "tools" for truly thinking differently when seeing the same thing as everyone else.

Tools of the Trade for the Ingredients

Curiosity (Why Are Things the Way They Are?).   Being curious about the things that go on around us is a very natural act. Just look at children as they are growing up. Why is the sky blue? Why can't I see air? Why does the ocean taste salty? And on and on and on …. As adults, we find being asked questions like this rather tedious and often respond with irritation and some answer just to get the young person to be quiet. After not too many of these types of answers, the one asking questions soon learns that being curious is apparently the wrong thing to do, and so the questions soon stop. When the questions stop, the mind stops being curious. But then, doesn't the same thing happen with us as adults? How many times have you been doing a job and asked your manager "Why do we do it this way?" and often get an answer that says, in effect, "Just do the job and shut-up." So even as adults, we soon learn not to ask questions and once again our curiosity is stifled.

Action: We need to, in spite of the pressure not to, continue to ask questions like: "Why is this the way it is?", "What caused this?", etc. If the environment that you live and work in is such that no one wants to hear the question, then ask it of yourselves. It is not just getting a specific answer that is important, it is getting the mind to think about the "why and what" that is important. We need to regain a child-like curiosity about things. One can never find creative ways to do things unless one first asks the questions of "why and what." Not only should we do this, but we should encourage others to do the same thing. You can practice this quite easily and you do not need to have a problem to work on. As an example, consider this scenario. You are driving somewhere and see a herd of cows bunched closely together and you notice that two of the animals are separated from the herd by some distance. Question: "Why is that so?" Of course you are not really trying to find the real reason, you just want to get the mind to start being curious about why things are the way they are. All around us are things like the cows that we can start to notice and ask questions about.

Risk Taking.   Again, I'm not talking about death-defying feats. What I am saying, though, is something as simple as this. Consider what almost everyone does when they ride in an elevator. You walk into the elevator, turn around, push the button for the floor you want, then either stare at the floor or at the changing floor numbers, and because of the "Rules of Elevator Riding," you will not say anything! Ever notice that? No one talks, even if you are riding with someone you know! Try this risk: Say something, anything. Perhaps, say, "I'm sure glad we are all going the same way," or "We are all sure quiet today," or, if it is crowded, ask someone to push the floor button for you. It is not that you are trying to be a stand-up, one-liner comic, but you are trying to use your risk muscle by doing something different, which is helping you think differently. If you really want to take a big risk, try standing with your back to the elevator door facing the other people in the elevator. Consider another situation. You are in a meeting (like a class, or a staff meeting) and someone says something that you don't understand. What do most people do? They don't say anything, hoping that someone else will, or maybe it will be explained later (usually this never happens).

Action: Take a risk and ask the question. In almost every case, you will find that there were others that had the same question as you had. This is especially difficult if the one who is talking is a senior (older, high-level) person. There is always some intimidation, even if unintentional, and it can be seen as a risk in that you might feel foolish (to ask such a question when no one else did), you might be subject to ridicule (all eyes turn to you as if to say, "How come you don't understand?", or any number of things that you could imagine). So with that fear, you just sit there quietly. The elevator, the meeting, and other similar situations gives us the opportunity to use our risk muscle, to think differently and then act differently. The act of thinking and acting go together, because the acting will strengthen the thinking.

When we find that our world doesn't come to an end when we do this then it will be easier the next time we take a risk. Soon, probably within 3 or 4 times, it becomes second nature.

Now you have started asking the why and the what (the curious part) and you are willing to act on this in meetings. When you are doing this on a regular basis and are feeling more comfortable doing so, then you are ready for the next step.

Paradigm Shifting (Getting Out of the Rut).   Start this step by identifying your paradigms. Don't get so deep that you are looking at paradigms that relate to the meaning of life or the like. One paradigm that has ruled industrial organizations for the last hundred years (and civilization for the last thousand years or so) is that organizations are structured vertically, with orders flowing down and results flowing up. If there is a need to communicate with other organizations, this is usually done through a manager, particularly if the communication has to do with making changes in the way results are achieved. This is how most businesses are run today, but there are some, very few, that don't run a business that way. The reason for this is that someone challenged the vertical organization paradigm. Someone got curious and thought: "Why does information have to flow up and down?" and "Why does a manager have to always be the one to make the changes?" Most likely this same person decided to take a risk and ask these questions in a staff meeting. Let's hope that the manager was inclined to think differently and said "Let's see if there is a better way." The point is not what the manager did or did not do, but that someone questioned the paradigm. The person that wants to think differently, to be creative, has to be curious, be willing to take a risk, and then start shifting from an existing paradigm to a different paradigm.

Action: Identifying your paradigms is the first thing to do. Some of these will be strictly yours, some will be shared with others, and some will be shared with society. Right now, though, identify those unique to you. For example, one might be the way you dress for the work you do. Is it always a suit and tie, a tailored business suit? A sport coat and slacks? A certain style of casual clothes? Another example is how you go to work (the route, by car, by other means). Do you always go the same way, at the same time? Admittedly, these may be in the trivial category, but looking at something simple and being able to do something about it is a good way to get used to shifting paradigms.

As you look at the examples I've mentioned or others that you have identified, ask yourself if it is really necessary to do what you are doing. What happens if you do it differently? Will it jeopardize your job (as in how you dress)? Will it cost or save you money (as in how you get to work), or take more time? I believe that the key to shifting any paradigm, large or small, is to first question it, then understand it, and then make a decision to either keep it or change it.

Continuous Exercise (The Mind Is a Muscle).   If you are now in the process of practicing (regularly) the first three ingredients of thinking differently, then you are already in the mode of continuous exercise. There is more, much more, that you can do to strengthen and improve your mind to the point where you easily think differently and are more creative. This idea of exercising the mind is not something

I discovered in some flash of insight, but it is something that I have come to appreciate more and more as the cheapest and easiest way to make thinking differently the rule and not the exception.

Action: So, just how do you exercise your mind? Simple, just play games! Games? How can something as serious as improving your mind benefit from playing games? Well, first of all, being creative is a fun thing to do and it is not all work and no play. Humor, play, fun, interesting, exciting, are words that aptly relate to the exercises that will help you on the journey toward being a person who can regularly think differently and therefore be more creative. The key, once again, is to play these games on a regular basis, the same way that you do exercise for the physical body. Just what are these games? The best book I have read for mind exercises is "Pumping Ions," by Tom Wujec, Doubleday and Company, Inc., 1988.

Let's take the time here to show the characteristics normally associated with each side of the mind and from this you can get a sense of the exercises you should use.

Left	Right
Verbal—words	Nonverbal—pictures
Analytic—step-by-step	Synthetic—holistic
Temporal—sequential	Non-temporal—nonsequential
Rational—reason, facts	Nonrational—no judging
Digital—use of numbers	Analog—relationships of all
Logical—order	Intuitive—insight, hunches
Linear—sequential	Holistic—patterns, wholeness
Vertical—narrow, sequential	Lateral—broad, many areas

Most of us have lived in an environment where the emphasis has been on the use of the left side, so this is the side that is more developed and easier for us to use. What we need to do is to balance our two sides a little more and to recognize when we need to use a particular characteristic; i.e., look at a situation from the aspect of the "other" side. Notice that each side has the opposite characteristic of the other side. Regular use of verbal and visual exercises will help us be more "whole brained."

Since I have mentioned *Pumping Ions* as the best source for mind exercises, let me also mention two other books that you should have and use. Roger von Oech has written two books that are probably the best for total creative thinking. The first is *A Whack On The Side Of The Head*, and the second is *A Kick In The Seat Of The Pants*. The first one, *Whack*, deals with the ten mental locks to being creative; i.e., being able to think differently. The second, *Kick*, deals with the roles you need to play when you think differently; i.e., first be an explorer (gather data), then be an artist ('sculpt' this data in many different ways), then be a judge (make a decision on the pros and cons of the 'sculpted' ideas), and last be a warrior (put a fire in the belly and a lion in the heart). With all that has been said before, this leads us to the final and most important of all elements.

Perseverance (All Good Things Take Time).   Of all the five elements, this one, perseverance, is the most necessary and the most difficult to master. If one does not develop this capability, having all the other four elements will be of limited value toward thinking differently and becoming creative. Thomas Alva Edison said (around 1932) that "Genius is one percent inspiration and ninety-nine percent perspiration." I believe that Edison would agree to the substitution of creativity for genius and not lose any meaning in his quote. If you truly believe in the mindsets shown in the section "Basis for Change," then being able to persevere, over long periods of time, can be accomplished.

Action: Start observing how you act on new ideas. Do you give up when someone tells you it won't work? Feeling discouraged when you receive negative remarks is perfectly normal, but the key is that you don't give up, but continue to think of ways to have your idea accepted and acted on. The Japanese phrase "Gambatte Kudasai" expresses this well. It means "Don't give up, carry on, persevere." Also observe how you react to the ideas of others. Are you positive and supportive, or are you negative and critical? Bob Conklin, is his course *Adventures in Attitudes* said, "To the extent that you give others what they want, they will give you what you want." So if you give positive support, you will get positive support.

Patience, persistence, perseverance—they all mean about the same, and they all are difficult to practice. Yet those that have been successful with new ideas, new businesses, new ways of doing a simple job or running an organization have had to have a large measure of patience, persistence, and perseverance.

The Next Step

We have been through the five major elements of the thinking differently process and let's say that you understand and accept them. Now what? Just what do you do to spread this to your work unit, perhaps to the whole organization? Consider these steps:

1. Make sure that you are convinced of the need for and the value of the thinking differently process. Ask yourself periodically (say, weekly, on a specific day) about how well you are doing on each element. Grade yourself by using a scale of 1 to 5 to indicate your strength in each element. If the number is low, you have more work to do. These five elements need to become the rule, not the exception, for how you work (and live, by the way).

2. Discuss with your work unit how thinking differently can enhance the team's capability for finding better ways to improve their work.

3. Ask the team to read this chapter and the books I have recommended (later in the chapter). Perhaps have one book assigned to each member of the team, then have them brief the others on the contents of the book. The team should particularly look

for problem-solving techniques; e.g., there are around 12 different ways to do brainstorming.

4. Discuss among yourselves the ways that the team will use the information gathered.

5. At the beginning of each team meeting there should be a period of time, say 15 minutes, devoted to creativity enhancement. This could be a review of one of the books, grading of the team's strength on the five elements, practice improving one of the five elements, use of one of the exercises from *Pumping Ions,* practice of a problem-solving technique, and so on. It is important to have the discipline to do this on a regular basis. You will not be able to improve your thinking differently process unless you practice again and again. Remember the fourth element: Continuous Exercise. There is no way around it if you truly want to see results.

You now have the foundation for thinking differently and the means to become more creative in your improvement journey. Remember, no one event (reading this chapter, a book, taking a class) will make you creative. Rather, it is a combination of all of these, plus your day-to-day use of the techniques we have discussed in this chapter that will lead you to the point where being creative is not a diversion from your job, but the normal way you think, do your job, and live your life.

■ Self-Managed Employees

The management process as we know it today has gone through many evolutions. The management process that lasted the longest was the tribal leader, or father, management process. This approach to coordinating activities within a group started back in the cave-dwelling era and is still used in many farming groups and subcontract situations in Asia. This approach eventually gave way to the guilds or craftsman/apprenticeship management process. As organization size outgrew this process, the hierarchical (pyramid) management process evolved. In recent years, this hierarchical management process has been modified in a number of different ways. Some organizations have organized their management processes to coincide with their business process flow. Others have tried to form matrix management processes. But at the grass roots level, these management processes all look alike. The employees have a manager that they report to who has been given the responsibility to accomplish a specific task, at minimum cost, and at the highest possible quality level, using the resources that have been assigned to him or her.

Little has changed in the first-level management structure since the cave dwellers first implemented it until about 1965, when some very advanced organizations started to empower their employees to take on more and more responsibilities. Out of this grew some major changes to the management structure. First, management focused on training the employees to understand the organization's goals, values, mission, and business plans. Then management explained how the team (department) contributes to the success of the business objectives. This logically led to establishing team measurements that reflected the team's customers and business plan requirements. Once the measurement system was established, the teams were provided with training on the organization's operating procedures, financial controls, problem-solving, and process measurement techniques.

The result of all this preparation was the beginning of *Self-Directed Work Teams*. The empowered Self-Directed Work Teams allowed the employees to get together and decide how best to do the tasks that they were assigned. As the teams became more effective at self controlling the tasks assigned to them, it was only natural to expand their empowerment to take over the total management role. As a result, Self-Managed Work Teams evolved. The Self-Managed Work Teams eliminated the need of a direct manager because the team took over activities like giving out increases, selecting team members, evaluating performance, discharging poor performers, and developing budgets.

In my article entitled "Worklife In The Year 2000" published March 19, 1990 in the *Journal of Quality and Participation*, I predicted that employees would become more than empowered in the twenty-first century. Their role would evolve to almost an independent contractor relationship in which employees would actually buy the rights to provide services to the organization. Well, that prediction is evolving faster than I thought possible. In China, employees who want to work in some organizations pay for the training that the organization will be providing before they are hired.

The evolution of Self-Managed Work Teams, network organizational structure (discussed in Chap. 14), and new communication systems that allow employees to work as effectively at home as they do in the office or plant, has resulted in freeing the employee from his or her bosses. The result is the *Self-Managed Employee*. The Self-Managed Employee process has at long last given the employee control over his or her life, and a new sense of freedom and dignity has resulted.

The Self-Managed Employee has specific measurements that relate directly to organization performance. Management monitors performance based upon the results achieved, not on the old type of performance indicators like attendance, hours worked, or items processed per period. Specific goals are set for the Self-Managed Employee and performance is measured based upon meeting these goals. How the employee meets these goals is left up to the individual as long as the approach used meets the ethical practices and values of the organization. This has resulted in a greatly expanded span of control for a manager. In the Self-Managed Employee environment, a manager can have 100 to 500 employees reporting to him or her and still do an excellent job.

Ernst & Young is one of the organizations that has led the movement to Self-Managed Employees. In 1990, Ernst & Young was a hierarchical organization organized around the partners. Each partner served as a manager of a group of consultants (6 to 18 employees), and partners were selected to manage groups of partners. The partner's job was to develop the consultants assigned to him or her and maximize the profits and customer satisfaction level generated by the group. Each employee had an office close to the partner who managed him or her, and when they were not at a client's office, they were expected to be in their own office.

Today, many Ernst & Young management consultants are operating in a much different role. A single manager is assigned to a large area (e.g., Northern California,

Washington, Oregon, and Nevada) and all of the employees in that area report to this manager, not the location office partner in the city they are assigned to. This has allowed Ernst & Young to establish large resource pools that the partners select from to meet the specific needs of the individual engagement. The result is that their clients are provided with better service because the partner can select just the right individual to meet the specific needs of the engagement.

In addition, there is no longer a need for the consultants to be in close proximity to a specific partner. This has allowed Ernst & Young to go to a process they call "Hoteling." In this process, only partners and principals have offices permanently assigned to them. When a consultant wants to come into work, he or she lets the secretary know that an office will be required, and one is assigned. The consultant's phone is also set up to ring within the assigned office. Prior to this change, many office areas were empty most of the time because the consultants were out with their clients. This consolidation has reduced the office area required by about 30 percent. In addition, the consultants have been provided with supporting portable tools that allow them to stay at home and work when they are not with a client. This reduces travel time, pollution, and allows the consultants who are away from their families much of the time to spend as much time as possible with their loved ones.

From the consultant's standpoint, goals are very clear. Each consultant has three primary measurements:

- Realization: The percentage of their chargeable rate per hour that is actually realized.
- Utilization: The percentage of time that is charged to a client.
- Customer Satisfaction Level

Each employee's salary relates directly to how much the client is billed for the consultant. It is easy therefore, for each consultant to multiply "realization" times "utilization" to determine if they are making a profit or loss for Ernst & Young. The employees have two factors to work with—realization and utilization. The clients that pay based upon a fixed-price contract or value-of-the-job contract provide the consultant with an opportunity to exceed 100 percent realization, which means that they could be profitable for the organization with a lower percentage of utilized hours. Of course the more profitable the consultant is to Ernst & Young, the more money Ernst & Young can pay the individual, so the consultants are highly motivated to keep realization and utilization high. The consultants are also motivated to expand their knowledge of methodologies by attending Ernst & Young schools so that they can be considered for more assignments, thereby keeping utilization high. The other factor that impacts utilization is how satisfied clients have been with their performance. Partners want consultants on their engagements who have proven high customer satisfaction ratings.

The consultant's performance is evaluated by each partner in charge of a project that the consultant is assigned to, as well as by the customer that the consultant works with. This means that most consultants are evaluated many times each year by different partners and customers. These accumulated evaluations, plus the consultant's self-performance evaluation, are used to define the total individual's performance.

To keep this process from being cold and impersonal, each partner and principal are assigned a group of employees for whom they will serve as career counselor and mentor. Whenever the employee has need of advice and/or guidance, they can meet with this individual to obtain the help that they need.

Another critical factor to the success of this new management process is excellent 5-way (star) communication. Effective use of voice mail, E mail, newsletters, business status reports, and social events is a very important part of enhancing the consultant's feeling of belonging.

A Self-Managed Employee process will not work without well-trained, knowledgeable employees and an environment in which there is a high degree of trust between both management and the employees. It is a process where bureaucracy has a hard time surviving.

Although there was a degree of concern at first, the pilot programs went very smoothly at Ernst & Young and the results obtained have been well worth the effort. Not only are the consultants pleased with the added freedom they have been given, but there has been a step function improvement in Management Consulting profitability in Ernst & Young, allowing them to hire more consultants. This is a good example of how creativity, organizational realignment, and technology can be combined to better serve the employee, the customer, management, the owners, and the community.

■ Summary

Excellence in an organization occurs when each employee goes home at night and looks into the mirror, thinking, "I did a great job today." For the organization to excel, the individual needs to excel. It is for this reason that management around the world are investing more and more money in developing their human resources. Technology, tools, and teams make the difference between failure and success, but it takes the individual's personal trust, commitment, and creativity to make the organization a world leader.

Now, let's discuss some of the ways different organizations look at individual employees:

Education and Training

Losers: Use on-the-job training only.

Survivors: Use on-the-job training and problem-solving training. Average about 15 to 20 hours per year in formal training classes.

Winners: Design training processes to meet the business and growth needs of the individual. Encourage individuals to take outside courses. Develop a learning environment where 40 to 60 hours per year is devoted to formal training of each employee, and the employees volunteer, on their own, an equivalent or greater amount of time.

Risk Taking

Losers: No room for risk taking at lower management levels or by nonmanagers.

Survivors: Risk-taking is a desired trait, but you had better be right.

Winners: Risk-taking is required, and the organization realizes that along with risks is an exposure to failure that serves as a learning experience.

Appraisals

Losers: Management gives appraisals. Most people are above average.

Survivors: Performance objectives are prepared jointly by management and employees. A formal appraisal is given once a year that documents performance to objectives.

Winners: Performance appraisals are linked to individual projects. The project objectives are used to evaluate the individual's performance. Input from the employee's customers has a major impact upon the evaluation. Many evaluations are conducted each year and are averaged out to define overall performance.

Career Planning

Losers: The organization is not involved in employee career planning.

Survivors: Career planning is used only with high potential employees.

Winners: All employees have a career plan that meets the organization's and the employee's needs and expectations. Career paths are well-defined and established.

Hiring

Losers: New employees are usually hired at the last minute when they are needed on a specific job. Little time is allocated to orientation and formal job training up front.

Survivors: Many applicants are considered and selection is based upon their ability to do the specific job. Heavy emphasis is placed upon grades and the quality of the school they attended.

Winners: Applicants are evaluated to ensure their personality and background fit with the organization's values and vision statements. Applicants are considered for a wide range of assignments and need to be acceptable for many of them. Total activities in school (social and scholastic) are considered and weighed. Candidates recommended by employees with good work ethics and demonstrated commitment to the organization, are given priority consideration.

Pay

Losers: Pay is based upon the job and time on the job.

Survivors: Pay is based upon the job and performance on the job.

Winners: Pay is based upon the individual's knowledge and total value to the organization. Management realizes that individual performance will vary based upon experience in the assignment and personality traits. Management feels a responsibility to have a good match between the work assignment and personality traits. When this is not possible, management does not penalize the employee, because management has chosen not to utilize the individual's skills and traits as well as they can be used.

Trust in Employees

Losers: Employees are viewed as individuals that will take advantage of the organization if they are not watched.

Survivors: Employees need to be watched because they make errors, and a small percentage of employees are dishonest.

Winners: The organization realizes that 99.9 percent of the employees will do the right thing if they are left to their own devices, so they should be trusted. The 0.1 percent of the people that cannot be trusted should be released from the organization.

Individual excellence results from people who take pride in their work and dedicate their physical and mental efforts to upgrading themselves, their jobs, and the organization. Their pride and self-esteem result from management's interest in them—in their creative solutions to today's problems, and their suggestions that prevent errors from occurring. The key elements that establish an environment where this behavior pattern flourishes are:

- Improvement ideas are a specified part of everyone's performance expectations.
- Rewards and recognition systems encourage creativity and risk-taking.
- Management balances the need to have an effective team with the individual's need to excel.
- Employees are empowered as their responsibilities and accountabilities are aligned.
- Management accepts the fact that they do not know all the problems or have all the answers.
- Employee careers are important to the organization.
- Both small and breakthrough improvements are important.

The organization can develop an environment where individual excellence can flourish, but as individuals, we need to take on the responsibility for our own work and the solutions/ideas we develop. To increase our ability to be more and more creative, we should continuously challenge our minds with things like:

- Take on a new assignment.
- Expand our self-image by taking on projects and setting goals for ourselves that are challenging and aggressive.
- Expand our view outside of the organization by increasing our professional activities.

- Learn a new language.

- Read job and nonjob related books, magazines, and articles.

- Listen to music to relax, but vary the type of music you listen to. Pleasant background music helps improve your thinking and creative skills.

- Set improvement goals for yourself and measure your progress against these goals.

- Play mind games with yourself and others.

- Train yourself to dream in color rather than black and white.

- Use meditation to clear your mind so that new thoughts can evolve.

- Always have something new that you are learning/studying.

- Exercise your body to help stimulate your mind.

- Measure your progress on a monthly basis. Set new monthly goals to use different tools to stimulate your creativity than you used the previous month.

■ Suggested Reading

There are many books available on every aspect of creativity. Let us offer some titles to you that we have found to be particularly helpful. We have put them in categories to help in your selection according to somewhat arbitrary levels of creativity. It is not necessary to follow the sequence exactly as we have shown.

Starter Set
  (Everyone should read.)

A WHACK ON THE SIDE OF THE HEAD
  Roger von Oech, Ph.D., Menlo Park, CA: Creative Think, 1983
  * A book on creativity written in a creative way, one of the best, easy to read.

Pundles
  Bruce Nash and Greg Nash. New York, NY: The Stonesong Press, 1979
  * Nice collection of mind exercises. Good warm-ups for the thinking process.

Intermediate Set
  (Everyone should be aware of contents, then read as appropriate.)

Brain Power
  Karl Albrecht. Englewood Cliffs, NJ: Prentice-Hall, Inc. 1980
  * Similar to "Brain Games," but more in-depth brain info. Discusses brain skills, patterns, how to make changes.

Serious Creativity
  Edward de Bono. New York, NY: Harper Collins Publishers, Inc., 1992
  * Updates and combines Lateral Thinking and Six Thinking Hats giving a step-by-step approach to creativity. The author is recognized as one of the modern-day "gurus" of creative thinking.

■ References

1. *Fortune,* March 22, 1993, p. 62.

2. *IBM Think*, Issue 1, 1992.

3. Richard Whiteley's book entitled, *The Customer-Driven Company; Moving from Talk to Action.*

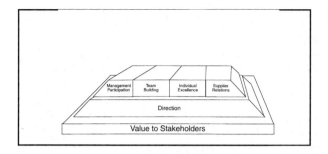

9

Supplier Relations: Developing a Supply Management Process

Charles Cheshire
Director of Quality, Dell Computer Corporation

and

Dr. H. James Harrington
Principal, Ernst & Young

Good suppliers can make even the poor-performing customers look good, but everyone looks bad if the suppliers are bad.   DR. H. JAMES HARRINGTON

■ Introduction

The incredible complexity of today's products, the drive to be first in the market, just-in-time manufacturing, and an increasingly quality-conscious global market place are causing all manufacturers to reevaluate the way they do business.

One clear trend has emerged—the top management of the leading-edge manufacturers have added Supply Management to their strategic initiatives. They have channeled major resources into their procurement organization with the mission to

313

manage and develop a supply base which delivers a competitive advantage in Availability, Quality, Delivery, and Total Cost improvements.

These leading-edge manufacturers' strategy is to get the materials and information moving faster, better and cheaper. These top managers are truly committed to achieving excellence; they realize that it cannot be achieved overnight nor can it be *bought*. Excellence must be developed from *within*.

This chapter discusses the "best practices" used by the leading practitioners of Supply Management. It defines strategies, tactics, methods, organization, and key performance measures which have proven to be effective. The key do's and don't's are based upon the success and failures of actual companies' efforts. This document is geared to the implementation of a Supply Management Process. The road map provided allows an organization to "jump start" and enhance some of the existing Supply Management efforts.

At the heart of this proposed supply management process is an attitude of teamwork and proactive prevention by an experienced cross-functional Commodity Team. The major areas of responsibilities, qualifications, and activities required have been clearly defined. The team performs their responsibilities by *linking* the organization's business objectives of Quality, Cycle-time, and Total Cost initiatives within the supply base daily activities. The linking mechanism is a partnering relationship attitude and a "creative purchase agreement" which defines the targets for improvement, the plan to achieve the targets, the supply/customer team, and the performance measure and review process.

■ Approach

The Executive Improvement Team should commission a task team to develop and implement a Supplier Management Plan in keeping with the three-year improvement plan. Typically this team would consist of people from Purchasing, Operations, Product Engineering, Manufacturing Engineering, and Quality Assurance.

■ Current State Assessment

The organization needs to do an unbiased review of the present Supply Management Process (SMP). This assessment will provide a picture of the way the SMP is working today. It is usually better if the assessment is performed by people who are not part of the present process, so that an objective picture is provided. It should probe into all sources that supply the organization with any item. To accomplish this task, the organization's budget is a good point for identifying all functions that purchase any item. A list of the functions that pay for services, materials, taxes, parts, assemblies, etc., should be prepared. Then a list of the items that will become part of the SMP can be defined. Very few purchased items should be excluded. Typical examples that may not be included: taxes, electric bills, national gas bills, etc. Once the items and the functions that will be included in the SMP are defined, the assessment team needs to understand how the different supply processes are performed. For example: Do the development lab, the product engineering function, and production control use the same process for procuring parts? If not, how do they differ?

The different processes should be evaluated for items like:

- How are suppliers selected?
- How are suppliers measured?
- Are good suppliers rewarded?
- How are suppliers involved in the design process?
- How much of the purchase budget goes to each major supplier?
- How well is the system documented?
- Is the Supply Management Process in keeping with ISO 9000?
- How good is the performance feedback process to the supplier?
- How good is the supplier history file?
- Are poor performing suppliers dropped?
- Who gets supplier interface training?
- What percentage of the items go through receiving inspection?
- How good is supplied equipment maintained?
- When and how were the suppliers certified, and how often are they recertified?
- Does the organization report cost to stock?
- How many suppliers are there per item?

This assessment should provide the management team with a view of today's process, its problems, and recommendations on how it should be improved.

■ Material Goals and Strategies

The results of a current state assessment of how supplier management is currently being performed within the organization provides a starting point for defining a set of material goals, objectives, and strategies. These goals, objectives and strategies must be in alignment and supportive of the organization's business strategic plans, visions, and improvement plans.

Typical material goals are:

1. Develop the most competitive world-wide supply base in the organization's market.
2. Perform material functions in the most effective and efficient manner.
3. Provide the best service to material customers, both internal and external.

4. Make the most effective use of the organization's material assets and working capital.

Typical strategies to accomplish these goals are:

- Manage the supply base as a valued resource and measure its quality, cost, delivery, technology, responsiveness, and business health performance levels.
- Develop a vertical integration/procurement strategy based on the organization's core competencies, available best-in-class suppliers, and strategic marketing relationships.
- Enhance the viability of end items by reducing variability in parts, suppliers, and processes.
- Institute an effective, efficient and adaptable material acquisition and management approach with emphasis on management of the total material cost, cycle-time, Quality and RISK.
- Achieve supplier and material involvement early in the design process and also in the proposal effort.
- Integrate commercial best practices.
- Develop and implement appropriate material systems.

■ Definition and Scope of Supply Management

The first and the key material strategies is "Manage the supply base as a valued resource and measure it to performance levels based on quality, cost, delivery, technology, responsiveness, and business health."

Although there are excellent examples of various elements of a supply management program within many organizations, seldom are they brought together into an effective total process that has the buy in by all functional groups.

To get this process started the organization needs to:

- Analyze the areas where the dollars are spent (purchase cost, inventory, supplier support, etc.) and identify both short- and long-term cost reductions.
- Design and model a Supply Management Process that can be integrated into the existing and future organization's activities.
- Benchmark "best" practices in purchasing.
- Educate and give example of implementation strategies, techniques, and pitfalls to avoid.
- Gain consensus and commit to a uniform approach to Supply Management by everyone at the organization.
- Outline the roles and responsibilities of a Commodity Team.

■ What Is Supply Management?

This section discusses the Benchmark studies and summarizes the approaches being applied to supplier management. It includes the underlying principles of operation, some of the key strategies for supplier management and a list of the important tools and techniques that must be incorporated if the program is to be successful.

History

Since the late 1960s, there has been a transition from the old concept of purchasing as a clerical appendage of management to a newer concept of material management that embraces inventory control, material logistics, distribution, and purchasing. Since the 1980s, this role has been greatly enhanced and labeled supply management. Companies that have organized for supply management include Motorola, Hewlett-Packard, GE, Xerox, IBM, Selectron, GM, Ford, Raytheon, and Rockwell. Most of these companies have been working on developing a Supply Management process and an integrated procurement system that uniquely fits their culture and business needs since the early 1980s. Most of the companies with advanced Supply Management Programs consider them as a strategic advantage and will disclose only the basic details. They closely guard their advanced tools and systems.

A key point is that it takes a significant amount of resources focused both internally and externally over a long period of time to achieve world-class status. All companies increased the Engineering resources in procurement, but total additional headcount decreased.

Benchmark Overview

Common elements found in the leading companies supply management program are:

- A totally new concept in management that involves Purchasing, Engineering, Supplier Quality Assurance, and the Supplier working together as one team early on, co-located, to foster mutually set goals.

- A long-term, win-win partnering for mutual growth and profits.

- A process of concrete, on-site, and frequent help to each other focused on New Product Introductions, quality, cycle-time, cost reductions, and co-training/learning sessions.

- The supplier is an internal partner including their chain of suppliers, i.e., early supplier involvement.

- All benchmarked companies state that Supply Management is a *Strategic Business Decision*.

- Trends are to centralized price negotiations and decentralized buying.

- Some manage production items and other commodities, while others also manage transportation.

- Single sourcing is OK, but most have two suppliers for capacity and risk reasons.

- Early supplier involvement is practiced carefully.
- Many invest capital into their supplier base.
- Suppliers are given firm, fixed orders covering a month's requirements: never canceled; with a 12-month rolling forecast.
- All have a credible Supplier Measurement System.
- All have "stretched" goals for improving quality, reducing cycle time, and an annual cost reduction target.
- Most have an advanced integrated procurement system.
- All have raised the profession levels of their purchasing staff-engineers, many with advanced business degrees.
- All have reduced their supplier base by as much as 50 percent.
- Most were strongly encouraged by their key customers or competitors. Few started supply management because it was "the thing to do."
- Most are developing partnership relationships.

Companies that are successful in Supply Management are clear that Supply Management is *not*:

Another fad that will fade away, if ignored

A more sophisticated tool to get suppliers to tow the line

Golden handcuffs for the supplier

Smoke and Mirror dance by top management based upon the same business-as-usual practices

A take or else position

In closing, all benchmark companies came to the following conclusions. Supply Management is not a technology issue—*it is a people and communication issue*—a mechanism for blending and coordinating the functional areas within the company and the supplier.

Perhaps the biggest problem in implementing Supply Management, once you obtain agreement to implement it, is that engineers in design, manufacturing, quality, and procurement do not speak the same language.

This is the main reason for developing Commodity or Supplier teams made up of members from each function. It forces communication and teamwork in order to get results. It is best to co-locate the teams next to their internal customers and each other.

■ Simple Classifications, Strategies, Tactics, Tools, and Techniques to Get Started

To get your Supply Management Process started, the organization needs to understand simple, but powerful implementation strategies and methods used to classify the entire supply base into manageable pieces. Most companies use more than one method or implementation strategy due to the diversity of needs and suppliers. The classification scheme is highly dependent on the procurement database system.

Excellent companies have an on-line database which has all information on past, present, and future requirements by part number, commodity, supplier, product, subassembly and user. Contracts are on-line and updated automatically on a quarterly basis. Quality records are on-line showing Incoming Inspection results, production fall out, field warranty, corrective action status, ECDs, and process change notices.

This allows the procurement function to analyze the total cost of doing business. If this database does not exist, early priority should be placed on developing it. Whatever classification scheme used most companies will have a subset of key suppliers in each classification. The following are the most commonly used methodologies.

Classification Schemes

1. Types of Relationships

There are four basic types of relationships, based upon the participation in designing, that have been identified by leading companies in defined their partnering with their suppliers.

- Design-build: Full design responsibility is placed with the supplier who will produce the part/assembly/system.
- Codesign and build: Buyer and supplier cooperatively develop design in joint responsibility and the supplier produces them.
- Build only: Buyer does the design and the supplier does not participate.
- Commodity: Catalog material which the supplier builds to their specifications.

2. Stage Classification

Stage 1: Confrontation with suppliers. This stage is one that is uncomfortable for both organizations. There is a lack of commitment on both sides. It is a stage that should be moved out of as soon as possible. It is characterized by long lead time and poor and yo-yo forecast.

Stage 2: Arms-length relationship. In this stage adversarial attitudes gradually give way to a cautious, tentative assessment of a working relationship. It is characterized by long production runs and long set-up times.

Stage 3: Congruence in mutual goals. This stage is the beginning of actual teamwork. The "We" attitude is prevalent. It is characterized by pull system, commitment to certification and small lot size.

Stage 4: Full-blown, open relationship. There is an open, frank relationship between all ranks of customer and supplier. Supplier becomes certified for direct shipment to stock. Partnering attitudes are developed. It is characterized by indirect labor productivity and near-instant customer delivery.

Stage 5: Strategic Alliance. Relationships that develop to this stage have a synergistic effect and a unique competitive advantage to both customer and supplier. Often formal partnership relationships are documented and signed. It is characterized by:

- Capital investments by the customer in the suppliers to improve their present performance and in support of future customer needs.
- Joint R&D, manufacturing, facilities, marketing/sales.
- Shared business plans.

3. Pareto Analysis

This Pareto Analysis is a quick and easy way to provide a road map for immediate action. It is based on three "critical" parameters: quality, logistics (cost/distance) and item dollar value (ABC). Therefore, concentrate on the "A" and "B" items, or high value ("A" items) versus low value ("B" items). Out source all "C" items.

Step 1: Quality Analysis. Benchmarks need to be made on the current quality performance of the suppliers. Generally this is the supplier who is certified (approved) versus non-certified (subject to approval). If the supplier is certified, the quality is considered high for the purpose of this analysis. There are many different types of certification plans. For example, 12 months with zero reject at incoming inspection, zero errors in paperwork, zero late-early delivery is high quality. The benchmark objective is to get products delivered to the line on-time with zero inspection.

		Quality	
		High	**Low**
Item Value	**High**	Move to JIT Check distance	Supplier quality program or new supplier
	Low	No inspection	Change supplier

All suppliers for each commodity purchased are evaluated using the matrix; high value (A items) are evaluated first. Suppliers of the A items are placed into the high quality or low quality box. The same is done for the low value items. Note: you can include Quantity × Item Value = Total $

Step 2: Logistic Cost/Distance Analysis. Given the completion of the quality analysis, the logistics (cost/distance) analysis is done. Both distance and cost are critical to Cycle Time (JIT). A guideline is developing sources as close to the using facility as possible. This has an impact on your international sourcing policy. Again, the same items above are evaluated and placed into the following 2×2 matrix.

Logistics Cost/Distance

	High (far)	Low (near)
High	Develop new supplier	Move to JIT High priority
Low	Focus on minimum deliveries	Key issue is order cost

Item Value

In this scheme, you need to establish a break point to determine high (FAR) distance from the low (NEAR) distance. This is a function of individual plant preference and conditions. A guideline might be any supplier that is within a 4 to 8 hour delivery window is considered low distance; all others are considered high.

Step 3: Commodity/Supplier Groupings. With the analysis of all suppliers completed, actions can be determined for the eight categories of suppliers. For example: High Quality—High Value—High Distance (H, H, H) Category

Tactical Actions: 3 possible scenarios
1. Get suppliers to move closer
2. Establish local warehousing
3. Other ways to reduce transportation and time factor (ask supplier for ideas)

Key: For JIT something needs to be done about the distance.

High Quality—Low Value—High Distance (H, L, H) Category

Tactical Plan:
1. Explore less frequent delivery
2. Product moves directly to line without inspection
3. Minimize handling / packing expenses

Step 4: Take Action. The eight preceding categories suggest a priority of action. The material team can quickly set a Supplier Program in place. Both high and low value items with high quality and low transportation cost can be delivered frequently.

This method helps the material team to evaluate and organize the suppliers into specific performance groups which can be effectively communicated to the internal organizations as well as the external suppliers. It reduces the complexity and confusion factors which arise due to the fit emphasis.

This technique can be done manually and then developed into an integrated data base information system. The key is, if not this; then develop an alternative, but do it now.

Strategies and Tactics

There are two main strategies that can be used to identify and develop world-class partners.

The Evolutionary Strategy.   In this approach, the customer and supplier simultaneously evolve a Continuous Measurable Improvement (CMI) effort. It the customer does not have an internal quality system, the customer starts its supplier improvement effort by initializing a pilot Statistical Process Control (SPC) project with one or more small groups of suppliers. The supplier provides the customer with SPC charts certifying that critical or designated product quality characteristics are monitored and controlled. Over time, more suppliers are asked and induced to adopt statistically-driven, company-wide prevention techniques and to improve at a faster rate each year or at the same rate as the customer.

The Step Progression Strategy.   This process can be characterized as a series of successively higher steps that a supplier must climb. At the highest level, a preferred supplier has an extensive continuous measurable improvement process, and has become a single source or one of several partners.

Suppliers are awarded a score based on the results of an audit. The score determines whether the supplier is a candidate, approved, or preferred supplier. If the specified level of performance has not been attained by a specific date, the supplier is often dropped. Each level indicates where the supplier's products will go, whether into inspection or dock to stock. At the highest level of certification, a supplier is considered a partner.

At the beginning of the partnering courtship, a customer may train or jointly train suppliers in Design of Experiment, Computer Integrated Manufacturing, Just in Time, Design for Manufacturability, and other higher-level tools.

Do Not Reinvent

The certification process is an excellent Progression Strategy. This process should be fully implemented into the supply base. The issue of effectiveness of the Certification is interesting. Again, there seems to be a lack of commonly agreed upon criteria for certifying suppliers in the United States. To resolve this issue we recommend the use of the American Society of Quality Control (ASQC) Standard.

The ASQC Customer-Supplier Committee has identified eight minimum certification criteria. They are:

1. No product related rejections over a period of time, usually twelve months.

2. No nonproduct related rejection for a period of time.

3. No production related negative incidents for a period of time, usually six months.

4. Successful passing of an on-site quality system evaluation (ISO 9000).

5. An agreed upon specification.

6. Documented process and quality system.

7. Ability to furnish timely copies of certificates of analysis, inspection, and test results.

8. Correlation and validation of laboratory results so that the customer can use supplier's results as it would its own certification of bulk suppliers.

The Ramp Up Strategy

This can be compared to a process of survival of the fittest. In this process, the supplier is expected to achieve a higher level of performance very quickly. If the supplier has a continuous improvement process, then it is easier for the supplier to achieve higher standards of quality quickly. Many companies that covet the Malcolm Baldrige National Quality Award are asking suppliers to apply for the award and be examined by the Award auditors. The supplier that ramps up to exceed the customer demand gets the total business and therefore becomes a partner.

No matter what strategy or tactical plan a customer might use in developing a supply base, the one major requirement is the ability to do a professional assessment of the supplier and your own internal organization.

Tools and Techniques

Assessment Techniques—There are a number of ways to assess and monitor suppliers. The choice of method is determined by the level of assurance the organization wants and the importance of the product. The higher the assurance, the higher the *cost and time* required to reach that level of assurance. Common assessment methods are:

- *Certifications:* Its major advantage is that a supplier must progressively pass more difficult tests to become a single source partner.

- *Surveys:* A survey is usually a short self-assessment that a candidate or existing supplier may be asked to conduct before the customer's audit team conducts and on-site survey.

- *Mechanical-chemical or similar tests:* A mechanical test, such as a tensile test, evaluates the strength of a product. Other tests may evaluate chemical, physical, or dimensional properties.

- *Reliability test:* This measures long-term product quality. These tests duplicate conditions in which the product is used.

- *First article inspection:* The first article from a process is examined to determine whether the product can be made to conform to all critical specifications.

- *Incoming material inspection:* Incoming statistical sampling and inspection at a certain quality level can verify the supplier's assertion of quality.

- *Failure mode and effects analysis (FMEA):* The major modes of failure of a product are identified and these are then eliminated or the product is made more robust at the points of possible failure.

- *Design of experiments (DOE):* This is a statistical technique to identify the major factors in a process that may cause variation.

- *Statistical process control:* SPC charts indicate whether a process is in control and capable of meeting specifications.

- *Capability studies:* A capability study indicates whether a product can be consistently produced by a manufacturing process.

- *Poor-Quality cost:* These indicate how costs are allocated in terms of internal errors, external errors, prevention, and appraisal.

- *Audits:* These can be used to evaluate and monitor supplier performance. An important use of an audit is to verify whether internal quality controls exist and work. An audit can utilize interviews, sampling data, and other tests to verify internal controls.

Assessment to Build the Partnership—Partnering is built on a proven, long-term, win-win relationship. To establish this, organizations must resolve *to trust but verify*. The verification may be a management audit, product performance check, inspection of a critical dimension, or a customer service evaluation. The ISO-9000 procedures covered in Chap. 5 of this book will reduce the cost of doing these assessments.

Both parties should view the Assessment process as the means to maintain and expand market share in existing products and to create new products for new markets by Continuous Process Improvements.

A periodic assessment also ensures that company wide, continuous improvements are being pursued. Continuous Improvement does not happen naturally; it requires a sense of market urgency.

All too often, managers dismiss quality partnering with a refrain of, "I have seen this before: It too, shall pass." Unfortunately, in the survival-conscious 1990s, such an attitude can only lead to complete loss of business. This process can be characterized as a series of successively higher steps that a supplier must climb. At the highest level, a preferred supplier has an extensive continuous improvement process, and has become a single source or one of several partners.

■ Supply Management Process

The purpose of this section is to model a generic Supply Management process and identify the actions that must be taken by the organization personnel, both within material and within other affected organizations.

Each major organization has its own definition of the Supply Management Process (SMP). We like to define the Supply Management Process as a business process that aligns the organization's business objectives within the supply base. The major focus is on customer satisfaction through continuous measurable improvement. Performance improvement is the number one priority.

Before embarking on the improvement journey with suppliers, the Purchasing organization should ask:

- Where are we going?
- What is the road map that will get us there?
- What the benchmarks of the journey?
- How are we going to get there?

- Who's in charge?
- What resources are needed to get there?
- How long will it take to get there?

These questions are fundamental to any business endeavor since they deal with strategy, objectives, plans, accountabilities, resources, costs, timeliness, and benchmarks. If these questions are not asked and answered, the commodity team will flounder. The supplier wanting to join in the improvement journey will receive mixed messages which may result in unsatisfactory product or service quality.

The organization's material management function must assure a coordinated and uniform approach is followed by the organization. There is, however, no set of right answers to these questions. Answers will vary according to industry, company culture, product line, supply base, technical capabilities, and human resources. Nonetheless, all continuous improvement efforts typically involved these steps.

■ Generic Supply Management Model—Ten Steps

A Supply Management Process can be developed using the following ten steps.

Step 1—Establish a team.

Step 2—Develop an action plan. A team should start with one supplier or a selected group of suppliers in a pilot effort.

Step 3—Develop specifications and standards. Continuous improvement can be achieved only if the customer sets and communicates explicit standards and measures for quality delivery, service, and cost.

Step 4—Prioritize product attributes. A product can have many quality attributes. Important product quality characteristics should be identified on the engineering print. Non-conforming critical characteristics can jeopardized health, safety, or welfare. Non-conforming major characteristics affect product function. A minor product non-conformance, a blemish, deals with product appearance.

Step 5—Determine process control and capability. Once a commonly understood measurement system has been devised and product service characteristics have been prioritized, the supplier is asked to submit a process flow diagram and specify the most suitable locations for tracking quality and testing products. The supplier establishes effective statistical process controls to the designated product quality characteristics.

Step 6—Measuring performance. The measurement system indicates how quickly improvements being pursued. It is a key measure of the commodity teams effectiveness.

Step 7—Improve continuously. Continuous improvement means that performance or specification targets have been set and over time, variation around these targets is gradually reduced.

Step 8—Take ownership. Once the customer has initiated the improvement effort, the supplier is encouraged to take ownership of the improvement effort and be responsible for it. *Only* a few suppliers will have the commitment and stamina for the long haul.

Step 9—Audit performance.

Step 10—Continuous improvement. Start at Step 1 and enhance the supply management process by using more sophisticated tools in each step.

■ The Commodity Team

Why the Word "Commodity"?

Webster's Definition:

Commodity: 1. Any useful thing 2. Anything bought and sold; any article of commerce

As we define the major areas of responsibility and tasks performed it will become apparent that the team's scope is broad in nature yet specific in terms of the commodity and expected benefits.

The Vision for a Commodity team might be:

Research the commodity so thoroughly and examine it from all angles then presenting the findings to our internal customers in a way that leads to the publication of a commodity plan and agreement to implement it.

The Commodity Team's focus is on the following factors for Total Improvement Management. They apply the General Rule equally to their internal organizations and their suppliers.

20% of profits
25% of assets
25% of people } as attributed to handling or coping with defective components/products and poorly defined business processes. (Source: H.P.)
40% of space
70% of inventory poor quality costs

Set in this context, a commodity concept and its proper implementation is a strategic business decision. It necessitates careful planning, proper resources, and time.

Therefore, one of the most important tasks of top management is the selection and nurturing commodity teams. These teams work in a high-risk environment, both internally and externally. Management must break any "barriers" that prevent the commodity teams from doing their job. The ability of top management to pay attention and address the "soft" or behavioral changes necessary from the management and white-collar support is another critical task to be done.

Major Areas of Responsibility

The Commodity Team is responsible for:

- Systematically obtain information throughout the organization, supply base, industry, and competition. This effort forms the basis for "management by fact."

- Develop and maintain a world-wide perspective.

- Identify short-term (<1 year) and long-term (3 years) projects which reduce total cost of ownership.

- Develop specific supplier quality improvements plans (i.e., suppliers, products, part numbers) which reduce total cost and lead to supplier *certification* and standardization.

- Develop manual or automated measurement systems to monitor supply base performance of price / industry trends, products, processes, and services.

- Provide written status reports to top management, internal customers, and suppliers which detail issues, benefits, and next actions.

- Integrate the supplier improvement plan with appropriate performance measures into corporate, group, or divisional agreements.

- Upgrade tools/techniques consistently by education / training of all personnel involved. Become a "Learning Organization."

A commodity team is in a sense a business team with the responsibility to achieve maximum return on investments, create a competitive supply base, and exceed customer expectations. It is a challenging and professional career demanding the best in engineering and business skills.

■ Application of the Supply Management Process (SMP) Leading to Certification

The Commodity Team must have flexibility in application of the SMP due to the complexity of the supply base, commodities, and today's business issues. Listed below is a scenario which, based on the author's experiences, is representative of implementing the SMP.

Supplier Surveys[1]

Supplier surveys have been one of the mainstays of supplier quality assurance programs for many years. As a result, a large number of proposed supplier survey check lists are available in various handbooks and standards publications. As would be expected, they are all quite similar, and both suppliers and customers have become quite comfortable with the supplier-survey ritual. The actual surveying practices of individual companies vary as to frequency and level of detail, but all too often surveys have become stiflingly routine.

A typical survey team consists of a customer's buyer, manufacturing engineer, and quality engineer. In some companies, two or more of these roles are performed by one individual, but any survey that does not cover all three disciplines is not adequate. Usually, each of these parties is armed with a check list unique to his or her area of expertise. Many of these check lists are "cookbook" affairs that make it easy for the surveyor to assign numerical ratings to each of dozens of aspects of a supplier's operation. Most modern check lists assign different weights to each questions, and some sort of weighted grand total can be calculated and compared to various threshold values, which correspond to general ratings, such as "acceptable," "conditionally acceptable," and "unacceptable."

If supplier surveys are flawed, it is usually not the fault of the check lists or the ranking schemes, which can often be described as elegant. The most common shortcoming of these surveys is that they often consist of a mass of detailed questioning, followed by a brief plant tour. Most are completed in a few hours, and they rarely last more than a full day. Long lists of questions and inadequate time allowances combine to make most surveys rather superficial. Add to this the fact that most of the check list questions have obvious right answers and the fact that most suppliers have become very adept at telling surveyors exactly what they want to hear, and the bottom line is that traditional supplier surveys can consume enormous amounts of time and accomplish very little.

This process can be made more efficient in a number of ways. First, lists of candidate suppliers should be established through disciplined searches of standard industrial registers, trade association membership lists, financial rating services, and the like, to pare down the survey candidate lists to only those companies that pass predetermined gross criteria, such as size, financial strength, location, range of services, and so forth. (Often organizations that are registered to a ISO-9000 standard need not be surveyed for their quality system.) Then a mail survey can be sent to the most promising candidates. This mail survey should cover the full range of questions asked on the traditional survey check lists, and should also include questions about the supplier's willingness to handle various possible new orders in certain relevant time frames.

These mail surveys dramatically improve the efficiency of the survey process. It will be possible to eliminate some suppliers on the basis of their answers to direct business and technical questions. Other very capable suppliers will clearly indicate that they cannot or do not wish to handle new orders of the type generally described in the questionnaire. Some will also express their disinterest merely by failing to respond to the survey, although each of these should be contacted by phone to be sure that is the intended conclusion.

The actual survey team can be dispatched to study only the most promising candidates. Since the majority of the check list questions will already have been covered in the mail survey, only a minimum of time need be spent on this material to clarify any confusing responses. The bulk of the on-site survey time should be spent instead on in-depth examinations of live evidence of the critical survey elements, such as testing the real strength and breadth of the supplier's in-house technical support staff through one-on-one interviews, examining actual process control charts on the shop floor, and seeing if the manufacturing operators are posting and interpreting the charts correctly. It is also very important to develop a good, albeit subjective, understanding of the supplier's commitment to achieving the desired levels of product quality, and the supplier's willingness to enter into a true business and technical partnership with the prospective customer.

In the case of surveys conducted at existing suppliers, an additional major factor is a key element in the survey process—supplier history. the survey team should be completely aware of the supplier's performance for at least the entire year preceding the survey, including the details of all corrective actions that have taken place if any material has been rejected during that time. If the history looks good, is it due to a low number of receipts, very lax incoming inspection, or a solid control system at the supplier? If, as would be hoped, it is the last, the surveyors may only need to assure themselves that the controls are still in place and show signs of improvement since the last survey. It is very important, though, not to avoid genuine critical examination just because past history looks good. What if the record is due to the influence of the original founder of the business, who is almost ready to retire? Perhaps it had a lot to do with very fine capital equipment, which is now beginning to show signs of wear. Or perhaps the company is beginning to implement a rapid expansion program, and many less experienced employees are being added to the work force.

Companies frequently become lax in keeping supplier surveys up to date. It is very easy to lose sight of the fact that a supplier has not been formally surveyed for a long time when there is frequent business and technical contact on a continuing basis. Since a formal survey is usually a very comprehensive exercise, it normally covers factors that aren't involved in the ongoing business exchanges. A good rule of thumb is to try to resurvey active suppliers at least annually.

Today supplier surveys are being driven by a new set of international standards called ISO 9000. In conjunction with this new standard is a 3rd party registration process that reduces the need of customers serving each supplier. This reduces both customer and supplier costs. As of January 1994, over 20,000 organizations worldwide had been registered to these document prepared by the International Standards Association in Geneva, Switzerland. In a survey conducted by Deloitte and Touche, 80 percent of the organizations contacted stated, "ISO 9000 registration status influences their selection of suppliers." To do business in Europe, ISO 9000 registration is a must. For more information on ISO 9000, see Chap. 5.

Initial Supplier Qualification[1]

Once a supplier has been selected as a possible production source, but before authorization is given to ship large quantities on a regular basis, a series of product qualification criteria must be satisfied. These vary widely, and the complexity of the qualification process depends on the complexity of the product, the newness of the product technology, the criticality of the product in the customer's intended application, and other similar factors.

The cycle begins with the submission of evaluation samples. These are usually few in number, and are subjected to a large number of physical, environmental, functional, and life tests, as well as being carefully measured for conformance to all applicable engineering drawing and specification requirements. The intent at this point is for the customer to develop only a gross confidence that the supplier is capable of producing an acceptable product at all. "Hard tooling," such as molds, dies, and machining fixtures, usually does not yet exist at this early stage. The evaluation samples may even be examples of existing production jobs, somewhat different from the new design, but accurately reflecting the supplier's production capability.

Once the evaluation samples have been approved, authorization is given to the supplier to start producing the "hard tooling" that will be used in making the production parts in high volume. Upon completion of this tooling, the supplier usually makes several short production runs (often as short as one piece, if the part is very complex or expensive), carefully checking the product after each run and making required modifications to the tooling and the process until an acceptable product is manufactured.

At this point, the supplier is ready to submit a "tool sample" for the customer's approval. Many serious problems have occurred in the past as a result of taking this step too lightly or doing it too hastily. Tool samples should be fairly large in quantity, and should be randomly selected from a production run that is truly representative of the anticipated full-volume production process. The degree to which these rather vague requirements are to be met is determined by experience and engineering judgment, and will vary due to factors such as cost, lead time, and the capacity of the process. An absolute minimum of five pieces would be needed before any possible estimate of the variability of the process could be determined, and samples of about twenty are preferable for many processes. The requirement that the tool sample run be representative of the ultimate process is often very hard to achieve, especially if the ultimate process will be run over two or three shifts by several different operators. It may only be economically practical to run the tool sample process for a few hours or days. In any case, the customer needs to be aware of the conditions under which the tool samples were run, and must be informed of significant deviations of those conditions from the ultimate process. The best way for the customer to obtain this information is for the responsible manufacturing engineer to be present at the time the tool samples are run, and to observe the process and randomly select the samples at first hand.

Once the tool sample is selected, it is sent to the customer, where it is subjected to very thorough inspection. The supplier's inspection results should be submitted with the sample parts, so that any inspection correlation problem between supplier and customer can be discovered and resolved at this time. There should be no excuse, other than a serious measurement correlation problem, for the supplier to submit tool samples that the customer finds deviant. But even if the customer finds all parameters to be within specifications, modifications to the tooling may still be required. This would be the case if all measurements of a given parameter were within specification, but uncomfortably close to one of the tolerance limits. Similarly, a change might be requested if the measurements of a particular parameter varied widely, and covered nearly the entire allowable tolerance band. It is also possible that it could be either expensive or risky to make tool modifications that appear desirable from analysis of the data. In this case, the more practical alternative may be for the customer's design engineers to consider a minor design change, widening a tolerance or altering a nominal specification if the function of the finished product would not be compromised. In any case, it should be clear that tool sample approval by the customer is far more than just a simple confirmation that all parameters meet their specifications. It involves detailed analysis of the data to confirm that the supplier is truly ready to begin volume production.

If the customer's product or manufacturing process is complex, the supplier's parts may have to be piloted in the customer's production line before volume shipments may commence. In this case, a substantial quantity (typically several dozen to several hundred) and are built into the customer's product on a very carefully controlled basis, and the finished products carefully observed in the final test stages to establish confidence that no degradation of the product has occurred. Pilot lots are also run through the process to be sure that the parts work well with the customer's assembly tooling, particularly if it is highly automated and potentially sensitive to small variations in component parts.

Some companies apply a final qualification criterion to suppliers after volume production shipments begin. These take the form of granting "qualification" to a supplier on a given part after a predetermined number of shipments have been received error-free.

Supplier Certification

The objectives of Supply Certification are to:

- Attain 100 percent confidence in your supplier's quality assurance, process controls, service performance, and management commitment. Zero inspection at your organization.
- Retain confidence in your certified suppliers' decisions to ship materials to you, providing evidence of process controls and program of periodic audits. (Dock to Stock; "0" defects)

The flowchart on Certification Strategy is jointly agreed upon, changed where appropriate, and managed by the supplier and the Commodity Team.

The certification process is very detailed and requires resources from suppliers and your organization. The focus is to work with key suppliers because of the cost and time involved. Most organizations certify six to seven suppliers per year. However, over the long term it will ensure a growing pool of highly competitive suppliers.

Task 1: Management sets organization, responsibilities and staffing resources.

Task 2: Commodity Team—Action Plan
- Review business volumes (past, present, and future).
- Reviews policies, procedures, system and practices.
- Performs business analysis using Activity Based Costing (ABC).

Task 3: Identify areas for improvement (supplier—commodity—items).

Task 4: Send off assessment/analysis report to:

Task 5: Send analysis/improvement plan to suppliers.

Task 6: Invites supplier(s) in for "expectation settings" work session where appropriate.

Task 7: Joint on-site audits (customer and supplier site).
- Use certification documents or standard procedures of either company.

Task 8: Jointly summarize audit findings.
- Develop presentation for both companies
- Recommend actions—certification
- Show dollar savings for both companies

Task 9: Document improvement plan.
- Incorporate into Corporate Purchase Agreements
- Letter of intent
- Other i.e.: Continuous Measurable Improvement Plan—Yearly
- Gain legal and management approval from both companies

Task 10: Begin joint improvement effort.
- Customer/supplier team, i.e., dual assignment:
  Detailed action plan
  Performance measures
  Specific owners for each task
  Tracking system
  Review system

Task 11: Update Commodity plan.

Task 12: Update supplier file.

Task 13: Supplier recognition.

Task 14: Develop new Commodity plan.

■ Commodity Team—Yearly Activities/ Responsibilities

Each supplier relationship requires a customized version of this suggested meeting format. These meetings are not to be used as negotiating meetings; they are working sessions.

Meeting				
Internal Preparation Meeting	Kickoff Meeting	Monthly Team Meetings	Quarterly Semiannual Management Reviews	Annual Management Review

Participants				
Customer & Supplier Team*	Customer & Supplier Team Exec. Partners (Suppliers/ Customers)	Purchasing Technical Quality Others	Purchasing Technical Quality Others (Suppliers/ Customers) Exec. Partner	Purchasing Technical Quality Exec.Partners (Suppliers/ Customers

| Internal Customers Meeting Meeting Purpose: Objectives Issues Participant Responsibilities | Introduce Program Obtain Mutual Agreement and Commitment Identify Teams Into/Suggest Exec.Partners Present and Discuss Customer, Supplier, Proposed, and Business Objectives | Establish/Update Mutual Key Results, Goals, Objectives, and Action Plans Discuss Issues Review/Discuss On-time Delivery, Quality, and Cycle-time, Corrective Action Plan, Certification and Progress | Major Issues Performance Review Objectives Expectations Actual Performance of Both Parties Technology Trends Business Trends Certification Program Effectiveness Modifications | At Supplier Location Tour Maintain Key Contacts Major Performance Review Set Next Year's Targets: Quality Cost Reduction Cycle-time Delivery Percent Share of Business Estimate |

*Team includes: Purchasing, quality, engineering, and others when needed.

■ Guidelines and Models for Implementation

Step 1—Executive directive. Obtain top operating officer approval and personal involvement for group-wide implementation. Select an individual (Champion) responsible for getting the Supply Management Process (SMP) accepted throughout the organization.

Step 2—Organize for success. Organize a Supply Management program to *fit* the organization's structure and culture. Following are three examples of successful models in use today:

Model 1: Supplier Manager (The Champion). Management appoints a Staff Manager to a full-time position for a minimum of 1 to 2 years.

Model 2: Supplier Council Team (Interdependencies). Management establishes a Supplier Council Team within the organization. The members own the resources required to support the SMP. The functions are the same for Model 1, but are assigned to several individuals.

Model 3: Supply Management Organization. Management creates a line organization, consisting of supply managers and staff that manages the selected supplier/products and supply management process described in Model 1. This organization receives support from the operational groups as required.

Step 3—Create commodity teams. A commodity team should be formed to evaluate and categorize potential key suppliers based on such factors as risk, criticality, importance to mission, and benefits.

Step 4—Gain management approval. The commodity team should get management approval for the following actions:

- Selection of key suppliers.
- Approval for proposed levels of involvement.
- Commodity Plan for each key supplier.
- Agenda for a supplier/customer kickoff meeting.
- Plans for managing the ongoing activities.
- Select an internal Executive Partner with critical suppliers.

Step 5—Select and notify suppliers. The commodity team should send letters to the organizations that were selected to be key suppliers, asking them to join in piloting a participative SMP for total improvement.

Step 6—Implementation of the pilot. After the suppliers understand the Supply Management Process and agree to participate in the pilot process, the commodity team should prepare an implementation plan and schedule in conjunction with each supplier. This will result in individual implementation plans for each of the pilot suppliers.

Step 7—Keep in tune with industry. As a group, the suppliers have a voice to be heard. The impact of the various Continuous Measurable Improvement (CMI)/Supply Management initiatives are very costly. The suppliers are very frustrated with their current situation. Some are selecting one or two major customers to work with while at best listening to the other customers.

The bottom line is that many suppliers are looking for other alternatives, just like the primes: also the suppliers are going to select a few partners to work with, just like the primes, so the future holds some interesting changes in the way business will be conducted for primes and subs.

Step 8—Reevaluate the overall pilot process. After the pilot process is complete, a review of the pilot process should be performed with the supplier(s), internal customers, and the commodity team.

Step 9—Expand SMP to other commodities. Using the improved SMP, the program is now expanded to all other appropriate commodities. At this point in the process, management should review the contributions made by the suppliers, internal customers, and commodity team, and reward them appropriately. It is also the time to develop and/or identify role models in all three areas.

■ Pitfalls to Avoid During Implementation

An examination of the need to truly recognize and understand what the company expect from SMP must be clearly stated and measured by the CEO.

SMP seems to be the "in" thing to do in the nineties. It had become such a universally accepted position that it is being broadly implemented without the benefit of full analysis. In fact, it had developed such a following that jumping on the bandwagon is not only the correct thing to do, but dangerous for anyone to suggest otherwise. The question to be explored—if the concept is judged excellent, if it is the correct approach, why have so many companies failed in their implementation? Perhaps even worse, why has the expense of SMP overshadowed the perceived ROI?

A company must recognize that the basic reason for Supply Management is that it is a *business strategy*. As such, it must be integrated into the company's overall business plan. There must be sound financial decisions made on SMP. There are up-front costs associated with it so there must be specific paybacks, gains, yields, etc.

How to Avoid the Pitfalls

Action 1: A company must approach SMP like any other business decision and not get enamored with the concept.

Action 2: Top management and staff must accept full responsibility for success or failure.

Action 3: Do not assume you can directly lift or apply someone else's program to your culture. It must be tailored to your needs.

Action 4: The work force—Engineering especially—must accept it as the next logical step.

Action 5: In the SMP plan, middle managers must know and have full reason to believe that their success in SMP implementation weighs heavily in their overall performance evaluation.

Action 6: Middle management must have the resources to perform.

Action 7: Establish an improved measurement system which is basic to SMP and the business tactical plan. Each measurement must be clear, simple, and understandable to all.

Action 8: Pilot program must be successful and a "role" model for others.

Action 9: Provide everyone with program awareness. Make training very specific. Identify and train on key SMP tools.

Action 10: Create the mechanism for incorporating new SMP ideas from all internal/external customers.

Action 11: Assign full-time SMP teams.

Action 12: Use Education/Benchmark to expand your frame of reference, fill in areas of weakness.

Action 13: Clearly define responsibilities and ownership for SMP.

Action 14: If we are to modify the company culture, then it must be part of each function's response to SMP.

Action 15: Set mutual Internal goals relating SMP to Engineering, QA, Mat'l, Purchasing, and Management functions. Set mutual goals with selected suppliers. There must be something in it for them.

Action 16: A company will never complete the effort because evaluation must always be against a continuously improved competitive business plan.

■ Supply Management—A New Competitive Advantage—Yes and Yes

The advanced companies are actively marketing and selling their Supply Management process as a competitive advantage to their customers and suppliers. The objective is clear: *"It is best to get the best first; leave the rest for our competitors."*

The Marketing/Selling process is based upon Partnering as a Team.

- Total business viewpoint
- We will shift the business paradigms
- We both will benefit
- We both will do business with the best
- Single sourcing is OK with only the 'vital' few
- Jointly we will develop and improve our business processes and products
- We will measure our success together and share the gains

■ Summary

The debate rages on. "Do you focus your early effort on improving your suppliers as they are the start of the supplier cycle." Or, "Do you improve your internal operations before you ask other organizations to do what you cannot." We see that there is only one correct answer to this question. Clean up your own house before you start throwing stones at others. The U.S. auto industry is a good example of how not to do it. Early in their improvement process they place requirements upon their suppliers that they could not meet internally. Although this approach got results it did not cement good relations between the customer and the suppliers. We have heard many auto part suppliers say words to this effect, *"They (Big 3) want us to make up for their mismanagement. Why don't they set the example by doing it first and showing us that it will work?"*

Set more stringent standards for yourself than for your suppliers. Don't expect your suppliers to be certified to ISO 9000 if you cannot pass the certification requirements.

Your supply management process must build a level of trust, understanding and cooperation between both parties. If something goes wrong always assume it is your fault unless you can prove otherwise. Don't put off notifying your supplier that you are having a problem, but let the supplier know that you are not sure that it isn't the result of something your organization is doing wrong.

The following shows the difference in thought patterns different types of organizations have related to supply management.

Who manages the supplier interface?

Losers: Everyone is empowered to discuss anything with the supplier.

Survivors: All contacts go through purchasing.

Winners: Employees that need to work with suppliers are trained in what commitments and information they can provide to the supplier. The materials group has a person assigned to each supplier that is notified either before or after the contact is made.

Attitude about their suppliers.

Losers: We pay them with good money—we want nothing but good parts from them—not excuses.

Survivors: If we hold the suppliers hand we can get them to provide us with good products most of the time.

Winners: The suppliers know the most about their products so they need to be part of our design process so that we can take advantage of their expertise.

Receiving inspection.

Losers: Receiving inspection keeps bad products out. It is our defense mechanism that protects us from our suppliers.

Survivors. It is expensive but necessary. We need to certify suppliers to reduce receiving inspection costs.

Winners: It is used for preproduction products to help us quantify our products/parts process. Something is wrong if we need to look at more than five lots. We still look at the first lot after a major engineering change and run a pilot sample through our manufacturing process to be sure that the integrity of our products/part process is maintained.

Supplier second-sources.

Losers: A supplier second-source is mandatory, four is even better. The more organizations competing the better price we can get.

Survivors: We have a few good suppliers that do not have unions that we use as sole sources, but on a whole we like to have a back up supplier in case something goes wrong.

Winners: We work with a few superior suppliers that really understand us and provide us with outstanding products and services. We will assign anything that they can provide to them even if it costs a little more. We invest in developing the best suppliers and leave the others to our competition.

Product development relationships with suppliers.

Losers: The organization designs the product and selects suppliers that can provide product to our specifications.

Survivors: Suppliers have many good ideas. We need to make them part of our product design team.

Winners: We invest in the supplier development process so that they can direct their efforts to be in line with our goals and projected future needs. Our suppliers play an active role in our product development and design teams.

■ References

1. *The Improvement Process*, published by McGraw-Hill.

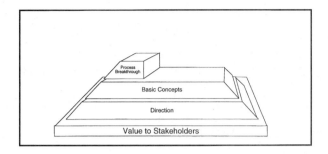

10

Process Breakthrough: Jump-Starting Your Process

You cannot win in today's marketplace using yesterday's processes.   DR. H. JAMES HARRINGTON

■ Introduction

Upper management provides the vision and direction, teams correct the problems, and individuals provide the creativity, but it is the processes within any organization that get things done. No matter how good your management and/or your employees are, your organization cannot be successful if it is using the same business processes it used in the 1980s. Process Breakthrough methodology provides a road map that will help your organization improve your critical processes as much as 1500 percent. This is accomplished by streamlining your critical business processes, using tools like process redesign, new process design, and benchmarking.

The question in everyone's mind today is: Should the organization concentrate on continuous improvement or on breakthrough methodologies (example: Process Redesign, Business Process Improvement, Process Reengineering, etc.) to be more competitive. The answer is that you must do both to survive.

Department improvement teams, natural work teams, task teams, self-managed work teams, statistical process control, quality function deployment, suggestion systems, etc., all have focused upon continuous improvement, and we do need continuous improvement. For example, one department improvement team in the IRS came up with an idea that nets the U.S. Government over $7 billion per year. But some parts of our business need to be jump-started. Many of the processes that we are using to manage the organization need to have their associated costs and cycle time cut by 50% within the next twelve months while improving the quality of their output. *The Process Breakthough methodology* (BPI) combines approaches like Benchmarking, Process Reengineering, Focused Improvement, New Process Design, Process Innovation, Activity Based Costing, and Big Picture Analysis, into one logical way of initiating drastic, rapid change in a single business process.

How to Improve Your Business Processes

The complexity of our business environment and the many organizations involved in the critical business processes make it necessary to develop a very formal approach to business process improvement. This methodology is conveniently divided into five subprocesses called phases that consist of 27 different activities.

Phase I	Organizing for Improvement	(7 activities)
Phase II	Understanding the Process	(6 activities)
Phase III	Streamlining the Process	(6 activities)
Phase IV	Implementation, Measurements, and Controls	(5 activities)
Phase V	Continuous Improvement	(3 activities)

Phase I—Organizing for Improvement

During Phase I, the upper management is trained on the BPI methodology, selects the critical processes and assigns someone (process owner) to each selected process. The process owner will be held responsible for improving the selected process' total performance, even though parts of the process include activities performed in more than one function. The process owner organizes a Process Improvement Team (PIT) that sets boundaries, establishes total process measurements, identifies process improvement objectives, and develops a project plan.

Activity 1—Defining Critical Business Processes

There are thousands of business processes going on all the time within most organizations. Management needs to select a few key processes for the organization to focus on. One to three critical processes are normally selected to start the improvement process.

To define the critical business processes the executive team should look at each macro process and define three to ten major processes included in each macro process. Once a list of 30 or more major processes are identified, the EIT needs to establish a prioritization matrix based upon the organization's business plan, competitive positioning, its core capabilities and competencies, and the key process issues like changeability, opportunity for improvement, impact on the external customer, etc. The executive team should then develop a prioritization index for each of the major processes, using real data if possible. First reduce the list to 8 to 15 critical processes that have the highest priority and then select one to three of these processes that will be started first.

Activity 2—Selecting Process Owners

It is the lack of ownership of the total businesses process that is creating the major problem. To offset this lack of process ownership, management should appoint a process owner for each selected process in Phase I, Activity 1. Someone whose salary and future growth will be based upon how well the total process performs. Management should select the process owner from the people who have a lot to gain from improving the total process performance. For example, if the process selected in Activity 1 was New Product Development, the Product Engineering manager might be selected as the process owner. This additional assignment will increase the process owner's workload for a period of 3 to 4 months but, in the long run, it will greatly reduce his or her total workload because the process itself will become more effective, efficient, and adaptable.

Activity 3—Defining Preliminary Boundaries

One of the first jobs the process owner undertakes is to define where the process starts and where it ends. This is harder than it sounds because different people involved with the same process see it in very different ways. It is important that the process owner defines the process broadly enough to resolve known problems, while being careful not to make it so large that it becomes unmanageable. The larger the process is defined, the more opportunity there is for improvement. Let's use the New Product Development process as an example. It could start when marketing conducts its surveys, or it could start when marketing delivers the product requirements to development engineering. It could end when the product specifications are released or when the first product is delivered to an external customer. It is always desirable to make the process as large as possible so that the opportunities for improvement are great, without making the Process Improvement Team (PIT) unwieldy. The PIT should have between 6 to 12 members. The PIT should also have a facilitator an data analyst assigned.

Activity 4—Forming and Training Process Improvement Teams

The process owner should then block diagram the process down to the department level. Each major department involved in the process should be represented on the PIT. If the department is significantly involved in the selected process, the department manager should assign a representative to serve on the PIT. The PIT will then be trained in basic team skills, because no team should meet until it has been trained in the basic team tools first.

Training for the following 10 Fundamental Business Process Improvement Tools[1] will be provided on a just-in-time basis.

- Business process improvement concepts
- Flowcharting
- Interviewing techniques
- BPI measurement methods
- No-value-added activity elimination methods
- Bureaucracy elimination
- Process and paperwork simplification techniques
- Simple language analysis and methods
- Process walk-through methods
- Cost and cycle time analysis (activity based costing)

Activity 5—Boxing in the Process

When the process is boxed-in, all the departments involved in the process, all the major inputs to the process, and all the major outputs are defined to the best of the PIT's knowledge. The PIT will now establish the finalized beginning and end boundaries plus upper and lower boundaries. The addition of the upper and lower boundaries completes boxing in the process. The upper boundary is used to define where inputs enter the process at points within the process. The lower boundary is used to define what leaves the process from points within the process.

Activity 6—Establishing Measurements

The PIT will now look at the total process to determine how it should be measured. Measurements for efficiency, effectiveness and frequently for adaptability need to be established at this point. Don't stop when you define your internal customer requirements. If you only look at your internal customer you may miss the real business objective. Once the desired measurements are defined, the PIT should establish a system for collecting the measurement data on an ongoing basis. Initial values should be established as soon as possible. It is important that not only average values be determined but that minimum and maximum values are also measured, since it is often the exception to the norm that loses a good customer.

Activity 7—Developing Project and Change Management Plans

The PIT now needs to prepare a project plan for the process under study. The project plan will include:

- The PIT mission
- The name of the project
- A list of key measurements and improvement goals
- Timetable for performing the analysis
- A change management plan to prepare the process stakeholders for the new process
- Resources required to complete Phases II and III

The importance of establishing a good process change plan and implementing it cannot be overemphasized. The objectives of the Change Management Plan should be to:

- Maximize the degree of commitment that the sponsors have to the change.
- Minimize the degree of resistance that the individuals that live in and with the process will have to the change.
- Maximize the effectiveness of the change implementation team.
- Minimize the time and resources required to implement the change.

■ Phase II—Understanding the Process

Unfortunately, most business processes are not documented, and often when they are documented, the processes are not followed. During this phase the PIT will draw a picture of the present process ("as-is" process), analyze compliance to present procedures, collect cost and cycle time data, and align the day-to-day activities with the procedures. There are six activities in this phase.

The purpose of Phase II is for the PIT to gain detailed knowledge of the process and its matrixes (cost, cycle time, processing time, error rates, etc.). The flowchart and simulation model of the present process (the "as-is" model of the process) will be used to improve the process during Phase III.

Activity 1—Flowcharting the Process

There are many different types of flowcharts that can be used to draw a picture of the business process. Some of the most common are:

- Block Diagram
- ANSI Standard
- Geographic
- Functional
- Data Flow

The PIT team now will block diagram the total process, following the process flow. Using the block diagram, start with the inputs to the process and develop the flowchart so that all activities within the process under study have been charted. Normally, flowcharts go down to the activity level only but often some important activities are flowcharted down to the task level. Once you have completed the flowchart, look at each block on the flowchart and estimate the following:

- Processing time
- Cycle time
- Cost per activity
- Percentage of items that go through that activity

At this point in time, the processing time, cycle time, and cost are based upon the best judgment of the PIT. Later on in this phase, actual data will be collected.

Activity 2—Preparing the Simulation Model

The data developed during the flowcharting activity is put into a computer along with other information like controlling documents, where the activity is performed, cost, cycle time, and processing time estimates. The computer program will define the process' critical path, overall processing time, cycle time, and cost per cycle. The computer program will also regroup the data so that it can be flowcharted in any one of the many flowchart conventions, providing the PIT with a number of ways to look at the process. There is a great deal of value in looking at the process in different ways. As actual data is collected, it will be used to update the simulation model. This model will also be used during Phase III to evaluate the impact of proposed changes to the total process.

The importance of developing a simulation model cannot be overemphasized. It is a monumental job to try to accumulate and manipulate the data that is involved in analyzing a business process by hand. There are a number of good software packages. Among them are: *Work Draw—Business Redesign Tool Kit* written by Edge Software Inc., Pleasanton, California; and *Envision* written by Future Tech Systems Inc., Auburn, Washington.

Activity 3—Conduct a Process Walk-Through

The PIT is now divided into two- or three-member teams. These walk-through teams personally observe each activity in the process and interview the employees performing the activity. Typical information collected related to each activity includes problems the operator is having, how the activity is performed, cycle time, costs, etc. This information is then used to update the simulation model.

Activity 4—Performing Process Cost and Cycle Time Analysis

Although the walk-through teams do their best to collect actual processing costs, processing time, and cycle time during the walk-through activity, there are often voids in the data and frequently, the total processing time exceeds the available rescources actually assigned to the activity. During this activity, validity checks are made on the basic information and voids are filled in.

Activity 5—Implementing Quick Fixes

By now the PIT should have identified many opportunities to improve the process. Many of them are things that can be done right away at little or no cost. These quick fixes often can save a lot of money and/or improve performance. As a rule of thumb, the change that has a three-month savings of three times the implementation cost should be implemented at this time. In this activity, the PIT eliminates the dumb things that are going on in the process.

Activity 6—Aligning the Process and the Procedures

The process walk-through usually reveals a number of activities that are not being performed in accordance with the documentation. If the employees have identified a better way of performing the activity, the procedure should be modified to reflect the present method. On the other hand, in cases where the present procedures are correct, employees who are not following the procedure should be retrained.

■ Phase III—Streamlining the Process

The streamlining phase of Business Process Improvement is the most critical and the most interesting. It is during this phase of the business process improvement methodology that the creative juices of the PIT members are really put into action. The streamlining phase consists of six activities (see Fig. 10.1).

During the streamlining activities, up to three very different options are available. They are: Process Redesign, New Process Design, and Benchmarking. The most frequently implemented streamlining option is process redesign (about 70 percent of the time), even though new process design provides greater levels of improvement. With the increased improvement comes increased implementation costs, risks, and time.

The PIT will review the process status and the improvement goals they have set for the process during Phase I. Based on this review they will select one, two or all of the streamlining options. Frequently, the PIT will start with process redesign and if that does not meet their goals, they will then do a new process design. Often, the PIT will benchmark to the point that they have the measurements from the best of the breed to compare their results against.

Activity 1—Process Redesign (Focused Improvement)

This approach takes the present process and removes waste while reducing cycle time and improving the process effectiveness. After the process is simplified, automation and information technology (IT) are applied, maximizing the process'

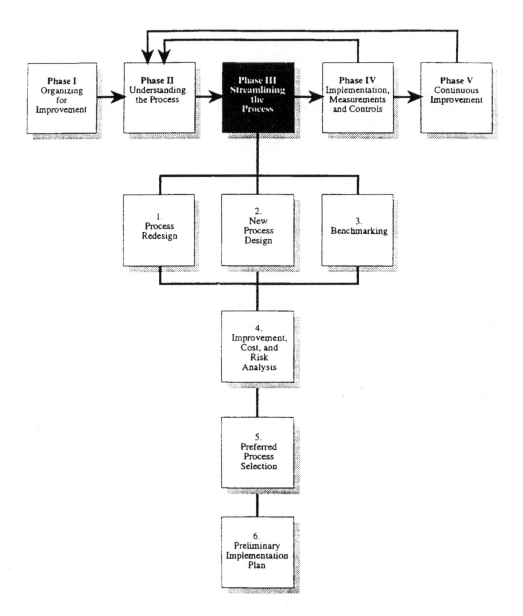

Figure 10.1. Phase III—Streamlining the process.

ability to improve effectiveness, efficiency and adaptability measurements. Process redesign is sometimes called focused improvement since it focuses its efforts on the present process. It results in improvements that range between 300 and 1000 percent, or reduces cost and cycle times by 40 to 60 percent. Process redesign is the one most frequently used because the risks are lower and the costs are usually less. This is the right answer for approximately 70 percent of the business processes.

The process redesign approach to streamlining the business process consists of eleven tasks. They are:

1. Bureaucracy Elimination
2. Value-Added Assessment
3. Duplication Elimination
4. Simplification
5. Cycle Time Reduction
6. Error Proofing
7. Process Upgrading
8. Simple Language
9. Standardization
10. Supplier Partnership
11. Automation, Mechanization, Computerization, and Information Technology

Each time an activity is considered to be removed or changed, the simulation model should be updated to determine the impact upon the total process. Frequently, an activity that is changed may have a positive impact on a specific part of the process, but a negative impact on the total process. The simulation model allows you to quickly determine total impact. In many other cases, an activity can be changed in a number of different ways. Each of the options needs to be evaluated from a total process standpoint before one is selected. Continue to look for the best option, not the first way of eliminating a root cause. Using process redesign, the PIT typically can complete Phases I through III in 90 days.

Activity 2—New Process Design (Process Reengineering, Process Innovation, Big Picture Analysis)

This approach takes a fresh look at the objectives of the process. It completely ignores the present process and organizational structure. The approach takes advantage of the latest mechanization, automation and information techniques that are available and improves upon them.

New process design, when applied correctly, can lead to improvements that range between 700 and 2000 percent, or reduce cost and cycle time between 60 to 90 percent. New process design is sometimes called "Process Innovation" because its success relies heavily on the PIT's innovation and creative abilities. Other organizations call it "Big Picture analysis" or "Process Reengineering."

The new process design approach provides the biggest improvement and requires more cost and time to implement than process redesign. It also has the highest degree of risk. Often the new process design approach includes organizational restructuring and is very disruptive to the organization. Most organizations can only effectively implement one change of this magnitude at a time.

The new process design approach to streamlining allows the PIT to develop a process that is as close to ideal as possible. The PIT now steps back and looks at the process with a fresh set of eyes, asking itself, how it would design this process if it had no restrictions. Often, this process stimulates the PIT to come up with a radical new process design that is truly a major breakthrough. New process design is often referred to as process reengineering, but we have difficulty thinking of this as a "re" anything. It is a new start. It is starting with a blank sheet of paper as you would if you were engineering the process for the first time.

The new process design approach to streamlining consists of five tasks, which are:

1. Big Picture Analysis

2. Theory of Ones

3. Automation, Mechanization, Computerization, and Information Technology

4. Organizational Restructuring

5. Process Simulation

Task 1: Big Picture Analysis. The PIT is now constrained in only very general ways. The results of the new process design must be in line with the corporate mission and strategy. It should also reinforce the organization's core capabilities and competencies. Before the PIT starts to design the new process, it needs to understand the direction the organization is going, how the process being evaluated supports the future business needs and what types of changes would provide the largest competitive advantage.

Once this is understood, the PIT can develop a vision statement of what the best process would look like and how it would function. In developing the vision statement, get the PIT to think outside of the normal routine that it lives with (think outside the box). Challenge all assumptions, challenge all constraints, go beyond the best, question the obvious, identify the technologies that are limiting the process, and define how they need to improve to help provide the best possible process. The vision statement defines only what must be done. For example: The organization does not require an invoice to pay a bill.

Task 2: Theory of Ones. Once the vision statement is complete, define what must be done within the process to take the input to the process and upgrade it to the point it can be delivered to the customer. Question why it cannot be done in one activity by one person in one place, or better still, at one time with no human intervention. Be a miser in adding activities and resources to the process.

Task 3: Automation, Mechanization, Computerization, and Information Technology. The PIT now looks at how the activities can be automated, mechanized, and/or computerized to perform the desired tasks at minimum cost and cycle time while providing error-free output. Set an objective of eliminating the need for any human being coming in contact with the process. Try to reduce cycle time to micro seconds in place of days.

Task 4: Organizational Restructuring. Now that the process activities and skills are defined, the PIT will evaluate how the process should be integrated into the business. Keep in mind that the chances of problems arising, increasing costs, and longer cycle time increase each time another department is involved in the process. It is usually best if the activity can be done by one natural work group (department), because the more departments that are involved, the greater the chance is that inefficiencies will develop in the hand-off activities. As you design the organizational structure that will manage the process, develop a justification for every hand-off.

Task 5: Process Simulation. When the PIT has completed the new process design, it should compare it to the vision statement to evaluate if the objectives set forth in the vision statement are met. When it meets the requirements defined in the vision statement, a simulation model is prepared and the measurement matrix is calculated.

Usually the cost and risk of implementing the new process design are much greater than those incurred in the process redesign approach, so the results must be much greater to justify the additional risk and expenditures. Using new process design, the PIT typically can complete Phases I through III in 9 to 12 months.

Activity 3—Benchmarking the Process

This is a very popular tool that compares the present process to the best similar processes available in the world. It may or may not compare processes that are from the same industry. Although benchmarking is usually not the approach normally selected it provides a proven performance measurement that can be used to evaluate the excellence of your other two alternatives. It also provides the process improvement team with many good ideas that are often improved upon and included in the other two alternatives. Approximately 10 percent of the time benchmarking is the right answer.

In all organizations that have operations at more than one location, benchmarking activities should start with internal benchmarking because of the ease of obtaining detailed data and good cooperation. This is often followed by external benchmarking activities. Often the external benchmarking process will focus on specific tasks or subprocesses that are defined as real-value-added or business-value-added during the streamlining process. An excellent computer program that will help with your benchmarking activities can be obtained from LearnerFirst, Inc.

Activity 4—Improvement, Cost, and Risk Analysis

After completing Activities 1, 2, and 3, the PIT can have a number of different process models available to them for analysis because each activity can define a number of options. To decide on which process is the correct one for the organization, the PIT needs to do an improvement, cost and risk analysis of each of the new processes. To accomplish this, the PIT needs to look at each process and estimate, with the help of the simulation model, its effectiveness, efficiency and adaptability. In addition, the PIT needs to estimate what the cost will be to implement the change, the length of time required to implement the change, the probability of success, and identify major problems in the implementation (see Fig. 10.2).

Activity 5—Preferred Process Selection

The analysis of alternatives should be presented to the Executive Improvement Team (EIT) along with the PIT's recommendations for the best alternative and how the changes should be implemented. The EIT must weigh the alternatives and make a decision on how the organization's resources will be invested. It is the duty of the EIT to identify from the alternatives the process that will be implemented and the implementation team.

Activity 6—Preliminary Implementation Plan

Now the PIT will prepare a preliminary implementation plan. The plan will include experiments and pilot runs to verify the performance estimates that were made in Activity 4.

■ Phase IV—Implementation, Measurements, and Controls

During this phase, an implementation team is pulled together to install the selected process, measurement systems, and control systems. The new in-process measurement and control systems will be designed to ensure that there is immediate feedback to the employees, enabling them to contain the gains that have been made and to improve the process further. This phase consists of five activities. They are:

1. Finalized implementation plan
2. New process implementation
3. In-process measurement systems
4. Feedback data systems
5. Poor-quality cost

PERFORMANCE ESTIMATE

	Original process	Benchmark process	Process redesign	New process design
Effectiveness (Quality)	.2	.02	.01	.009
Efficiency (Productivity)	12.9 hrs./cycle	7.5 hrs./cycle	6.3 hrs./cycle	5.3 hrs./cycle
Adaptability	25%	Not measured	80%	65%
Cycle Time	305 hrs.	105 hrs.	105 hrs.	85 hrs.
Cost/Cycle	$605	Not measured	$410	$380

IMPLEMENTATION ESTIMATE

	Benchmark process	Process redesign	New process design
Cost	$1,300,000	$20,000	$280,000
Implementation Cycle Time	24 months	6 months	15 months
Probability of Success	50%	95%	85%
Major Problems	Need more data	Training time	New organizational structure

Figure 10.2. Improvement, cost, risk analysis.

Activity 1—Finalized Implementation Plan

An implementation team is formed to prepare a detailed implementation plan and coordinate the changes. It may or may not include all the members of the original PIT. Often, Department Improvement Teams (DITs) become part of the implementation plan so that the teams within the functions that will be impacted by the change are part of the group that plan and implement the change. Sometimes the implementation team is divided into sub-teams (example: information system teams). The implementation plan usually is divided into three parts:

1. Short-term changes—Changes that can be done in 30 days
2. Mid-term changes—Changes that can be done in 90 days
3. Long-term changes—Changes that require more than 90 days to implement

An implementation plan will be prepared for each change.

Activity 2—New Process Implementation

The implementation plan and the change management plan are now united to bring about an effective overall implementation of the new process. The implementation team will maintain close control over each change to be sure that it is implemented correctly. Often, complex changes will go through a series of modeling and/or prototyping cycles to prove out the concept and to ensure smooth implementation. After each change is installed, its impact is measured to ensure it accomplishes its intent and has a positive impact upon the total process. As the change is implemented, the simulation model is updated so that it always reflects the present process.

Activity 3—In-Process Measurements

Before you can design a measurement system, you need to define requirements. Each activity on the final flowchart should be analyzed to define the customer requirements and how compliance to these requirements can be effectively evaluated. You will note that up to this point, the measurement system focused on the total process. Now the task is to develop measurements and controls for each major activity within the process.

A good measurement and feedback system is one in which the measurements are made as close to the activity as possible. Self-measurement is best because there is no delay in corrective action. Often, though, self-measurements are not practical and/or possible.

Activity 4—Feedback Systems

Measurement without feedback to the person performing the task is just another no-value-added activity. Feedback always comes before improvement. In most organizations, too much data is collected and too little is used. Employees need ongoing positive and negative feedback about their output.

Although we need ongoing feedback to the employees involved in the process, we also need summary reports for the same people and for management. The summary reports should be exception reports so that masses of data do not waste management's and the employee's time. Exception reporting allows everyone to focus in on where improvements can be made.

Activity 5—Poor-Quality Cost

Waste costs money. In many business processes, poor-quality costs run as high as 80 percent of budget. Poor-quality costs of 50 percent or more are common in business processes before BPI is applied to them. If that is cut by 50 percent through the use of BPI, the process is still wasting 25 percent of the organization's budget and is a gold mine for future continuous improvement. More information on Poor-Quality Cost can be found in the book, "Poor-Quality Cost," by H. J. Harrington, published by Marcel Dekker, Inc., New York City, 1987.

■ Phase V—Continuous Improvement

Now that the process has undergone a major breakthrough in performance, you cannot stop improving. This is not the end of the improvement activities; it is just the beginning. Now the process must continue to improve, usually at a much slower rate (10 to 20 percent per year), but it must continue to improve. The Natural Work Teams or Department Improvement Teams now take over.

■ Does BPI Work?

Does business process improvement work? Just ask any of the organizations that have tried it—Ford, Boeing, IBM, 3M, Corning, Nutrasweet, McDonnell Douglas—and you will get a resounding yes. The following are some typical examples of results:

McDonnell Douglas

- 20 to 40 percent overhead reduction

- 30 to 70 percent inventory reduction

- 5 to 25 percent material cost reduction

- 60 to 90 percent quality improvement

- 20 to 40 percent administrative cost reduction

Federal Mogul

- Reduced development process cycle time from 20 weeks to 20 business days, resulting in a 75 percent reduction in throughput time

Aetna Life and Casualty Co.

- Reduced Information Technology workload by 750 employee years per year
- Consolidated 65 property casualty claim offices to 23.
- Net income rose 50 percent to $207.2 million.

■ Summary

The process and the system which controls it represents the real problem facing business today, not the people who work within the boundaries set for them by management. Employees must work within the process and management must work on the process. The improvement efforts and their supporting systems must be directed at the process and not at the individual. This means that all functions must work together to optimize the efficiency, effectiveness, and adaptability of the total process. This can best be accomplished when one person is held accountable for the performance of the total process and is given the authority to bring together members from all the individual functions involved within the process, with the objective of maximizing its total performance.

Now let's look at how different types of organizations feel about their business processes.

Losers

- The employees are the problem. They need to be motivated.
- If everyone does their job, things will get done.
- Employees and management cannot be trusted. There needs to be a lot of checks and balances.
- Downsizing is equally distributed across all functions. That's the fair way to do it.
- Business processes, on an average, have 15 percent real-value-added content.

Survivors

- If each natural work team improves its part of the process, everything will be OK. It is the high number of errors that are the problem.
- Bureaucracy is something that we have to live with.
- They decide to use either continuous improvement or breakthrough methodologies.
- Computerization of their business processes is used to decrease cycle time and costs.
- Business process improvement activities are directed at reducing costs.
- Process changes that will reduce headcount are kept confidential until the last minute.
- Business processes, on an average, have 30% real-value-added content.

Winners

- The primary focus of BPI is on things that impact the external customer (example: quality, cycle time), not on headcount or cost reduction.

- Any bureaucracy is bad.

- When downsizing is necessary, they remove the no-value-added activities from the processes. Unnecessary work is eliminated and the associated resources are removed from their budgets.

- They never computerize until the process has been streamlined.

- They prioritize their investment in business process improvement activities based upon their business plan, competitive position, core competencies and core capabilities.

- Business processes, on an average, have 50 percent real-value-added content.

For more detailed information on this paper, see *Business Process Improvement*, by H. J. Harrington, published by McGraw-Hill, New York City, 1991, and *Process Innovation* by Thomas H. Davenport, published by Harvard Business School Press, Boston, 1993.

■ Reference

1. Harrington, H. J., *Business Process Improvement*, McGraw-Hill, NY, 1991.

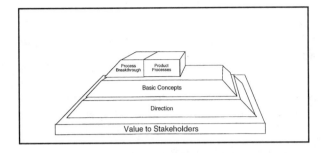

11

Product Process Excellence: The Production Side of All Organizations

Jose R. Rodriguez-Soria
President, Q-2000 Group

and

H. James Harrington
Principal, Ernst & Young

We must save our smokestack and manufacturing industries. For no nation can defend itself with even the very best service.   DR. H. JAMES HARRINGTON

■ Introduction

In today's complex business world, it is often increasingly difficult to separate product and process design activities. The need for ever increased quality and reducing the time to get the product to market have made it imperative to integrate the Product and Process Development efforts to a degree never seen before. Most organizations, particularly in the computer and related fields, are bringing out products in extremely short development cycles. AMBRA Corporation, owned by IBM, claims it will redesign each product to reflect the most recent advances 3 or 4 times a year. To effectively compete, organizations must bring into play a number of new techniques and methods like Computer Aided Design (CAD), Computer Aided Manufacturing (CAM), Concurrent Engineering (CE), Design For Manufacturability (DFM), Just In Time (JIT), and Failure Mode and Effect Analysis (FMEA). These methods, properly integrated, will allow organizations to bring high quality products faster to market.

While this chapter is written with respect to manufacturing and production processes that produce product, the same principles and many of the techniques apply to the service industries equally as well. It is important to understand that all organizations are made up of production activities (example: paying bills) and service activities. This means that it is very important today for all organizations to look at themselves to identify which activities should be classified under the service category and which are really production activities. This is important because different improvement approaches need to be used in each case. This realization has changed the way the best organizations think about themselves and has resulted in a major focus on exchanging data between the service and industrial sectors.

■ Product Processes

The Product Processes can be viewed as consisting primarily of two phases: the design of the product, and the design and implementation of the production process. The product design phase can include the following steps:

- Concept generation
- Concept selection
- Product design
- Design tests
- Final product design
- Prototyping

The production process phase covers the following:

- Process concept
- Process design
- Prototyping the process
- Pilot production
- Process installation
- Process qualification
- Full production

■ Product Development Phase

Product Processes start with "market-in," meaning to design and manufacture products the customer wants. In order to effectively accomplish "market-in," we need products that meet customer requirements and are the result of the focused application of the organization's core capabilities. For products to be successful, they must deliver value over competing products, and/or a radically new creation, not just an improvement, and provide high quality as viewed by the customer and/or consumer of the output.

Core Competence

Core Competence or Capability is defined as the unique combination of resources that provides each organization with its position in the marketplace. When speaking of resources, we are referring to such things as technology, personnel, and finances. A much needed Core Competence today is the ability to develop products quickly. Time-to-Market is declining in industry after industry, from automobiles to mortgages. For an organization to attain success in today's competitive climate, Time-to-Market needs to be shorter and shorter. Ford has cut the product development cycle to three years from the previous five, and is currently aiming for two. Honeywell has cut new product development time to less than 12 months. Hewlett-Packard derives more than 50 percent of sales from products introduced in the previous five years.

Another dimension of core capability is the understanding of customer needs, and how the new products in the development process pipeline meet those needs.

Quality Function Deployment (QFD)

Quality Function Deployment (QFD) is the most comprehensive technique developed to date to ensure that the customer requirements are defined and met. QFD not only sets the requirements, but brings the *voice of the customer* impact to processes, subprocesses, and activities, while deploying quality, capabilities, costs, and reliability. QFD assures that customer requirements will be met by the product and processes developed. It builds a "natural" quality system for the organization to follow, that quality is built into the product. Quality Function Deployment integrates the different Product and Service Design Steps into a single process.

Quality Function Deployment was developed in actual practice by Japanese organizations such as Bridgestone, Matsushita, Kayaba, and Toyota. Some of the benefits of implementing QFD are prevention of defects, faster development time, fewer exchanges in design, greater knowledge about customer requirements, and improved customer satisfaction.

What is Quality Function Deployment? Basically, QFD connects the needs of the customer, or "What's" to "How" those needs will be met by the product, the production process, and the organization.

QFD starts with a detailed study of the customer wants (the "needs" and the "seeds"), then in a structured and disciplined manner, identifies and carries the voice of the customer through each stage of product development and implementation.

Focus groups are more subjective. They consist of groups of people led by a moderator and following a previously prepared discussion guide that usually talks about a product or service. While focus groups will not usually generate a new product idea, they do provide "seeds" that might go on to grow into new products or new approaches to a service. Focus groups' research gives a direction, an insight, by probing the mind of the customer.

In the author's experience, the most difficult, time-consuming, and critical aspect of QFD is to obtain the voice of the customer.

It is highly recommended that members of the cross-functional QFD Team participate in the actual gathering of information from customers. These observational studies are quite effective in gathering useful information as to what the customers need. There are valuable insights to be gained. For example, during one study a customer was asked to open a package. When asked for his comments on the package, he stated that it was "easy to open." Yet he was observed using his teeth to open the wrapper after a few frustrating tries.

The data collected from the various market research studies is organized in a "Customer's Needs Chart," and then translated into "Required Quality" elements.

The House of Quality. The House of Quality, or Product Planning Matrix (Fig. 11.1), portrays the relationship between the customer requirements and the Quality characteristics. It is also a way to show the actual performance, the competitor's performance, the organization's plan, and potential sales points. It allows the formulation of design priorities based on the voice of the customer. The House of Quality can be used not only for individual product planning, but also in the planning of product families. In this application, QFD can greatly contribute to reducing the Time To Market by eliminating duplications of efforts and providing focus.

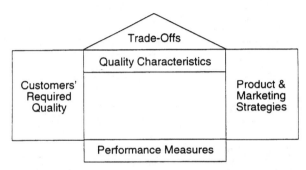

Figure 11.1. Product planning matrix or house of quality.

At this point, we prepare a matrix (Fig. 11.1) by listing the "Required Quality" elements down the left side of the matrix. The upper portion of the matrix lists the "Quality Characteristics," which are inferred from the Customer's "Required Quality." We then determine the relationship between "Required Quality" and "Quality Characteristics." This allows us to set the "Performance Measures" we need to achieve the customer's required quality. At the top of the "Quality Characteristics," (sometimes called the "roof" of the House of Quality), is an area where trade-offs are looked into and as a result, design changes are made as needed. The area of the matrix to the furthest right is where the Product and Marketing strategies are compared. The critical customer requirements are then determined with reference to competitors and the organization's own plans. The determinations of critical customer requirements allow the organization to focus on what is really important to the customer. It allows everyone in the organization to concentrate on those few things that will really deliver a true competitive advantage with reference to the customer.

Reliability

Reliability studies started with machinery and electrical systems. A simple definition of reliability is that the product functions in the manner intended for its designed life. Therefore, reliability must be built in. The design must take into consideration every potential failure that could occur during the life and use of the product. In simple products, reliability is measured in mean time to failure. For complex products, reliability is measured in mean time between failures. Usually reliability only applies to a product's performance after it has been accepted by the customer. As a result, reliability is the most important customer-related performance measurement. Products that have poor reliability often cost the customers more to own the product than to purchase it.

The reliability at the design stage can be improved by using Fault Tree Analysis (FTA) and Failure Mode and Effect Analysis (FMEA). Fault Tree Analysis (FTA) is a Tree Diagram that shows failures and/or defects in increasing levels of detail. It helps to narrow the root-causes and focus on prevention. See Fig. 11.2 for an example

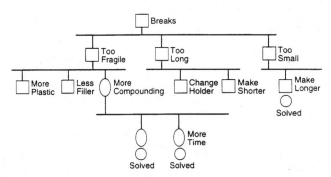

Figure 11.2. Fault tree analysis.

of the FTA. A closely related technique is the Process Decision Program Chart (PDPC), one of the Seven Management and Planning Tools. The PDPC not only attempts to prevent failures, but assists in formulating contingencies.

Failure Mode and Effect Analysis (FMEA) can be applied to a product or a process, just as the FTA can. In applying FMEA to the product design, the team looks at problems or likelihood of failures due to the design. When the FMEA approach is applied to the process used to produce the product, the team looks at the capability of processes to manufacture the parts. It assumes that the design was correctly recreated. Both methods, the Design FMEA and the Process FMEA, center around the determination of the probability of occurrence, the severity of the failure, and the likelihood of detecting failure prior to shipping. By analyzing the information provided by the FMEAs, the product designs can be made more "robust" and processes can be improved.

QFD helps prioritize failure modes for the design and manufacturing. FTAs and FMEAs allow the organization to focus on the critical performance issues. FTA looks at product and process failure modes and prioritizes them by comparing them with:

- Customer demands
- Functions
- Products
- Processes
- Parts
- Materials

Product reliability requirements should always be part of the product specification developed by Marketing and should be periodically reconfirmed with sample customers throughout the design cycle. Component parts life testing is used to develop reliability figures at the component level. This data is used to analyze designs to predict product reliability performance even before the first unit is built. The reliability of the product is determined by the way the components are connected, the percentage of total power rating that is used, the environmental conditions the components are subjected to, etc. To improve reliability, parallel backup circuits are often used.

In addition, completed units, assemblies and total products are life tested to ensure that the reliability requirements are met before the product is delivered to a customer. High customer satisfaction and high reliability go hand-in-hand.

Design for Manufacturability and Assembly (DFM/A)

While DFM/A had its beginnings with parts simplification of assembly type products, the general concepts can be applied to many manufacturing processes. The general concepts of eliminating complicated processes and products are key to improved performance and error prevention. The general rules for DFM/A are:

- Use minimum number of components
- Avoid fasteners
- Consider ease of assembly
- Use modular designs

These rules apply whether the assembly process is manual or automatic. Motorola is a good example. When they were building the Quasar TV sets just outside of Chicago, they designed the product with "The Works in a Drawer." This made it very easy to repair and the repairmen loved them, not just because the sets were easier to repair, but also because the sets required frequent repairs. This approach to improving ease of repair in contrast to eliminating the need for repair through improved reliability put Motorola out of the TV business, even though they had many of the basic patents.

The key to DFM/A is to understand the process and its costs. In addition, the product structure needs to be analyzed, and improvements identified, not only in structure, but also in design, resulting in easier assembly and lowered costs. Figure 11.3 shows some of the benefits that can be obtained through DFM/A.

There are a number of techniques and software packages available to assist in DFM/A analysis. Some of the software packagers are Boothroyd/ Dewhurst & Component Analysis System (CAS), PPS and PERL.

DFM/A for greatest benefit, must be part of a Concurrent Engineering effort. The U.S. industries that have most extensively used this technique are the automotive and electronics industries. DFM/A has been instrumental in reducing Time To Market (TTM).

Concurrent Engineering

Concurrent Engineering or Simultaneous Engineering involves Manufacturing, Quality, Marketing, and all other functions in the design cycle, as early as possible. Concurrent Engineering (CE) is a systematic approach to product design that considers all factors of the product cycle. The system uses cross-functional teams to develop the product and process design simultaneously. This results in a shorter

Company	Product	Results
DEC	Mouse	Reduced Parts: 50 percent Time Saved: 53 percent
NCR	Point of Sale Eqpt.	Reduced Parts: 80 percent Time Saved: 75 percent
TI	Reticule	Reduced Parts: 67 percent Time Saved: 84 percent

Figure 11.3. Benefits of DFM/A

development cycle and fewer engineering and process changes during the production stage of the product cycle.

Concurrent Engineering utilizes the capabilities of Computer Aided Engineering (CAE), Computer Aided Design (CAD), and Computer Aided Manufacturing (CAM) to the fullest. The advantage is best realized when all of these tools are integrated into a network that allows for the downloading of the designs to the specific piece of manufacturing equipment. This cuts the Time to Market significantly. In addition it saves a great deal of money, since fewer trial and bad products are built. Computerization improves quality by reducing the opportunity for error, particularly since it helps to bring together the traditional enemies in engineering: design and manufacturing.

Concurrent Engineering works perfectly with QFD, Design of Experiments, DFM/A, and other TQM techniques. It helps to create a natural quality system by building quality into the product and process by concurrently designing both. Concurrent Engineering procedures will greatly enhance compliance to ISO-9000 Regulations.

Concurrent Engineering is not a new concept. For more information about how it has been applied, contact IBM for a copy of Technical Report #TR02.901 by H. James Harrington, September 1980.

Innovation and Time To Market

Faster development of new products, processes, and services is essential to maintain an edge in today's global markets. Manufacturing cycle times must also get shorter and shorter. This new type of Innovation strategy has been highly successful whenever it has been applied.

Innovation means growth and survival. The most successful organizations such as 3M (which expects less than 25 percent of sales to come from products developed in the last 5 years), Sony, and Honda set the pace in their industries, not content to merely match competitors. They want to be the ones with the most successful new products to market. For a period, Sony was introducing a new Walkman model each day. They soon totaled over 200. In the automotive industry, Chrysler and Ford have reduced the Time To Market for the latest models to well below 3 years.

In the pharmaceutical industry, products have a very long development cycle, mostly due to pharmaceuticals being a highly regulated industry. This cautious approach has cost hundreds of thousands of American lives and untold agony that could have been avoided if the development cycle had been reduced. Nevertheless, recently, the government has fast-tracked drugs so that the public can benefit from their effects sooner.

Slow to market has a significantly more negative effect on profitability than exceeding product development budgets, the traditional way to measure success. Faster Time To Market allows organizations to increase the variety of products. Conventional practices make operations more difficult and costly. A way around this dilemma is to apply Just-In-Time (JIT) principles, not only to manufacturing, but to other corporate functions as well.

Process Design and Innovation

The objectives of Innovation Process Design are to:

- Increase volume and speed of production
- Improve manufacturing capabilities
- Use new technologies
- Reduce time to market
- Increase quality

This puts the organization in a much more competitive position.

The competitive advantage of many German and Japanese organizations lies in the increase in quality and lowered costs made possible by their process strategies. Japanese organizations are noted for their extensive use of Flexible Manufacturing Systems (FMS) that make extensive use of robots. This allows organizations like Hitachi, Sony, and Lyon to provide a much faster response when making product changes.

The Just-In-Time concept is designed to have suppliers submit just enough inventory to keep their customer operating until the next order is submitted. Instead of storing suppliers' products before they are used, they go directly to the product process. This approach greatly reduces the storage areas needed to support product build. Toyota has its suppliers deliver products every 4 hours in some cases. Other organizations plan on 24, 48, or 168 hour order cycles.

Zero stock applies Just-In-Time concepts to the internal manufacturing process. In these cases, manufacturing engineering designs equipment to minimize set-up time. This approach has reduced die set-up times from 6 hours to as short as 10 minutes. This allows very small lots to be manufactured at one time, which again reduces stocking area and inventory costs. The most advanced concept along this line is called single unit build. Organizations that are using the single unit build concept actually process a lot size of one unit or item.

These concepts reduce lead times, improve responsiveness and promote continuous product flow. Zero stock can be a low investment strategy since it is based on people rather than on extensive use of automation. Zero stock and JIT promote continuous improvement and quality performance.

Manufacturing cells (which can be an FMS) are the result of combining zero stock and automation. Local Area Networks (LAN's) Technology handle most of the information integration needs. Fully automated cells are basically the same as FMS cells except that the degree of computerization is much higher.

Benchmarking is a process that can hasten process innovation by measuring where you are in product and process technologies versus your competitors and world-class organizations. Benchmarking begins by understanding your processes; the objective is to cause improvement. Product and product process benchmarking has been an effective improvement approach since the early 1950s. Ford, GE, IBM, and HP have used product and product process benchmarking for more than 30 years. Xerox credits benchmarking as a major factor in the organization's turnaround back in the early 1980s. Benchmarking can be of great help in selecting process alternatives for greater improvement.

Process Capability

Process Capability measures the variation from all sources in a process as they affect the product. Once a process is stable (it shows variation only due to common causes), it is in statistical control and the capability can be measured and expressed in number known as the capability index. The index compares the process variation to the specification spread. In a way, it measures the balance of the goodness of the product design and compares it to the variation that exists in the manufacturing process. More and more process capability target setting is done prior to the start of the design stage in order to achieve the specifications and variability targets needed by the customers.

Motorola has targeted for what they call a Six-Sigma performance standard. Unfortunately, Motorola is having a hard time obtaining this objective, even though the way they are calculating their process capability is much less stringent than the normally accepted approach. George Fisher, Motorola's Chairman of the Board and CEO stated, "Furthermore, improvements in product quality seem to be stalled at a 5-5.2 sigma 'wall.'"

The standard definition of process capability is the reproducibility over a long period of time with normal changes in workers, material, and other process conditions (Reference page 16.15, Juran's Quality Control Handbook). Using this definition, Motorola's 6 sigma program is designed to give the organization a process capability of 4.5 sigma, which is just a little better than the normally accepted criteria of 4.0 sigma.

■ Process Qualification

Process Qualification is a key step in process development. If this step is performed correctly, it will measure how well the preceding activities were performed and set the stage for future improvements. A qualified process is one that has demonstrated that all the necessary procedures, training, documentation, measurements, controls, and checks and balances are in place to ensure that the process can produce high-quality output even under stress conditions. When this level of performance is demonstrated, the product and process design are complete and the program can go into a maintenance and improvement mode.

This does not mean that the process has reached its optimum performance or error levels. Quite the contrary, it is the start of the improvement process and the beginning of the voyage to error-free performance. One of the best ways to help an organization move from an appraisal-type philosophy into a preventive-type philosophy is by implementing a systematic process qualification activity.

The complexity of a qualification activity varies, based on the complexity of the process itself. To provide as thorough a picture as possible, let's examine what is required to qualify a complex thin-film, electronics-component process.

Qualification Plan

At this point, let me provide two definitions:

Certification applies to a single operation or piece of equipment. When an acceptable level of confidence has been developed that proves the operation and/or equipment is producing products to the required level when the documentation is followed, that item is then certified. Typically this will require that the process capability is a minimum of plus or minus 4.0 sigma.

Qualification applies to a complete process consisting of many operations all of which have already been individually certified. For a process to be qualified, each of the operations and all the equipment used in the process must be certified. In addition, the total process must have demonstrated that it can repeatedly produce high-quality products or services that meet customer expectations.

In the manufacturing process areas, quality assurance normally is responsible for process qualification. In the non-manufacturing areas (business process areas), process qualification is the responsibility of the process improvement team and probably will not go through all the stages indicated in this example.

For a complex process, there often are three separate production lines, which are:

1. *Development Line*—The development laboratories produce crude models, using complex laboratory equipment, to prove out a theory or concept.

2. *Pilot Line*—A model pilot line is set up to manufacture larger quantities of product for internal evaluation.

3. *Production Line*—The production line is established to produce finished product for the customer.

Now let's look at how the four levels of process qualification are applied to the 3 different production processes.

Level I qualification evaluates the acceptability of the development process. It takes place in the development lab during the early phases of the program. In this evaluation, it is important to establish some basic controls, collect pertinent data, and study manufacturability without interfering with the creative nature of the work environment.

Level II qualification is directed at the pilot process used to produce product for internal evaluation and specification preparation. It is very important to have a good understanding of the pilot process before these critical internal tests are run and specifications prepared. Level II qualification evaluates this new process to ensure it meets both customer's and the organization's expectations.

Experimental lots must give way to mass production quantities, and hours of processing time must be reduced to a minimum to meet cost targets. The manual development processes must give way to automation. The equipment must undergo major changes to allow manufacturing operators to replace highly skilled development technicians. These activities present a major set of challenges to manufacturing engineering that result in drastic differences in the facilities. Manufacturability, as well as the ability to meet committed yields, must be demonstrated for the first time.

In most cases, Level III qualification applies to a single-stream product line that has limited ability to produce the customer-shippable products. Once this line is established, the process continues to expand, adding equipment to the manufacturing facilities.

Up to this point, the process qualification activities have been directed at bringing the process under control before we start shipping products to our customers. When Level IV is granted, the floodgates are opened, so every control must be in place before this point in the program is reached. The purpose of Level IV is to characterize the process to ensure it is under control and has the ability to consistently produce output that meets customer expectations.

To get a better understanding of process qualification at each of the four levels, let's take a look at the three major activities that go on during the Level III qualification study:

1. Certifying each operation in the process

2. Process qualification lots

3. Independent program audit

Certification.   Certification activity looks at the four facets of each operation:

1. Documentation

2. Test and process equipment

3. Operational requirements

4. Output acceptability

Documentation is an extremely important part of any process because it allows the experience and knowledge of previous activities to be transmitted to the individual currently performing the job. An activity left to chance always has the chance of producing errors.

Test equipment and *process equipment* (fixtures, tools, dies, etc.) can have a major impact on the quality and productivity of the area. Certification of the equipment determines if it is capable of doing the assigned job and being maintained properly. Some of the activities appropriate to this facet of certification are as follows:

1. Accuracy and repeatability assessments, including operator variation, must be conducted.

2. Calibration and preventive maintenance procedures and intervals must be evaluated.

3. The equipment must be evaluated from a safety standpoint.

4. The adequacy of standards controls must be evaluated.

5. Correlation studies between similar pieces of equipment and other locations should be conducted.

6. Long-term drift studies need to be performed.

7. Extremes of setting must be evaluated and taken into consideration when the process window is established.

8. The acceptance criteria must taken into consideration. If the inaccuracies in the measurement equipment is more than 5 percent of the specification limits, working limits need to be established. These new working limits ("guard bands") protect the customer from receiving out-of-specification output.

Operational requirements also need to be evaluated. At this point in the certification activities, we are evaluating the process, not the employees. It is the process that causes errors and places individuals in a position where they cannot fulfill their God-given right of error-free performance.

The first three facets of the certification activity have all been building up to the point that we can ensure *output acceptability* on a continuous basis. To evaluate output acceptability for certification, the following activities and requirements are typical:

1. Projected yields for this point in the program must be met for four consecutive weeks.

2. Daily going-rate projections must be met and must be following the learning curve.

3. Stress tests must be in place to evaluate product reliability at the lowest possible level.

4. Component traceability must be established.

5. Correlation studies between suppliers and receiving inspection must be completed.

6. The plus or minus 3-sigma variation (99.7 percent of the product) must be less than 75 percent of the engineering specification tolerance. By Level IV this variation should be reduced to 50 percent of the specification tolerance (a Cp of 2). Note: This was IBM's requirement in the late 1970s, which was adopted by Motorola in the late 1980s.

Qualification Lots. Once each activity in the process has been certified, qualification lots are processed to measure the effectiveness of the process design. The purpose of a qualification lot is to evaluate the continuity of the total process, to measure process yields, and to identify process volume limitations under controlled conditions. A typical Level III process qualification design experiment would be conducted over a five-week period; one of the five weeks would be used to measure equipment capacity and throughput volumes. A minimum of five separate lots would be processed through the manufacturing operations.

Independent Process Audit.   The next step in the qualification process is a detailed process audit. The audit team is headed by Process Quality Assurance and consists of an independent group not assigned to the process. For example, you might have representatives from Product Engineering, Development Engineering, Manufacturing Engineering, Product Assurance, and Sales. Some of the things the audit team would evaluate are:

1. Has product manufacturability been proven?
2. Has the design considered and corrected problems existing in similar products?
3. Does the performance specification represent an improvement in reliability and quality performance compared to products that it will replace?
4. Are there any major technical exposures to the program or the supporting technologies?
5. Have the certification and qualification activities been implemented as required and have all major exposures been satisfactorily addressed?
6. Is there a good correlation between today's customer expectations and the engineering specification?

As soon as the audit team members have completed their assignments, they will meet with the concurrent Engineering team and report their findings. These findings will be subsequently documented in an audit team report. The concurrent Engineering team will generate corrective actions to solve each of the problems described in this report before the process is qualified at Level III.

The qualification process just described was in use within IBM in the late 1970s and still is the best practices for the evaluation of production processes. For more detailed information, contact IBM to request a copy of their Technical Report TR 02.901 entitled, "Process Qualification—Manufacturing's Insurance Policy" dated March 15, 1980.

■ Product and Process Design and Innovation

Manufacturing Process Design Concepts

In the past, almost all of our research and development budget has been dedicated to the products we produce. Very little has been set aside to improve the manufacturing processes. Today, this is changing as management begins to realize that the manufacturing process design is equally as important as the product design. In fact, often the competitive advantage is based upon the excellence and innovative nature of the manufacturing process.

Continuous Flow.   For 100 years we believed that large lots of parts processed together maximizes our use of resources. The error that we made is that we forgot to consider all of the costs (example: storage space, inventory, scrap, etc.). Today our focus is on designing processes that have the capability to build one unit at a time (single unit build). This has changed our manufacturing process layout significantly. In the past, we centralized all of our major resources of capital equipment in similar locations (example: one area set aside for milling equipment, another area set aside for grinding equipment). This allowed us to maximize the use of these costly pieces of equipment. On the other hand, if built in a large amount of additional cost in storage, movement, production controls, inventory, and scrap. IBM first used this concept in 1948 in Endicott, New York to produce their 077 machine. But somewhere in time, they forgot the lessons they learned and slipped back into Henry Ford's mass production concepts until late in the 1980s.

With today's focus on single unit build processes, these departments have been divided up and the equipment is grouped together so that the unit can be serially processed through a sequence of operation without leaving the manufacturing cell. To accomplish this, many things needed to be changed. Among them are:

- Supplier Just-in-Time programs need to be implemented.
- Employees need to be trained on a wide variety of activities.
- Set up time needs to be reduced from hours to minutes.
- Plant lay-out needs to be designed in parallel with the product design.
- Simplify production control system needs to be developed.
- The manufacturing line needs to be balanced.

This new application of manufacturing process design has changed the way we measure the effectiveness of manufacturing. For example, Harley Davidson uses the following measurements:

- Productivity improvement
- Schedule attainment
- Quality levels
- Maintaining requirements
- Conversion costs
- Manufacturing cycle time
- Inventory levels
- Scrap and rework
- Overtime required
- Materials cost variance

One Minute Change-Over.   Three keys unlock the door to continuous flow manufacturing. They are:

- High quality components
- Highly trained workforce
- Ability to change the production line from one product to another rapidly

Of these three items, the last one, being able to convert the manufacturing line fast has become the biggest barrier to implementing continuous flow. Manufacturing engineering reasoned that big lots allowed the hours of machine set-up time to be distributed across many units thereby keeping the total cost down. All of a sudden, in the late 1960s, we began to challenge the need for long set-up time. Slowly, at first, we began to reduce set-up time as we began to feel the increasing pressure to convert lines faster in order to meet the customer's changing demands. Toyota was among the first organizations that broke away from the old traditional model. Their engineering team focused upon designing jigs and fixtures that would reduce conversion time. This led to a major breakthrough in fixture design and production strategy. For example, using the old approach, Toyota required approximately four hours to convert a punchpress from one part to another. Using the new concept, this six-hour time cycle was reduced to less than 10 minutes. These concepts now have become commonplace in most advanced organizations and are an absolute critical factor that must be considered if you are going to compete on the international market.

Error-Proofing.   As we began to look closer at the tooling, in order to improve its changeability we realized that a great deal of scrap was being produced because the tooling was set up wrong or the materials were located in the tooling wrong during the production cycle. To eliminate these possibilities, the concept of error-proofing was developed. Essentially, when error-proofing is applied, the equipment or tools are modified so that parts cannot be installed or used incorrectly. For example, if a flat bar of steel that had a notch in one side was being milled to length, a pin would be added to the fixture that would protrude into the notch when the part was installed correctly. If the part was reversed, the pin would keep the part from fitting into the fixture. Shigeo Shingo was the leader in this approach. For more information about error-proofing and continuous flow manufacturing read his books published by Productivity Incorporated entitled, "Zero Quality Control: Source Inspection and the Poka-Hoke System" and "Nonstock Production."

Production Control Systems.   Simplicity of the production control system is a requirement for a successful single unit build process with no "just in case" stock. We cannot afford to wait for a computer program to analyze manufacturing status and tell us what to do and when to change to a different product. Here again, Japan made some major breakthroughs. While the East was trying to use information technology to control their manufacturing processes and its scheduling, Japanese organizations simplified the processes and obtained superior results. Japanese organizations simply told their workers to build parts to fill the empty containers that had been returned to them. Other organizations approached it by putting a card on each unit built that was returned to the operator when the product was used.

This instructed the operator to build a replacement part. Often, information technology is the answer to data control. But on occasion, it is better to simplify what you are doing to maximize the effectiveness of the total operations.

■ Production Phase

Once the process is qualified, the flood gates are open and large quantities of product start to ship to external customers. During this phase of a product's life cycle, great care must be exercised to ensure that the basic product design and production process are not compromised as a result of fluctuations in materials, processes, or people.

The Process Control Cycle

The objective of the early-entry activities is to bring the process under control, then implement improvement activities. A typical process evolves through four different phases during its life cycle (Fig. 11.4):

Phase A: Out of Control

Phase B: Stable

Phase C: Step-by-Step Improvement

Phase D: Customer Ship (Continuous Improvement)

Phase A: Out of Control. During this phase of its evolution, the process frequently goes out of control. The first priority of the process improvement team is to understand the true cause of the out-of-control conditions and to implement control over the process elements that cause them so that the out-of-control conditions are eliminated.

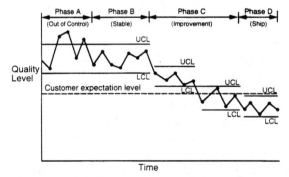

Figure 11.4. The process control cycle.

Phase B: Stable.   During this phase, the process performance is statistically stable, but it is not producing output that meets customer expectations. It is a major breakthrough to bring the process to a stable state, but it is not an answer in itself. A statistically stable process that does not produce products that meet customer expectations requires 100 percent screening and is very expensive to operate. As a result, the improvement team must move the process quickly into Phase C.

Phase C: Step-by-Step Improvement.   Once the process is under control, we need to carefully analyze the output of the process that does not meet customer requirements to define the root cause of the deviation. Priorities should be given to the potential improvements, based on cost, ease of implementation, and estimated impact. A plan should then be developed that allows the potential improvements to be implemented and evaluated one at a time. After each change, the process is given time to stabilize so that the change's impact on the total process can be evaluated. This corrective action cycle should continue until a process capability of 1.4 is reached. At this time, 100% inspection of the product can be dropped.

Phase D: Customer Ship.   We now have statistical evidence that the output meets customer expectations, and we can start delivery to our customers with a high degree of confidence that they will be satisfied with the output.

Process Control

Once the process is under control, ongoing measurements must be made and evaluated to keep it under control. These are several easy-to-use methods that exist for identifying, measuring, and displaying defects:

- *Check Sheets:* Used for data collection and organization.
- *Line Graphs:* Used to display data. There are several kinds, including line graphs, bar graphs, and "pie" graphs.
- *Histograms:* A type of bar graph, used to display distribution of whatever is being measured.
- *Pareto Diagrams:* Another type of bar graph, showing data classifications in descending order from left to right.

Measuring Processes.   To measure the effectiveness of a process or how it is meeting requirements, it is necessary to use statistical methods. Brief descriptions of three such methods follow:

1. *Sampling:* A method used to obtain information from a portion of a larger population when it is too expensive or time-consuming to measure the total population.
2. *Data Collection:* A method with three purposes:
   a. Analyze a process

    b. Determine if a process is in control

    c. Accept or reject a product

3. *Stratification:* A special sampling technique utilizing information from subgroups (strata) of a larger population. Used in conjunction with histograms that indicate an abnormal distribution.

Process Evolution.   Once the measurement characteristics are defined and the measurement system is in place, the process improvement team should make a thorough analysis of the total process to identify characteristics that lend themselves to statistical process control methods. This would normally be completed before Level III Qualification would be granted to a process.

Process Control Charts.   A great deal of time and effort was spent to evaluate and bring the process under control as it was qualified. The problem now is how to keep the process from degrading and going out of control. One of the best ways to accomplish this is through the use of control charts.

    The process control chart theory is based on the fact that all things have variation between different entities. As the process improvement team and process employees become more and more familiar with the process and become very adept at continually reducing variation, a very interesting pattern will begin to show up in the control charts. The charts will literally show that the sample data are becoming tightly clustered about the desired nominal condition. This is the point at which new "natural" limits for the process need to be calculated and placed on the control charts, to be sure that the process does not drift away from this improved state. This iteration can go on continuously, and this is the state referred to as "never-ending process improvement." It is an extremely desirable state, and organizations that reach it find themselves in strong competitive positions, continually able to beat their cost objectives and bid aggressively for new work.

    We are often asked, "Why try to reduce variation below the specification requirements? Why not use up the total specification range?" There are two good answers to these questions. They are:

1. As the output variation is reduced within the product specification, there is less need to inspect output to be sure it is good.

2. Experience has proven that the closer to the nominal value the 3 sigma values are, the better the end product will perform. We know of examples where product performance has improved over 100 percent by having a process capability of 2 compared to the same product that was 100 percent to specifications.

    Figure 11.5 shows a typical process improvement cycle. During Phase A the process is unstable and the output should be inspected 100 percent. In Phase B the process is able to produce output to specification, and output only needs to be measured on an occasional basis to prove there are no negative trends. During Phase C the specification limits are easily met and measurements are made to ensure the process does not degrade.

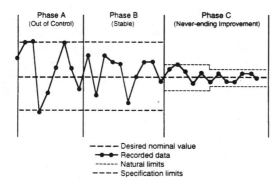

Figure 11.5. Process improvement cycle.

Design of Experiments

In process capability studies, one factor in the process is examined. A Designed Experiment includes many factors that are looked at once. The data is then examined in a number of ways, facilitated by the manner by which the data was obtained. There are two approaches to design of experiments: the classical approach, and Taguchi methods. Tools used are fraction factorials in the classical approach, and orthogonal arrays in the Taguchi methods. Classical design looks for the optimum, while Taguchi looks at variability. Design of Experiments is a more powerful method to reduce variation than statistical process control.

■ Information Technology

Information Systems

The trend toward more sophisticated computer-based information systems in the design and control of processes continues. There are a number of tools in wide use today, such as:

- Computer-Aided Process Planning (CAPP)
- Computer-Aided Manufacturing (CAM)
- Computer-Aided Acquisition and Logistics Support (CALS)
- Electronic Data Entry (EDI)

CAPP interacts with CAD/CAM systems to develop the process plans. CAM is defined as: "the effective utilization of computers in the management, control, and operations of manufacturing and processes" (adapted from CAM-I Arlington, Texas). CAM is used to control operations including robots. CALS is a DOD (Department of Defense) initiative that focuses on the integration and exchange of data not only internally, such as between design and manufacturing, but also between suppliers and customers. One of the benefits of CALS is a tremendous reduction in paperwork and an increase in the reliability of the data.

EDI focuses on the data exchange between organizations, particularly in the interface between customers and suppliers. The auto organizations make great use of this capability in the planning and control of production, particularly in the coordination of supplies to assembly plants. EDI systems have allowed "Wal-Mart" to deal directly with manufacturers, by-passing distributors and giving the consumer lower prices.

Measurements

The change revolution has arrived to the measurement field. New techniques such as Benchmarking and Activity Based Costing are providing new perspectives. Activity Based Costing (ABC) reduces the inaccuracies brought about by the treatment of overhead costs in standard cost systems. Today many organizations are distributing overhead based on time and materials as opposed to direct labor.

Traditional measurements such as machine utilization, labor utilization, and cost variances have virtually disappeared. The new measurements are:

- Lead Times
- Inventory
- Process Availability
- Defect Rates
- Material Fields
- Capability Indexes
- Mean Time Between Failures
- Variability
- Time To Market
- Time To Customer

The new measures are centered on Quality, Costs, Delivery, and Time

■ How Motorola Woke Up

In May 1974, Motorola sold its television holdings to Matsushita Electric Industries of Japan (MEI) for $108 million dollars. To manage this new holding, Matsushita Electric Industries formed Matsushita Industrial Company (MIC) as their United States manufacturing branch. With a Japanese advisory team that never exceeded 30 members, MIC went to work immediately, restructuring its newly acquired business. Motorola, at the time of the sale, had three plants in the Chicago area producing television sets and a sizable multi-layer management system guiding the operation with vice-presidents heading up each function and directors of each sub function unit (a total of 18 officers). MIC took several significant steps to improve the quality of the product and increase profits:

1. Closed down two of the three plants.

2. Transferred product development to Japan where development work was already underway, eliminating the need for two separate development activities.

3. Eliminated levels of management and unnecessary functional activities.

4. Spent 9 million dollars for new manufacturing and materials handling equipment.

5. Started purchasing 85 percent of their components from Japanese suppliers whose output quality was more stable than U.S. vendors.

6. Changed the production line from a continuous moving conveyor belt to a "free-flow" (buffer conveyor) line.

7. Initiated a concerted program to make the employees feel that management is interested in them and supportive of their needs.

8. Redesigned the products, reducing the number of complaints significantly and making it more manufacturable.

The results achieved were truly amazing:

1. Morale improved greatly.

2. Indirect personnel dropped almost 40 percent.

3. Warranty costs dropped from 22 million dollars a year to 3.2 million dollars.

4. The line reject rate dropped from 140 to less than 6 percent.

5. Individual worker productivity improved by over 30 percent.

6. Total plant output increased by more that two times (through design, process and sourcing changes).

7. Rework and scrap decreased by 75 percent.

As we toured the plant and talked with their management and the employees, we recognized an air of confidence, commitment and an attitude of personal excellence in every member of the MIC team. They display a personal pride in their work, a team atmosphere, a team commitment and pride in the product they produce.

This was a rude awakening for a proud giant that had been one of the world leaders in the early development of the TV industry.

■ Summary

Banks, hotels, and auto manufacturing organizations all have production processes. It is important that all of these organizations identify these critical production processes that impact their external customers so greatly and continuously upgrade them. Manufacturing-type organizations have historically focused much of their effort on these production-type processes. But the service industry has been slow to realize that these processes can, should and must be controlled if the organization is going to compete in the twenty-first century.

The following shows how the three types of organizations approach product and process design and implementation.

New Product Cycle

- *Losers:* A serial activity with one organization completing their task and handing it off to the next.

- *Survivors:* Manufacturing people participate in the design process to ensure manufacturability.

- *Winners:* Seamless process with the same people doing product and process design. The team that starts the product process remains with the task until the product is qualified in the consumer's environment.

Process Qualification

- *Losers:* No formal process.

- *Survivors:* Primary focus is on process capability analysis. If the process is not able to reach a process capability of 1.0, inspection operations are added so that the product can be shipped to the customer.

- *Winners:* Worldwide process compatibility is obtained through a thorough certification of all elements within the process and qualification runs. Ongoing, continuous comparisons between locations ensures complete interchangability of product throughout the process.

Process Controls

- *Losers:* Gate inspections are the primary means of controlling quality. Normally, these gate inspection points are performed by Quality Control.

- *Survivors:* Process control charts are used by management and engineering to determine trends. Data collected by Quality Control or Manufacturing operators.

- *Winners:* Employees use statistical process control to identify and correct trends. They are empowered to solve problems as they arise and to involve management and engineering, if necessary.

Quality Control

- *Losers:* 10 percent of the manufacturing direct labor force are Quality Control inspectors. Gate inspection is the primary means of ensuring quality.

- *Survivors:* Manufacturing employees are empowered to do their own inspections and are held accountable for the quality of work they perform.

- *Winners:* Quality assurance is held responsible for conducting audits of the organization at all levels. They are held accountable for enforcing compliance to procedures.

Automation

- *Losers:* Little or no automation.
- *Survivors:* Automation designed to save labor costs.
- *Winners:* Automation designed to improve product quality and reduce job monotony. Emphasis is also placed upon quick set-up.

Production Process Design

- *Losers:* High volume, serial design.
- *Survivors:* Just in time, lot design.
- *Winners: Continuous pull (single unit build) design.*

Characteristics of the Winners

- Winners implement the right degree of technology and automation.
- Winners perform concurrent engineering.
- Winners have process capabilities of 1.4 min.
- Winners pursue innovation.
- Winners improve the reliability of their products and services.

   WINNERS: DO IT!

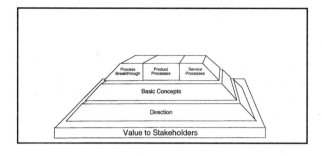

12

Service Process Excellence: How to Best Serve Your Customers

Craig A. Anderson

Chuck Bayless

James S. Harrington

M. Melanie Polack

and

Dr. William D. Wilsted

Organizations are three times more likely to lose a customer over poor service than poor products.
DR. H. JAMES HARRINGTON

■ Introduction

Well, the only important thing is to make the children happy. Whether Macy's or somebody else sells the toy it doesn't make any difference. Don't you feel that way too?   SANTA   CLAUS, 1947

Every Christmas season, viewers across the world are treated to an excellent example of quality in the service sector. They find it not in seminars, classrooms, books, or business conference rooms, but in repertory film theaters and late night television. This fine example of quality service is the George Seaton directed film, "Miracle on 34th Street," starring Maureen O'Hara, John Payne, Natalie Wood, and Edmund Gwenn as the unforgettable Kris Kringle.

The interesting part of this story is that in the movie, Macy's hires the real Santa Claus to play Santa Claus in their busy New York store. When parents tell their shopping problems to Santa Claus, he provides them with solutions—even if it means going to another store! This extra concern for the needs of the customer might seem counter-productive, but it turns out to be a great asset for Macy's. By valuing their customers, the store gained their respect and trust, and therefore their repeat business.

Granted this example is strictly out of the minds of Hollywood. This Santa Claus was an actor hired by a studio, not by Macy's. But there are real-world displays of organizations that realize that the customer is definitely the most important person in the business world.

Perhaps Macy's should have more closely followed this celluloid version of themselves, because it is this type of allegiance to the customer that has put Nordstrom, and other select service organizations, into the forefront of their industries.

■ What Is a Service Industry?

All organizations provide service to their customers, even such "hard goods" organizations as U.S. Steel, GM, Ford, and Boeing. These organizations are not in business to produce products. They are in business to deliver a service to their customer. Ford does not exist to make Mustangs. They exist to service their customers by providing them with a means to move from one place to another while expending a minimum amount of energy. Every product manufacturer has at least some pure service processes that do not deliver tangible products to their customers, probably more than each of them realize. It is for this reason that all organizations can benefit from applying the things that the best service organizations are doing to satisfy their customers.

In their ninth new collegiate dictionary, Webster defines service as "contribution to the welfare of others" and "useful labor that does not provide a tangible commodity." Using those two definitions, one can see that the service industry is a major participant in the everyday businesses of the United States and the world.

For the purpose of this chapter, we need to go beyond the previously listed definitions. Our definition is as follows:

A service process is one whose main contribution to the welfare of others is providing an intangible commodity.

This would include such organizations as Nordstrom, who, even though they do provide a tangible commodity (clothing), gained their main distinction in the industry for delivering outstanding service to their customers.

It is this intangible transfer between the organization and the customer that mainly separates the service from manufacturing processes. When we begin to look at the issues involved with service processes, we see that there are many differences between the two types of processes.

■ Importance of Service Industries to the U.S. Economy

It has been said time and time again, that the United States is in a time of transition. We are transcending from the manufacturing age to the service age, much like we did before, going from a farming to a manufacturing economy. A simple look at basic data shows that the United States is not in a period of transition, but has already transcended to the service age.

Service industries employ, by far, the largest percentage of the working population of the United States. Currently, 75 percent of the jobs in the United States are in the service industry, and that figure is projected to keep growing until it reaches the staggering figure of 88 percent in 1999.

The largest single work classification is government: national, state, county, and local levels. And the government is, of course, a service organization. The next largest job classification is agriculture, followed by another service industry: health care.

With service encompassing two of the three biggest activities, it becomes obvious that service already is the dominant force in today's economy. This dominance of the economy is not just seen in the United States, but also in almost all parts of the world. The service industry has become the major employer of the world's population. For example, in the United States and Japan, service industries account for 60 percent of the gross domestic product (GDP).

■ Service Is America's Number One Problem

As we continue to lose the "smokestack" and manufacturing industries, more and more individuals will count on service organizations for their futures. We need to focus the same effort in improving quality and productivity in the service industry, as we have in the manufacturing area.

It has been in vogue for economic analysts to criticize the manufacturing industries, pointing out that such American industries as automobile and steel, have not kept pace in productivity and quality issues with their counterparts in places like Japan. The truth of the matter is that the worst Ford assembly worker's output error rate is much better than most economists'.

While it is easy for someone to notice a problem with a product, it is much more difficult to pinpoint a service problem. For instance, when the specifications on an assembly line at Ford are askew, the result is obvious, because the parts suddenly stop fitting together. But how do you pinpoint the exact cause of the bad service in a hospital? In fact, how does the average patient even know if they are being serviced correctly?

This differing nature between service and product activities, has been the reason why the majority of improvement processes have been aimed at the products. Manufacturing activities have improved greatly over the past 15 years. In the 1980s, productivity in the industrial sector of the United States rose 34 percent. Unfortunately, during the same period, productivity in the service areas has remained flat.

If the service industry had kept pace with the improvement shown by manufacturing in the 1980s, the additional revenue generated by taxes would completely pay for the national debt. This is a phenomenal figure, but totally believable when one thinks of what a 34 percent increase in productivity would mean to a sector that includes over two-thirds of our businesses.

To highlight what has happened, consider the following example of the banking industry. At the end of the 1970s, America had six of the top 10 banks in the world. Today the first U.S. bank is ranked number 25. If you consider assets control as being the primary measure of a bank, 8 out of 10 of the largest banks are Japanese. In fact, the most productive bank in the United States is Fuji Bank and Trust.

While manufacturing firms have been actively pursuing such goals as "zero defects", service firms have been turning out hideous quality percentages. Government organizations provide a wealth of prime examples. The IRS phone lines give wrong answers 38 percent of the time. The U.S. census missed 2 million people. As a result of including an incorrect class of personnel in a computer program, the Federal Financial Help Program in one city paid out an additional $18 million.

Why Do the Service Processes Need to Be Different?

There are certain innate properties involved with service processes that make them distinctly different from that of manufacturing processes. These properties are as follows:

1. Services are intangible. They cannot be measured, tested or verified in advance of delivery.
2. Services reflect the behavior of the provider.
3. Once the opportunity is missed, there is no second chance.
4. The customer is part of the process.
5. The service cannot be recalled, like a manufacturing product.
6. Quality assurance activities must be completed before the product is produced. You cannot inspect quality into a service.
7. The more people the customer interacts with, the less likely he or she is to be satisfied.

Service Processes Major Dilemma

In service processes, unlike production processes, quality and productivity often are in conflict. People tend to relate personal attention to the quality of service, which often distracts from productivity. The desire to provide the customers with outstanding service while trying to service more customers per employee, puts service industries in a "Catch 22" situation. For it seems reasonable to believe that high quality and high productivity of service are in contention with each other, once you have eliminated all errors, but this is not true.

Having been an employee at Waldenbooks, I was given the chance to experience this "Catch 22" firsthand. The manager presented each employee with two directions. The first direction is to be totally responsive and helpful to the customer's needs. The second direction is to keep the book shelves stacked at all times. In other words, the first direction was quality, and the second was productivity.

What I, and other employees, found out was that if we obeyed the first direction, we could spend 30 minutes helping someone find the right zoology book, causing us to fall behind in our other duties, resulting in a loss of productivity. On the other hand, when we found ourselves behind in our stocking rotations, and needed to increase productivity, many customers were left with unsatisfactory service.

Not just at Waldenbooks, but at service organizations throughout the world, the issue of simultaneously obtaining high productivity and quality will become the major concern. No matter what the service industry is, there is a manufacturing or assembly line facet to each industry's business. The manufacturing activity at Waldenbooks is stacking books, at Chase-Manhattan it is processing checks, on British Airways it is loading luggage. There is always an assembly line function to any business.

This assembly line function is what causes problems, because customers are people, not products, and they need to be treated as such. Customers want both quality and productivity. The problem is that they sometimes want it at different times. You hope that the airline reservation clerk will take the time to help you with all your needs and questions (quality), but you also want him or her to speed through the customers in line in front of you (productivity).

In a different circumstance, you may even want different degrees of quality and productivity with regard to yourself. At Victor's in San Francisco, I want a nice leisurely evening of dining. At Jack in the Box in Los Gatos, I'm really disturbed if I have to spend more than 10 minutes waiting in line for my order.

We see from these examples that dilemmas in service industries arise from the difficulties intrinsically associated with the areas of quality and productivity. The successful service organizations will be the ones that can achieve the right balance between these two areas for their particular customers.

■ Characteristics Involved in Service

Given that there are so many different types of service industries, there is bound to be an almost limitless degree of issues to be dealt with, related to improving service. For each unique service area, and indeed each individual, the service firm will have at least a few small differentiating facets of their business that they will need to deal with to be competitive. The issues concerning Delta or American Airlines are not the exact same ones that will concern Safeway or Wal-mart.

Understanding this we still can see that there are at least five main issues that need to be addressed by all organizations that desire quality service. These areas are:

1. Having user-friendly interfaces—Bank of America customers want an easy-to-use Versatel machine that does not take much time to learn to use. People want their services to be as simple and as easy to use as possible.

2. Providing job-related training—Customers do not want to deal with someone that is not knowledgeable about the service that the customer is trying to acquire.

3. Developing and maintaining an external customer-driven and focused organization—This means that all employees must not only give their utmost to do and provide for the customer, but also that the employees must realize that they are ambassadors of their organization.

4. Shortening response time—To keep customers, you need to respond to their needs as quickly as possible.

5. Developing empowered employees—Organizations need to design their structure so that the front-line people can take care of problems when they arise.

■ Major Classifications for Service Organizations

Although our wide, generous, encompassing definition of a service organization makes it hard to find classifications that can accommodate all service organizations, a vast majority can be labeled into the following eight categories. These eight categories often are broken down into several subclassifications.

- Travel
- Health Care
- Financial
- Restaurant
- Retailing
- Entertainment
- Utilities
- Government

Although all eight sectors are vitally important to the economic health of the United States, we have chosen for the purposes of brevity to highlight only three very different classifications:

From these three comprehensive examples, we will see how best to improve our service processes.

■ Service Quality in the Banking/Financial Industry by Dr. William D. Wilsted

Introduction

The banking/financial services industry is an important one in the United States economy by virtue of its sheer size and the central role it plays in the working of our national economy as a custodian of the payment system. Total assets of the 11,360 commercial banks in the United States totaled $3.477 trillion at the end of 1992.[1] Further, the services of banks and other financial institutions are used by nearly all adult individuals and nearly 100 percent of all business and organizations in the country. For these reasons, a review of improvement opportunities and practices in banking and financial institutions is important to any discussion of total improvement management in the United States.

Major Problems Facing the Banking/Financial Industry

Today the banking and financial services industry segments are experiencing historic challenges. Commercial banks have been through their most threatening years since the Great Depression, and now are emerging from a decade in the eighties of falling earnings, record numbers of closures and the radical changes in industry economics brought about by "deregulation and reregulation." The plight of the Savings and Loan industry has been a national scandal. Brokerage companies had their share of difficulties prior to banking, resulting in numerous acquisitions, mergers, combinations and failures. And today, many experts believe that the insurance industry is headed for their own "shakeout," with similar, if not even more massive threats to financial soundness than banking and brokerage experienced.

While many different factors account for the difficulties of each of these financial services industry sectors, some things are not common to all. These include "uneven regulation," chaos in traditionally stable revenue/cost relationships, high-cost distribution systems that cannot compete with lower cost competitors from other sectors, and massive over-capacity leading to cut-throat price competition and its concomitant downward spiraling of earnings. Indeed, there is a critical need for improvement in every aspect of operations in these industries, from customer interface processes right along the whole length of the value-chain into the bowels of back room operations.

Comparison to Foreign Industries in the Same Field

While financial services markets have traditionally enjoyed some protection from competitors through regulation, these barriers are falling. Today, more that 25 percent of bank assets in California are held by foreign owned banks, and global competition is entering our traditionally "protected" financial services markets at a accelerating rate. Foreign competitors often have advantages over U.S. companies in U.S. markets. These include easier access for foreign companies to enter U.S. markets than vice versa, often a lower cost of funds for foreign competitors, and cooperative arrangements between foreign companies and their government, suppliers and customers that favorably affect their operations in the U.S. marketplace. Structural differences also account for significant competitive disadvantages for U.S. companies. For example, many foreign banks can sell insurance or brokerage services directly and hold equity in companies they lend to. Deutche Bank is reported to hold 28 percent of the shares of Daimler Benz. These activities are prohibited for U.S. banks.

Quality and Productivity Measurements and Trends

Productivity and Quality are important competitive factors in banking and financial services. These industries have a tradition of accuracy and reliability. But, historically, they have lagged manufacturing in developing measures and techniques for improvement. Productivity measures have been relatively crude, and driven by operational concerns—not centered on customer needs and expectations. For the majority of banks, productivity measures seldom go little beyond such general measures as the dollar amount of assets or deposits per employee, the number of customer relationships per employee, or the number of processing errors per thousand items. And until recently, quality initiatives primarily were based upon inspection and reducing errors in back room operations areas.

Status of Improvement Activities

But this relatively low emphasis on quality, improvement and customer satisfaction is now beginning to change dramatically.

Research by this author, A Survey of Quality Practices in Banking[2] indicated that

a. Less than 20 percent of the nation's then more that 12,000 banks had initiated a formal Quality or Customer Satisfaction program.

b. The majority of the top 100 banks had done so, but their programs varied widely in breadth, depth and sophistication.

c. Where most early bank quality programs started in the back room, the focus was shifting to customer contact areas.

d. Some had well developed measurement programs, or customer complaint processes, or other limited initiatives.

e. Less than two dozen banks had comprehensive TQM programs.

f. A handful had well developed, comprehensive TQM programs and ambitions to compete for the Malcolm Baldrige National Quality Award.

Eight Quality Strategies

Another study by this author, Eight Quality Strategies[3] identified eight separately identifiable quality strategies in use in the banking industry. The implementation of these strategies ranged from superficial to comprehensive, but all quality initiatives found to be in use by banks fell into one of the eight basic categories. They are:

1. High Rev—"Rev up" the employees with highly motivational speakers. This approach to Quality can have a quick, but short lasting effect as a stand alone tactic, but is an essential element of almost all successful long-term Quality efforts we have seen.

2. Claim Jumper—Lay claim to Quality through advertising, without making any other quality effort. Surprisingly, research indicates organizations employing this strategy unanimously claim success.

3. Technocrat—Use technology, job and product-line simplification, scripting, and highly routinized procedures to sharply reduce the chance of human error.

4. Keep Them Calm and Happy—Reduce or eliminate negatives of business that are the target of many complaints, i.e.: reduce wait lines, extend hours, implement telephone service, express lines, provide a place to sit, etc.

5. Complaint Solver—Focus on solving customer complaints rather than on eliminating errors or poor service.

6. Quality Guarantee—Promote service Quality through a Quality guarantee of customer satisfaction. Waiver of fee or cash back are both used increasingly. Research in banking indicates only 0.5 percent of accounts are refunded at banks guaranteeing customer satisfaction.

7. Quality Segmentation—Recognizing the 20-80 Rule, any or all of the above are implemented only for selected customer segments (starting with upscale).

8. Quality Culture—The most complete and effective approach to Quality. Encompasses all seven of the above, with infrastructure to support and sustain.

Update of Eight Quality Strategies

In a 1990 Update of Eight Quality Strategies[3], some equally interesting, but not surprising findings were reported.

1. The Quality Culture Strategy (No. 8) is now commonly referred to as *Total Quality Management* (TQM) among quality practitioners.

2. While early success was claimed in 1987 by organizations using one or more of the eight strategies, there was no evidence that early benefits of implementing strategies 1 through 7 was sustained over the three-year period.

3. Only *Total Quality Management* (Strategy 8) showed sustainable competitive advantage over the three-year period.

The findings of this Update of Eight Quality Strategies support the proposition of Total Improvement Management.

Data from The International Quality Study (IQS)[4] indicates that only a small percentage of the participating banks (fifty-four leading U.S. money center, super-regional and regional banks) made use of formal quality tools.

It is believed that other sectors in the financial services industries are at about the same place in adopting formal improvement programs—varying from a majority doing little or nothing to a handful in each sector making world-class progress.

How Organizations Are Improving

But the low emphasis on quality and improvement in banking in the late eighties is now changing dramatically. Evidence of how banks are improving is provided for the consumer banking line of business by The Banking Industry Report[5] of The International Quality Study. IQS collected data to attest to the strength of banks' intentions to change during the next three years. The following are the highlights of these projections.

- A greatly enhanced strategic emphasis on quality will be taking priority over cost cutting or meeting schedules.
- The emphasis on reliability, performance, convenience, responsiveness and adaptability of all products and services will approximately double.
- Banks believe that quality will become the most important element of reputation.
- The ability to meet major goals such as profitability and loyalty is expected to double through the increased emphasis on quality.
- Quality will increase more than any other factor in influencing banks' strategic plans.
- Even banks' staff groups (e.g., legal) will double their quality efforts.

The International Quality Study further reveals some exciting trends in banks' move toward more customer-driven products:[5]

- "Banks expect to double the rate at which they use modern 'consumer goods' marketing tools such as focus groups or market analysis.
- Departments not involved in new product development in the past will now be an integral part of the product development process.
- Customers will be added to internal product design and product review teams.
- Customers' expectations will be identified beforehand, through market analysis, instead of afterwards, through complaint analysis.
- "The customer is always right"—while not practical in all cases (e.g. in granting credit), will become about twice as common as it is today as a fundamental approach to business.
- Every channel available to listen to the customer will be used, from surveys and 800 numbers to benchmarking and statistical analysis and visiting customers in person."

Who Are the Customers and What Are Their Expectations?

Bank customers include almost the entire adult population and nearly every business and other organization in the country. Both businesses and organizations are becoming more sophisticated in money management, the understanding of the time-value of money, and awareness of the ever-increasing number of options they have in financial products, services and providers. But in other research by this author, Two Sides to Quality[6], it was determined that bankers (and other sellers of products and services) often fail to understand their customers' or potential customers' expectations. In fact, the data revealed that sellers' perceptions of how customers defined quality was often the mirror image of their customers' perceptions. (See Figure 12.1.) These findings were of sufficient interest to help motivate the Cover Story of the June 4, 1990 issue of *Fortune.*[7]

Interestingly, while bankers (and sellers from manufacturing and high tech industries) placed greatest emphasis on the "effective" dimensions of how well the product or service performed in tangible ways, customers consistently demonstrated greater concern about the "commitment" dimensions—the commitment they felt from the bank to their own personal, unique financial needs and to themselves as individual customers.

Indeed, the evidence is strong that there are many special dimensions to banking services. Figure 12.2 indicates some of these.

These differences along with others have important ramifications for improvement efforts. In one sense, they make improvement through standardization more difficult. Second, they make sensitivity to individual customer needs even more important than in the case of manufactured products.

What Improvement Tools Are Being Used Today and What Are Future Trends?

The most comprehensive review of the use of improvement tools and trends is (again) The International Quality Study. In the area of operations improvements, the IQS data[9] indicates:

- Designing and building quality into services instead of "fixing it in" will increase significantly.

- Technology is expected to be used as much to improve quality as to cut costs or increase revenue.

Sellers' perceptions of quality		Buyers' perceptions of quality
± 70%	Effective	± 10%
± 20%	Responsive	± 20%
± 10%	Committed	± 70%

Figure 12.1. The two sides of quality (Research by Dr. William D. Wilsted).

Manufactured products	Bank services	Consequences for banking services
Tangible	Intangible	Difficult to compare from bank to bank
Durable	Not durable	Proliferation of product/service combinations and permutations
Inventoryable	Consumed at delivery	Capacity must cover peak demand
Impersonal	Highly personal	Each transaction is somewhat unique
Easily standardized	More difficult to standardize	More training required for delivery
Quality less affected by customer	Quality (Interpersonal transaction) highly affected	Customer is both consumer and influencer of how well transaction goes
Lower emotional tie	High emotional tie (It's my money you're dealing with)	Greater customer sensitivity to variation from expectation

Figure 12.2. A Comparison of Product and Service Attributes (by Dr. William D. Wilsted.)

- Most of the considerable future growth in technology will go toward improving delivery systems and paper productions processes—not core accounting systems.

- The rate at which business process improvement practices are used is predicted to quadruple.

- The amount of management information devoted to quality is expected to triple.

- About 20 to 40 percent of employees will use formal quality tools in the future as opposed to about 5 percent in the past.

The IQS[9] also documents banks' expectations in the area of people improvements:

- Greater understanding of the strategic plan at all levels—including employees, suppliers and investors.

- Dramatically greater use of relationship training—by a factor of five.

- Close to 100 percent empowerment of employees to fulfill the requirement of their jobs.

- Use of employee teams by virtually all banks to solve quality problems.

- Significant change in assessment of management to rely on quality, profitability, and team performance.

Outline of a Typical Improvement Process In Banking

The International Quality Study showed that banks intend a sharp increase in the use of quality and improvement tools and paradigms. Of particular interest are two that are increasingly employed in banking: outsourcing and strategic alliances. There is a major trend in outsourcing nonstrategic operations in banking today. It is increasingly common for banks to outsource complete functional areas to achieve better performance, lower costs and other advantages.

Another improvement trend of note in banking is the recent move by three of the nation's largest super-regional banks to combine efforts to develop the next generation of processing hardware and software. By bringing together sufficient combined volume and resources they plan to exploit the next generation of processing technology on a level that no single competitor could achieve alone. Only a few years ago such cooperative efforts would have been looked upon by banks, their competitors, and regulators as suspicious.

The International Quality Study data suggests a valuable set of prescriptions, depending on whether your present performance is lower, medium or higher, as indicated in Fig. 12.3.[10]

Banking Industry Summary

Banking has a long tradition of accuracy and reliability in the information processing function of its operation. But as recently as the late eighties, use of quality and improvement tools was limited to a small minority. Empirical evidence now shows that banks are embracing quality and improvement methods and paradigms at an impressive and accelerating rate. Certainly the expressed intent and recent resource commitment by banks lead to three conclusions:

1. Quality and improvement initiatives are imperatives—not options.
2. Competitive success will require significant further improvements in costs and the ability to achieve superior customer satisfaction.
3. Total Improvement Management (TIM) provides the tools and paradigms to do both.

Lower Performing Banks (Losers)	Use basic tools to correct fundamental process problems in serving customers, and avoid "trendy" tools which don't address your bank's problems.
Medium Performing Banks (Survivors)	Use process improvement tools to tune your processes and leverage what you do well.
Higher Performing Banks (Winners)	Use sophisticated processes to "fine tune" your processes and anticipate changing customer needs.

Figure 12.3. Summary of IQS prescriptions by performance level.

■ Service Quality in the Health Care Industry (by M. Melanie Polack & Craig A. Anderson)

Health Care Industry Introduction

One may wonder why this particular book includes a section on health care quality. The answer is simple, clear and direct. United States health care represents a major portion of our economy and impacts everyone; individuals, corporations, State and Federal budgets. In 1992, U.S. Federal Medicare Expenditures alone totaled $138 billion. This figure is equivalent to the total Domestic Expenditures of the entire country of Finland, which has the fifteenth largest total Domestic Expenditures of any country worldwide. The United States spends more in total on health care than any other country ($751 billion in 1991), with Japan ($223 billion) and Germany ($133 billion) in second and third places, respectively.

The rising cost of health care in America is posing an interesting challenge to our competitiveness in a global marketplace. At approximately 13.9 percent of the 1992 U.S. Gross Domestic Product, our health care costs are more than twice the cost of health care for the countries whose products and services we compete against. (See Fig. 12.4.)

As Fig. 12.4 illustrates, the current problem is expected to get worse as the U.S. cost of health care continues to escalate, while competing countries hold their costs in control. This fact is the primary reason the U.S. health care system is under a microscope, and also why federally sponsored Health Reform (The Clinton Plan) is receiving national attention.

The most important reason for a section on health care is that we all can and must work toward a higher quality, lower cost system. The principles of Quality discussed elsewhere in this book are directly applicable to the delivery of patient care services. This section will explore and explain how health care may be unique, but is also a prime candidate for continuous improvement. It should be noted that if and when Health Care Reform is adopted, many of the incentives and issues discussed may dramatically change.

Are Quality Solutions for Health Care Really Different from Other Organizations?

A great debate is ensuing as to whether the total quality management approaches heralded in manufacturing organizations can be translated to work in health care

	1960	1970	1980	1990
U.S.	5.4	7.5	9.3	12.5
Canada	5.4	7.3	7.4	8.2
Germany	4.8	6.1	8.6	8.2
Japan	3.0	4.6	6.5	6.7

Figure 12.4. National healthcare expenditures as a % of GDP.

as well as other service organizations. The health care industry is split on this point. Some health care organizations consider it a passing fad, some have tried to implement total quality and had a negative experience, and some are staking their entire future on the implementation of TQM with their delivery systems. There are multiple elements in health care that need to be clearly understood as we examine this question.

The health care industry is composed of numerous unique entities including hospitals, nursing homes, physician groups, home health agencies, equipment suppliers, health maintenance organizations, and others. Within the hospital category are multiple delivery components—acute care hospitals, specialty hospitals, teaching hospitals, not-for-profit hospitals, for-profit hospitals and many more. Let's examine characteristics of the not-for-profit acute care hospital to demonstrate how health care organizations differ from other businesses.

First.   The not-for-profit status creates an interesting mindset. Until recently, hospitals received payment for their costs regardless of what those costs were. Although the payment rules have changed, the competitive requirements for efficiency and effectiveness have only recently been felt. Clearly the economic drivers which direct for-profit entities have been missing in health care.

Second.   Quality patient care is for the most part intangible. While the automotive industry has J.D. Powers to provide quality score cards and many other manufacturing industries have Consumer Report type publications, health care is provided and selected based on intangible criteria. Patients most often rely on their confidence in their doctor as the only criterion for evaluating quality.

Third.   The purchasing agent in the hospital, the physician, is not an employee. Doctors order the tests, determine services needed, and in many cases determine how long the patient stays as an inpatient. However, the doctors have no economic interest in the efficiency of the hospital or resources used. They are independent parties. Consequently, budgeting and managing resources is dictated by outside organizations, such as insurance organizations.

Fourth.   The individual consuming the products and services of the hospital, the patient, is doing so at a personal time of crisis. They are physically ill and their health is more important to them than anything else. They are worried, often afraid and extremely uncertain about making decisions. Contrast that mindset with the purchaser of most goods and services who make a choice when and where they chose to act. Buying a car, television, clothes or even food is not a life and death situation but usually a pleasurable planned for event.

Fifth.   Hospital care has been for the most part a local market service, usually with few if any competitors and no competitor that can easily or quickly move into your market. Individuals desire to be hospitalized in close proximity to their home and family. They have difficulty differentiating care provided by different hospitals in the same town and usually rely on the advice of physicians. As a result, until recently there has been little pressure for hospitals to compete against one another on the basis of cost or value. Unlike fast food restaurants who open new franchises over night, or hotel chains who differentiate themselves through National marketing campaigns, patients rely on the friendly blue and white Hospital signs to lead them to their hospital of choice.

We could go on with several other elements that distinguish hospitals from other businesses, but in the end all we could say is that hospitals are unique, just like any other businesses. If Cadillac Motor Company and the Ritz Carlton organization can be so different yet both win the Baldrige Award, why can't hospitals? An examination of several quality principles will help answer that question.

Who Are the Customers in Health Care?

Most quality thought leaders would agree that meeting or exceeding customer expectations is a primary quality goal. The concept of customer usually strikes a clear vision in most minds, someone who purchases a good or service provided by or sold by an organization. The concept of customer in health care is more confusing. First of all, the individual deciding what services to purchase is not the recipient of those goods. The physician, as the decision maker, dictates which hospital to use, when to use it, and what combination of tests and services is needed. But we have already said that in many cases the physician is an independent agent who does not directly benefit from the quality or cost of these services. In fact, the legal environment has put the physician in an unfortunate position where he often orders more tests than may be needed to avoid a malpractice claim. The physician does not, however, have to suffer the consequences of these choices. He or she is not the one who must lie in the hospital bed, get stuck by additional needles, or have the privilege of consuming the most unusual hospital cuisine for a few extra days.

Maybe, then, the patient is more of a customer. Certainly the patient is the recipient of hospital services. But the patient has almost no ability to evaluate whether he is receiving good or bad services. In crisis, the patient accepts whatever is ordered, and meeting or exceeding his or her expectations is often irrelevant. Additionally, patients do not directly pay for what they receive. In most everyday transactions, customers part with personal wealth to obtain goods or services. Hospital patients, however, usually have some form of insurance that they rely on to pay for what they receive.

Perhaps then the insurance organization is the customer. They ultimately pay the hospital bill. Insurance organizations have been significantly removed from the patient care process. This fact is changing however, and numerous health theorists speculate that managed care organizations combining insurance and care delivery will be the future patient care models of choice.

These multiple customer groups increase the complexity of designing a quality effort for a hospital based care setting. Even when individual expectations can be clearly identified, they are often counter effective. Accordingly, the central focus of most quality driven businesses, the voice of the customer, is most challenged in the health care industry.

Measuring Quality in Health Care

Another critical component of quality driven organizations is the process of measurement. World class organizations measure quality at all levels and on a continuous basis. And measurement usually starts with the quality of the output. We have already alluded to the fact that the automobile industry has J.D. Powers to thank for comparable measurements of consumer purchased automobiles. Outcome data has been one of the more significant deterrents to the adoption of TQM in health care. Health care leaders profess that providing patient care is not like building automobiles. Every patient is different, they contended, and the success of the patient care process could not be subjected to traditional means of output measurement. Most of this rhetoric occurred at a time when hospitals still received cost based reimbursement for their services. There was no incentive to define or measure outcomes.

In the late 1980s, organizations began inventing new and more sophisticated mechanisms to measure the effectiveness of patient care processes. Many of these measurement applications were associated with inspection and external certification agencies. In particular, The Joint Commission on Accreditation of Health Care Organizations created a complex inspection based approach which hospitals have adopted to measure, monitor and report on quality. This system, which has been evolving since the 1950s, created a "bad apples" approach to measurement—finding the individual who had performed outside the norm and changing his or her behavior.

Only recently have health care organizations begun to see the merits of total quality oriented measurement. And in doing so they are developing a richer appreciation for measuring quality through cycle time reduction, customer satisfaction improvement indices, error reduction and other quality driven measurement criteria. Quality is achieving a broader definition and, in turn, measurement is becoming an essential element of the health care delivery process.

Leadership Skills Development

Most successful quality oriented organizations can point to organizational leaders who have led the process through the recognition that their role as a leader is not to catch people doing things wrong, but to create an environment for people to become heroes. These leaders have had both the responsibility and the authority to make things happen. And these leaders usually have an active Board of Trustees who take a serious role in looking out for stockholders' interests.

Contrast this scenario with a typical hospital. The average tenure of a hospital CEO in the United States is less than four years. While they have the responsibility to run an effective hospital, they often have little authority over the medical staff who ultimately determine resource consumption, cycle time, and selection of supplier products. In addition, the Board of Trustees are frequently composed of individuals who are not compensated for their time, do not report to shareholders, and consist of community interested individuals who do not understand the health care delivery process and are certainly not going to challenge the process of medical care.

Thus the environment for leadership development is less intense than in industry. Hospital middle management is also stuck in historical management/control paradigms. Their performance has been evaluated the same way for years—how effective they are at producing defined levels of budget driven production quotas (i.e., lab tests per FTE) and how little controversy they create with the medical staff. As health care learns new leadership patterns, where individuals are evaluated based on their role in advancing defined quality, adding to the effectiveness of teams and creating an environment for individuals to become heroes, the quality of patient care will surely improve.

Technology Applications

World class organizations depend on the synergy of people, processes and technology to interface in improved services and products. This objective is challenged in industries where technology is advancing faster than organizations can build cultures supportive of the human side of quality and focused on process driven business objectives. No industry experiences a greater challenge with this issue than health care. Patient care driven technology is changing faster than organizations can economically afford or assimilate the technology into their care processes. Often such technology is mandated by physicians who, if not satisfied, will take their patients to other hospitals. Concurrently, these technology advances are demanded by patients, already worried about life threatening issues, who hear from friends or through public communication that such technologies are critical to the "best" care processes.

The movement in health care to define and measure patient outcomes is also putting new pressures on information technology. Hospital based systems have grown through band aid based architecture, largely because many hospitals have not understood the power of information technology and are waiting for clear requirement definitions from external financing agencies before they make major hardware and software purchasing decisions.

In industry, new technology has been driven by customer expectation, whereas in health care, technology has taken a stronger scientific and research direction where it is invented, marketed and "sold" to the industry.

Rapid advances in technology will continue to reshape how we think and provide services or make products. But health care has two unique issues that will impact its direction. First, the patient care process is still driven based on emotional issues that are often life and death in nature. This creates a situation where market demand is easy to create and justify. At the same time, however, the increasing use of sophisticated technology is one of the more significant drivers of the cost of health care. Technology helps keep low birth weight infants and the frail elderly alive for longer periods. As a society, we hold dearly onto the extension of life at all possible costs, yet technology is driving our costs beyond our reach. As much as 25 percent of all health care costs are consumed in the last six months of life. At some point, society will have to choose between unlimited technology and limited financial resources, and that time is coming soon.

The Buying Cycle as it Relates to Value

When health care needs do arise, our purchasing decision is largely based on the advice of our doctor—advice we seldom understand and almost never question. We consume or "buy" what we are told to buy, when we are told to buy it, and we rarely know whether what we are buying is good, fair, or poor in quality, cost, or value (we almost always assume it is good). This lack of a rational buying process creates an environment where quality is misunderstood and organizations are less concerned about being able to communicate to the marketplace the value of their products and services.

Healthcare Summary

Several sections of this book have focused their summaries on strategies for losers, survivors and winners. We shall take a similar approach in this chapter, with one exception. Because of the lesser competitive environment in health care, "loser" organizations will probably not be forced out of business in the short run if they fail to modify their strategies. In the long run, however, these organizations will most likely not survive. Therefore, the strategies for losers need to be articulated to help them get to "survivor" status and survivors to the "winner" status. At the same time, failure to implement these strategies may send your organization in the other direction. The following strategies are based on results of the International Quality Study.[4.]

Leadership

Coaches who can outline plays on a blackboard are a dime a dozen—The truly great coaches get inside their players and motivate."   VINCE LOMBARDI

The single most important quality issue facing health care organizations is turning managers into leaders. Whether right or wrong, health care has been stuck in a patient care paradigm that is about to change and change very rapidly. This change will require a new health care model designed around customer expectations—patients, physicians, insurance companies, and government. With the average tenure of a hospital CEO less than four years, there is far too little sustained innovative leadership to champion quality initiatives. Mid-level managers (department heads) will also need to transition from good managers to great leaders. This issue will clearly separate the winners from the losers.

Losers: With the pace and degree of change likely to accelerate in health care, organizations with traditional management styles will surely fail.

Survivors: Survivors will learn to create quality based cultures which are flexible to new delivery models. These organizations will turn out to be the foundation of the emerging American health care system.

Winners: The future health care leaders will be innovative, redesigning old systems into new integrated delivery models. These organizations will be saturated with energized leaders.

Training

Anyone who stops learning is old, whether at twenty or eighty. Anyone who keeps learning stays young. The greatest thing in life is to keep your mind young.
                                HENRY FORD

The total quality movement in health care is focused far too heavily on training and far too little on leadership and culture development. More specifically, the health care industry is focusing an unbalanced amount of energy on training associates in the tools and techniques of quality. It is almost always assumed that training is and will be synonymous with improvement.

Losers: Almost all types of training will help lower performing hospitals (losers) improve to the next level. Customer focused training deserves the highest quality priority. But the real priority for these lower performing organizations should be organizational alignment and leadership formation. Training alone will never dig them out of their last place position.

Survivors: Need to more carefully focus their training plans around structured curriculum. Fewer elective courses and more courses focused on leadership development will be most successful. Continuous training will be required to build the culture necessary to move into a "winner" category.

Winners: Training hours should be limited for top management with a very structured curriculum to achieve specific objectives. These organizations have achieved reinforcing alignment of organization enablers.

Teams

Build for your team a feeling of oneness, of dependence on one another and of strength to be derived by unity.
                                VINCE LOMBARDI

Most hospitals in the improvement movement place a heavy reliance on teams to solve business problems. Teams have been, are, and will always be critical to effectively solve business problems. There are many different types of teams, however, and each type can produce a different result.

Losers: Departmental and Cross-Functional teams are found to be helpful. Departmental teams help focus on basic business needs while cross-functional teams being attention to mission critical processes. Again, however, losers believe teams alone are the answer, and alone results are never implemented.

Survivors: Should focus their teams on customer driven issues that are deemed to be critical to quality patient care.

Winners: Should avoid departmental teams. Departmental teams tend to reinforce barriers between departments and take the organization's focus off fundamental patient care processes.

World-Class Benchmarking

*Everything should be made as simple as possible, but
not simpler.*   ALBERT EINSTEIN

Benchmarking is frequently touted as a critical strategy for all successful organizations. As health care organizations strive to simplify processes, searching for appropriate benchmarks will be critical. However, benchmarking can actually prove harmful in organizations that don't understand the cultural implications of searching for excellence.

Losers: Losers usually are unable to realize benefits from world class benchmarking because their cultures and processes are in need of such significant modifications and improvement.

Survivors: Survivors are more prepared for challenges to stretch their organizations and benchmarking is critical to create this growth. Again, however, benchmarking outside the industry is usually not mission critical, although an increasing number of benchmarks for middle-of-the-road hospitals will come from outside the health care industry.

Winners: Winners must base their benchmarking targets on best performers, and these targets almost always come from outside one's own industry. Accordingly, hospitals in this category will have to challenge their practices to stretch into the world of innovation, creating services for customers who don't even exist today.

Empowerment

*Never tell people how to do things. Tell them what to do
and they will surprise you with their integrity.*
                                                    GEORGE PATTON

Many quality experts claim that creating an environment where employees are empowered will be critical to achieving excellence. While this may be ultimately true in a mature organization, empowering employees may require careful consideration in hospital based care.

Losers: Empowering associates can actually be detrimental for losers. Empowering individuals in organizations that are not organized around processes may allow caregivers to take out of control processes even more out of control.

Survivors: Survivors are more likely to gain benefit from empowerment, but only in limited, clearly defined care processes. Life threatening processes require structured approaches and empowerment may not be best suited for these activities.

Winners: Winners should expect significant benefits from empowering associates. In order to create new modalities of customer focused care, empowered caregivers will be required to bring about innovation.

As can be seen from the previous comparisons of losing, surviving, and winning strategies for selected quality practices, there is no "one size fits all" approach to a quality strategy in health care. While total improvement does hold great promise for the delivery of the highest quality care at the lowest possible cost, the strategy to get there clearly depends on a hospital's starting position.

■ Service Quality in the Utilities Industry (by Chuck Bayless)

Introduction to the Utilities Industry

Not since Thomas Edison first invented the electric light bulb and gas lamps graced the corners of America's downtown quarters, has the electric and gas utility industry experienced such a dramatic and seemingly life threatening period of change. The recent legislative and regulatory changes that have flooded the industry with new competitors combined with increased environmental pressures are altering the very structure of the industry. To be successful in this changing, competitive environment, utilities must achieve two critical objectives. In the short term, they must make improvements in productivity and performance that result in greater value for their customers, and in the long-term, they must change their very cultures to create the opportunity to thrive in their new environment. Utilities are pursuing a number of strategies to meet these challenges, the most successful being those strategies that address both productivity and culture change. Their success at change is heavily influenced by the history and culture of the industry as well as the particular strategy chosen.

Industry Characteristics That Impede Improvement Efforts

Historically, electric and gas utilities operated within their service territories under a virtual monopoly and had an "obligation to serve" all customers in that territory. In exchange for this monopoly status, they were regulated by state public service commissions (PSCs), which authorized them to receive a specific rate of return, and by federal governmental entities including the Federal Energy Regulatory Commission (FERC) and the Nuclear Regulatory Commission (NRC). Utilities traditionally served captive protected markets, with the regulatory focus differing by state and circumstance. Some states emphasized low rates in order to attract industry. Other states emphasized reliability of service. Because their monopoly status protected them from effective competition, utilities often focused on satisfying their regulatory commissions and avoiding adverse publicity. In this noncompetitive rate making environment (where prices are set by regulatory agencies rather than the market), utilities could make improvements in service by increasing expenditures. As long as the expenditures appeared reasonable to the regulator and as long as there was no strong basis for determining what the real (i.e. competitive) cost of the improvement ought to be, utility companies could improve service results by increasing the resources used rather than fundamentally examining and changing the way they did business.

The regulatory process itself was in many ways flawed in that there was a strong disincentive to be different from others in the industry because the associated costs of those differences might be disallowed because they were not "normal" or "standard" for the industry. In addition, every effort to improve performance has an associated risk of failure. The costs of failure were certain to be disallowed whereas the potential improvements (to the extent that they exceeded allowed rates of return) might not be recouped. With a built in disincentive to take risks and with reduced potential to benefit from improvements, the regulatory process shaped an industry that emphasized stability and low risk taking.

There are relatively few utilities that hold their managers accountable for process results. Most managers are still measured and rewarded primarily on how closely they adhere to their budgets. Productivity measurement also varies significantly across the industry. A few companies use relatively rigorous workforce management mechanisms with good tracking of activities and detailed reporting of productivity. In general though, this has been the exception rather than the norm.

In part, this lack of process performance and productivity measures may be attributable to the fact that profits have been essentially regulated with an allowed rate of return determined by the PSC. To maintain this security and the reliability of dividends, the industry has been characterized by periodic temporary belt tightenings. A mild winter or summer with low energy demand often led to year-end belt tightening as budgets were cut to maintain the dividend. Significantly overrun capital projects also often led to temporary budget cutbacks. By and large though, growth was predictable enough and commissions compliant enough, that utilities did not have to control their costs rigorously. In addition, since the shareholder base of utilities was comprised mainly of small investors who sought long-term stable growth, security, and reliable dividends, and as long as the economy was growing and interest rates stable or declining, the stock market was kind. Such is no longer the case. Utilities are struggling to reduce their costs and improve productivity to meet the demands of their changing shareholder base.

What Do Customers Want?

As competition increases in the utility industry, it is increasingly critical that electric and gas utilities continuously strive to meet or exceed customer expectations. Some of the industry's key customer requirements are as follows:

- Supply safe gas and electric service
- Establish reasonable rates
- Respond quickly to emergencies
- Maintain reliability
- Restore service promptly if interrupted
- Send accurate and easy to understand bills
- Respect customers' property
- Deliver consistent quality of electricity
- Listen to customers

These requirements are not particularly different from those five, ten and twenty years ago. What has changed is the customers demand for improved on each of these requirements. With competitors in the market now, they have a baseline for comparing performance that is not frequently beneficial to traditional utility companies.

Culture Change and Performance Improvement

One of industry's major strengths prior to competition was a loyal workforce, a strong culture oriented to service and consistency, and a significant profile in the community. Its drawbacks were lower productivity than in competitive industries and a culture that discouraged change and risk taking. With the introduction of competition, utilities are faced with a difficult balancing act. In the short term, to meet the competition, utilities must improve productivity, which is an issue of skills and techniques. But to succeed in the long term they must change their culture to become more adaptive, risk taking, and responsive to changing targets. This involves questioning the assumptions and behaviors that form the foundation of the industry.

The dilemma facing utility executives is how to increase productivity, increase responsiveness, and increase risk taking without losing the strengths they had, such as employee loyalty, service focus and community service strengths. Utilities have undertaken different types of improvement ranging from doing more of what they have done in the past (such as budget cutting), to more wrenching efforts such as some quick fix downsizing, to newer more venturesome efforts (such as process benchmarking, Total Quality, problem solving, and Business Process Improvement). Efforts at improving performance in the industry have fallen into four camps (with a fair degree of overlap):

- Improving performance and changing the culture simultaneously.
- Improving performance through culture change.
- Improving performance using traditional centralized decision making and focused projects.

- Improving performance through traditional budget management and downsizing techniques.

Each of these approaches is examined in the following.

Continuous Improvement

The regulatory changes affecting utilities in the late 1980s occurred just when the concept of Total Quality began to come into fashion. It was, therefore, only natural that executives would look to Total Quality as a means of addressing the consequences of deregulation. Because Total Quality focuses on both performance and cultural issues, it seemed the approach that offered the utility industry the most long-term benefit. It is also the most difficult to accomplish because the industry's culture and history have not provided it with the necessary skill and behavior base. Several utilities have undertaken total quality efforts with varying levels of success.

Public Service Electric & Gas (PSE&G) of Newark, New Jersey, has implemented an aggressive Total Quality program. "I believe TQ is the key to positioning ourselves for the future," said Lawrence R. Codey, president and chief operating officer. "It will enable us to implement the level of change that we must achieve in our business if we are to successfully meet the challenges of a changing and increasingly competitive industry. Total Quality was once a preferred strategy; now it's an absolute necessity."

As part of PSE&G's Total Quality program, the utility has:

- Developed a state-of-the-art system to continuously monitor and measure customer satisfaction.
- Moved to align its strategies, business plans, and individual goals to meet long-term stretch targets.
- Developed the work environment and systems that support quality initiatives and introduced the training, education, and reinforcement needed to build the necessary quality skills within the organization.
- Identified key competitive processes and established the foundation needed to continuously improve business processes.
- Expanded the use of benchmarking and began to assess the organization against the criteria for the Malcolm Baldrige National Quality Award.
- Targeted all employees for involvement in improving performance.

In the future, Codey said, "PSE&G will be doing significant work on our quality process improvement: how we do our work; how we relate to one another; how we relate to the external business environment. And we will develop and display the analytical tools of quality to help us achieve our objective of continuous improvement."

For six years, Illinois Power (IP) has been pursuing its Quality and Productivity initiative, which is closely linked to culture change. "We have strengthened our efforts to foster a 'winning team' attitude among our 4500 employees. It is a deliberate change in our corporate culture," said Larry Haab, chairman, president, and chief executive officer of IP. "A winning team isn't complex," he continued. "To make it work, we must create an environment in which employees at all levels are treated with respect, encouraged to contribute individually to the achievement of organization goals, and fairly recognized and compensated for their work." The utility's efforts have been so successful that in 1992 IP won the electric industry's Edison Award, presented each year to the utility whose operation, customer service, and innovation set the pace for the rest of the electric industry. Some of IP's Quality and Productivity measures include:

- Centralizing customer service operations and making representatives available to customers 24 hours a day, 365 days a year via a toll free number.

- Installing the Trouble Outage System, a computer program that pinpoints damage to the transmission and distribution system and then prioritizes a response plan.

- Creating a Customer Assistance Advisory network to help residential customers with special needs.

- Launching a program in which over 500 employee teams proposed more than 5000 ways to improve customer service and profitability.

- Including all employees in an incentive compensation plan.

- Implementing an employee Career Development Program.

- Establishing a program to reengineer business processes.

Florida Power & Light's (FP&L) experience with Total Quality highlights some of the pitfalls that a utility may encounter when pursuing a quality program. The perceived shortcomings of FP&L's effort are the subject of potentially lengthy articles, but some commonly derived lessons are that the organization focused too much on the objective of winning the Deming Award and the creation of improvement teams and too little on the nature of the improvements being sought and obtained.

Culture Change in the Utilities Industry

A number of companies have recognized that the key to successful change is to change the way they manage the business. This involves changing managers' assumptions, their behaviors, and probably, as critical, the way personnel are appraised and rewarded. This approach is probably one of the most alien to the industry because it grapples with softer management issues and is counter to the more structured, engineering view typical of many utilities. This approach is also difficult to sustain because it often puts changing the organization's culture at least in parity with, if not ahead of, improving results.

The rational for using this approach is two-fold. First is the view that the traditional micro-managing, activity, and budget focus of utilities is so strong that it is not possible to successfully transition to process and performance management without completely changing the view point and reward structure of personnel. The argument is that any focus on results will drive personnel back to micro-management and defeat the change. The second reason for using this approach is that it is much more people focused and therefore leverages off of the organization's traditional strength in terms of employee loyalty. Citizen's Gas has used this approach, and the successful results and experience are recounted in the book *Implementing Quality with a Customer Focus* by David Griffiths.

Centralized Improvement Efforts

Utilities have traditionally had a strong centralized decision making apparatus that decides on and initiates changes that are then implemented in a decentralized command and control fashion. Some utilities approach improvement in this fashion, but realize little success in implementing changes because there is little effort at obtaining organizational consensus. Without consensus, decentralized functional groups can undermine the effort.

To mitigate this tendency, companies are using more external consultants to assist in their change efforts and are relying on more sophisticated and disciplined improvement methodologies such as business process improvement based improvement, benchmarking and other techniques. Their focus with these efforts is to achieve the results necessary to meet competition, not to change the culture. Changing the organization's culture is most often a secondary objective—if it is an objective at all.

The utility industry has done a good deal of key measurement comparisons within the industry for much longer and more comprehensively than most other industries. Since utilities in the past faced no serious competition, they were relatively open about their operating practices. The American Gas Association, Edison Electric Institute, and Electric Power Research Institute have long served as clearinghouses of research on utility management issues.

Process benchmarking looks not only at differences in process performance measures but also at process customer requirements and process technology and management. Though still rarely used, process benchmarking is becoming more common in the industry as companies strive to meet competition. The industry is also becoming much more willing to look and learn from experiences outside the industry. For example, the budgeting and planning benchmarking study mentioned previously included roughly equal numbers of participants from within and outside the industry.

Many companies have also attempted to use problem solving teams and other such disciplined process based improvement methodologies. While this has generated many improvements, the companies have often encountered two sets of barriers, 1) a skills barrier particularly when attempting improvements through some of the more sophisticated approaches such as business process improvement and 2) a behaviors barrier (poor commitment and follow-through, poor management of change, etc.).

Downsizing

The last strategy still used by the industry to achieve change is budget cutting and downsizing. This approach has a number of drawbacks. First, the changes in the industry are so fundamental that they demand a change in the way business is actually done—utility business processes need to be redesigned or reengineered. Simply removing resources alone without changing the underlying process usually is self-defeating. A second drawback is that downsizing is counter to the companies' culture of long-term employment and therefore has a negative and significant impact on employee morale.

As a result downsizing has often resulted in unexpected and unplanned for behavioral reactions that have negated expected downsizing benefits. Since utilities did not redesign the processes that they downsized, there was no corresponding reductions in workload. The downsizing benefits were typically only temporary, and headcount often rose again after a year or two as permanent and contract personnel were added to meet the unchanged workload. Downsizing continues to be an instinctive response to lost or threatened market share, but many utilities are beginning to realize that it is an inadequate response to their changing environment.

Utilities Industry Summary

The gap between current performance and management practices and the performance and management practices necessary for a competitive environment is so great that any effort will be a significant challenge and subject to high risk. Utility experience with Total Quality is mixed, and, with some significant exceptions, many companies have difficulty achieving the levels of improvement that they anticipate and that they see in other industries. It is clear, however, that relying on downsizing and contraction alone is unsuccessful.

The utilities with the greatest potential to "win" the improvement race are those that are aggressively pursuing a portfolio of process improvement strategies (rather than functional or departmental improvement efforts) that change both the way they do business and the culture of their companies. The key factor behind these winning efforts is a cadre of leaders that is not only committed to change, but willing to potentially risk their jobs to support it.

These successful companies create a portfolio of improvements based on an accurate and shared vision of their market and its future changes. Based on their future process requirements and their current performance, they align their improvement efforts with those processes that have the largest performance gaps and which are the key drivers to meeting future requirements.

Implementation is based on a balanced approach of fast and measured improvement efforts and the relative importance of the processes selected for improvement. Successful utility companies seek to achieve performance improvement in key processes quickly and at the same time pursue culture change which involves more decentralized decision making, improvement efforts and employee involvement. A second important component to successful implementation is the companies' willingness and ability to manage the consequences of the improvement effort and to measure improvement not just in terms of inputs but in terms of process measures and outputs.

Finally the successful companies are marked by their ability to manage change. Many utility companies have a long and distinguished track record of initiating major projects and not completing them. One of the reasons for the frequent failure to implement projects has been poor change management skills. This is often demonstrated by lack of coordination, poor communication, not involving key implementing individuals and groups, and poor consequence management. This element is especially critical to utilities because their long term success depends as much on their changing their basic culture as on their ability to achieve big improvements in the short-term.

Utilities that could be characterized as "holding their own" against the tide of change are probably those pursuing centralized improvement strategies that strive for performance improvements without seeking culture change, and those who will probably lose the battle are pursing traditional downsizing and budget cutting strategies.

◼ Overview of Improvement in the Service Industry

From these three in-depth reports on the service sectors, you can clearly see why it is so crucial for organizations to apply the concepts of improvement to their services. As an overview, I would like to go over a few examples that show what to do and what not to do in order to achieve and maintain service quality excellence.

Putting Their Money Where Their Mouth Is

Everyone expects improvement from banks, hospitals, computer manufacturers, but few nonprofit making organizations have done much about improving. This includes the U.S. government. The American Society for Quality Control (ASQC), a nonprofit making organization by design, preaches that through improved quality, America will be at its most competitive state. ASQC is in fact a service organization who serves its members and society as a whole, providing them with numerous ways to learn about improvement methodologies. Often those who preach quality service do not practice it, but that is not the case with ASQC.

ASQC's improvement process formally started in the late 1980s, and their efforts are now beginning to pay off. Executive Director Paul E. Borawski remembers how it was before the improvement efforts: "I can remember the afternoon an employee walked into my office with a fistful of checks wondering what to do with them. It turned out those checks represented $500,000, and some of the checks were three months old. I can remember learning that we had 7000 address changes waiting to be inputted. Some of them were six months old. I can remember days when the membership department staff went to customer service to steal membership applications so that they could mail membership materials, while the applications waited another 8 to 10 weeks to be entered into the database."

Although ASQC's improvement process cannot be given total credit for the following business performance gains, the process did have a major impact on these business results:

Measures of Business Performance

	Year 1	Year 2
Growth rate	25 percent	30+ percent
Operating surplus	$ 0.18	$2.5 million
Member retention rate	77.2 percent	83 percent
Complaint level	27 percent	7 percent
Service rating "better" or "best"	65 percent	85 percent
Overtime and temporary help	heavy	appropriate
Staff turnover	34 percent	8 percent
Board of Directors	rumbling	cheers

As Borawski puts it, "There could be no more appropriate sign of our commitment to our customers than our service center and quality improvement process. These efforts have contributed significantly to our performance improvements over the last 5 years."

Through an incentive program that they have in place, the staff earns financial bonuses if they accomplish certain service-oriented objectives.

In one quarter in 1993, the following accomplishments were achieved by the staff:

- ASQC staff met all established deadlines, achieving 100 percent compliance.

- ASQC product development and promotion staff were able to increase non-dues target revenue by 25 percent above last year.

- ASQC customer service representatives answered over 96 percent of all calls within three rings.

- Satisfaction with ASQC courses and certification has increased due primarily to the efforts of staff to improve information, cycle times, service, content and instructors.

And believe it or not, this was considered a less-than-adequate quarter for ASQC. This particular quarter was the first one since the incentive program began in 1991 that no incentive payout will be made. Truly, we are looking at an organization that has put their money where their mouth is.

Heir Apparent—Michael Eisner at Disney

Perhaps there is no better example of an organization that has been focused on the needs of the customer than Disney over the last ten years. Since Michael Eisner arrived at the organization, Disney has gone from just a theme park to an entertainment giant. Earnings had fallen for three straight years before Eisner arrived, but have increased eightfold under his direction, going from a $1.4 billion organization to that of $5.8 billion a year.

This success was due in great part to Eisner's uncanny ability to know not only what the customers wanted, but also in many cases to exceed their expectations. Eisner expanded the organization into new areas, bringing that same Disney feel and service to every place he went. He had the formula for success, and it would work wherever he pointed his finger—the Disney Stores, blockbuster movies, new rides—he just could not lose. Or could he?

Enter into the picture: EuroDisney. And proof that even "Golden Boys" can make mistakes. What is the old adage about, "give a person enough rope …"? Eisner saw a need. He defined that need as a large population of individuals that did not have easy access to a Disney Theme Park, but had the monetary resources to afford to go. Where Eisner went wrong was that he defined the need as Disney-specific, instead of just a theme park that would be best suited to the location, people, climate and nature.

Eisner rolled out EuroDisney with the exact same ingredients that had made the United States and Japanese locations successful, without modifying those ingredients to suit the needs and wants of the French and other European communities. A major discrepancy was that the patented Disney wholesomeness did not call for wine to be served at the park. (And for a Frenchman, on a scale of 1 to 10, with 1 being utter annihilation of the universe and 10 being oxygen, wine has got to be a 9.) Another problem was that all the food service was directed toward a quick, fast-food, American style, while Europeans are much more apt to want a longer, slow-paced meal.

Eisner has set about correcting these problems. Whether or not he will be able to recover from this costly error in judgment and turn it around, will be one of the interesting case studies of the twentieth century. One thing is certain; it won't happen without a proper understanding of the customer's needs in Europe and how best to service those needs.

Service Organization Examples—Enlightened or in the Dark?

We will close this chapter with a few examples of two types of service organizations—"enlightened" service organizations and "in-the-dark" service organizations. Enlightened organizations are ones that have addressed the problems within their organizations and have set about improving the quality of their service. They may not be there yet, but at least they have begun addressing the need. In-the-dark service organizations are ones that are just trudging along, providing bad service and firefighting problems as they arise.

Enlightened.   The following organizations and individuals have shown to be enlightened toward providing service excellence:

AFCO Financial Services Co. of Canada—AFCO started their improvement process in the 1980s. Eighteen months after they started their improvement process, the following results were recorded:

- Income was up 51 percent
- Profits were up 200 percent

- ROI was up 26 percent
- Plans for 20 new offices
- Went from a distant No. 2 in personal loans to a runaway No. 1

Oklahoma City Parks and Recreation Dept.—An automated system to provide more efficient maintenance of municipal property and more relevant productivity data has been developed by the Oklahoma City Parks and Recreation department. The planned maintenance system (PMS) allows for computer-generated work schedules of maintenance activities. Reports can then be generated to provide a statistical analysis of completion and cost data. PMS has enabled the parks and recreation department to complete maintenance of municipal property in a more timely and orderly manner. This has been demonstrated through a 69 percent reduction in the number of complaints from citizens received concerning park maintenance.

Vice-President Albert Gore—The vice-president has committed 50 percent of his time to reengineering the government. He hopes to cut down on procedures and regulations that take up a large portion of United States Government resources (i.e., time, money, etc.). The question is, "Is this just a political commitment that will never be lived up to, or does Al Gore really mean what he says?" Will history remember Mr. Gore as an enlightened vice-president, or as one who was in the dark.

In the Dark.   The following organizations and individuals have shown to be "in the dark" when it comes to providing service excellence. To coin a phrase, "The actual names of the organizations and individuals have been concealed in order to protect the guilty."

Foreign Health Care—A leading London teaching hospital is unable to find the patient's case notes 30 percent of the time. As a result:

- The patient has a wasted journey, because the doctor will not see him or her.
- The patient's history has to be recorded again, wasting clerical time.
- Investigations often have to be duplicated.
- The operational costs of the hospital are increased.
  Note: One patient's notes were lost 29 times.

Recent U.S. President—In 1987, the President of the United States issued an Executive Order to increase productivity in government by 20 percent. While the President was "enlightened" to the need to increase productivity, no improvement came about. (In fact, the percentage of the GNP consumed by government has continuously worsened, resulting in the largest governmental debt ever recorded.) He seemed "in the dark" about the need to follow up on executive orders and staying committed. Remember, it all starts with top management commitment and it requires ongoing top management interest and follow-up to ensure it is effectively implemented.

General Marketing Departments—Some marketing departments spend 70 percent of their time chasing problems of late deliveries.

The Best Service Organizations

Raymond J. Larking of American Express states that their performance standard is "not just better than our competitors or in line with customer expectations—but noticeably superior to the competition and above customer expectations." This sets a very challenging requirement on every employee within the organization—a standard that requires everyone to continually improve.

You know the best service organizations when you see them and deal with them. They are the ones that you are likely to do business with again and again. You may not always get the best price at Nordstrom, and yet Nordstrom is always busy—why? In a word: service. Organizations such as Nordstrom, American Express, British Airways, Federal Express, Disney, and Ritz-Carlton Hotel, know how the customer defines service quality, and what their expectations are based on. These organizations have set the standards for a world-class service organization. The key characteristics of a true world-class service organization are:

- Well-defined service strategy—It highlights the real priorities of the customer.

- Front-line people that are interested in and like people—They make a good assessment of the customer's current situation, frame of mind, and needs.

- Customer-friendly systems—They are designed for the customer rather than the organization.

■ Service Sector Summary

In our MBA classes we learned that the four P's—product, price, place, and promotion—were the four keys to successful selling. Though a good product, innovatively marketed, may give an organization an initial edge, sustained success only comes by delighting the customer, by providing him or her with a surprisingly good experience. They have to walk away from the contact thinking, "How smart I am for doing business with this organization." It is the total experience that the successful service organizations manage. What do our customers look at when they measure the interface experience? The following five things can win or lose you points:

- *Output*—Do you have or can you create what the customer wants?

- *Knowledge*—Do you understand what the output will do, how it does it, and what its limitations are? The customer always wants to do business with people who are well informed, who will make recommendations of other output that can do the job better at less cost.

- *Timeliness*—The customer does not want to be pressured, but they also do not want to be ignored. They want someone there to answer questions, to give them alternatives, and to take their money. (Never keep people waiting to give you money.) They want you to act like you perceive them as important people whose time is valuable, without making them feel that you are pushing them to buy something they do not need and/or want.

- *Reliability*—Customers want to be able to rely on you to be open the hours you advertise, to provide the service you promise, to live up to your word, and to be honest even if it means losing a sale.

- *Tangibles*—The days of the little repair shops with dirt all around is gone. Customers want the physical conditions to be excellent since they reflect the attitude of management and the work force. They do not expect the machinist to be covered with grease and dirt. They expect everyone to look clean and neat. They turn away from a window that contains a cluttered display. The store does not have to be a showplace with soft music in the background like Nordstrom. Take for example: Wal-Mart.

> Wal-Mart has a deceptively simple plan to sell name-brand, good-quality, fairly-priced merchandise in a no-frills setting that focuses on family shopping. Walk into a Wal-Mart and you will hear no fancy string quartet serenading you, no attentive sales representative looking like they stepped off the cover of *GQ*. Everything is simple and basic. Is that excellent service? Definitely. If done well, a self-service store can be every bit as satisfying to individuals as the high-maintenance service of a Nordstrom.

The lessons that you have learned from this chapter will help enable you to provide excellent service to your customers, whether you consider your organization to be part of the service sector or not—because every organization has some part of its operation that is service. Every time someone from your organization comes into contact with an external customer, you had better consider yourself a service organization, because that is one customer that you could lose.

The banking/financial, health care, and utilities industries are some of the most important sectors of business and life in the United States. The examples of these case studies offer excellent guidance for any organization. Key differences between different types of service processes follow:

Customer Requirements

Losers: We know what they need.

Survivors: Surveys are used to obtain customers' thoughts.

Winners: We talk routinely to customers in order to understand their real requirements and find out what they will need in the future as their requirements change. Customers become part of the design process.

Employees

Losers: Employees are a process input that need to be motivated.

Survivors: Employees need to be trained.

Winners: Employees are not the problem but can be part of the solution.

Management Style

Losers: Control through hierarchical organization.

Survivors: First-line and middle managers use participative management.

Winners: Informed and empowered team.

Primary Business Focus

Losers: Quarterly profits.

Survivors: Meeting customer expectations.

Winners: Building a strong and responsive organization.

Business Planning

Losers: Done by top management and kept confidential so that competition is unaware of plans. Focus tends to be on budgets.

Survivors: Done by top management and partially communicated to employees.

Winners: Everyone is involved in the planning process and their personal goals are directly related to the business plan. Focus is on results.

Communications

Losers: Inconsistent communication of organization objectives. Give orders, little upward flow of information.

Survivors: Good up and down communication but communications are not always consistent with actions.

Winners: Five-way communication and actions are understood in the context of organization objectives.

Reactions to Business Downturn

Losers: Cut discretionary spending, cut back overtime, reduce contractors.

Survivors: Early retirement programs.

Winners: Plan for business down-turns so that most of the impact is on temporary employees. Increase training of regular employees and increase investment in improvement activities.

Work Environment

Losers: Complacency and little accountability. Individuals do not understand the effect of their actions on the long term health of the organization.

Survivors: Decisions made by consensus. Teams used throughout the organization. Sense of urgency but little coordination.

Winners: Need for action is understood and accepted and organization functions as a coordinated team to achieve results. Risk taking is encouraged along with accountability for results.

Top Management Leadership

Losers: Support the process by assigning resources.

Survivors: Involved in the process and show their interest by attending meetings and talking about how important improvement is to the organization's success.

Winners: Excited about the process and totally dedicated to it. Aggressively search for ways to further improve the process. Recognized internally and externally as the leader of the improvement activities within the organization. Willing to take risks with their career in order to improve the organization's performance.

Measurements

Losers: Unintegrated measurement efforts focusing on activities and inputs (the budget).

Survivors: Integrated business measurements such as Activity Based Costing.

Winners: Measurements integrate both the strategic plan and business plans. They focus on the whole picture, including business processes and the stakeholders' priority interests.

■ Banking Source References

1. The Federal Reserve Bank of Chicago.
2. "A Survey of Quality Practices in Banking," a proprietary, unpublished survey, William D. Wilsted, October 1987.
3. "Eight Quality Strategies," a proprietary, copyrighted report, William D. Wilsted, December 1987.
4. "The International Quality Study," The American Quality Foundation and Ernst & Young, Cleveland, 1992.
5. "The Banking Industry Report" of The International Quality Study, The American Quality Foundation and Ernst & Young, Cleveland, 1992.
6. "Two Sides of Quality," a proprietary, copyrighted report, William D. Wilsted, May 1988.
7. "What The Customer Really Wants," *Fortune*, June 4, 1989 issue, pp. 58-68.
8. "A Comparison of Product and Service Attributes," William D. Wilsted, April 1993.
9. "Banking Industry Report," The International Quality Study, The American Quality Foundation and Ernst & Young, Cleveland, 1992.
10. "Best Practices Report," The International Quality Study, The American Quality Foundation and Ernst & Young, Cleveland, 1992.

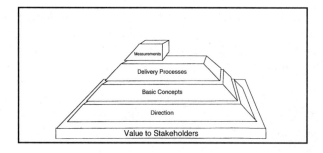

13

The Measurement Process: The Balanced Score Card

Dr. H. James Harrington
Principal, Ernst & Young
International Quality Advisor

and

Dorsey J. (Jim) Talley
President, Tally-Ho Enterprises

To measure is to understand, to understand is to gain knowledge, to have knowledge is to have power. Since the beginning of time, the thing that sets humans apart from the other animals is our ability to observe, measure, analyze, and use this information to bring about change. DR. H. JAMES HARRINGTON

■ Introduction

There is a very simple, fast way to establish the personality of an organization—any organization. All you have to do is look at the way it measures itself and the priorities that are set on each measurement. To this end, try to determine which measurement drives management bonuses. A financially driven organization will have measures that focus on things like production costs versus target, after-sales service costs, profit, compliance to budget, etc. An organization that is quality driven will focus on measurements like customer satisfaction level, poor-quality costs, first time yields, installation and early life defects, customer complaints per unit, etc. A resource oriented organization will measure things like inventory turns, value-added per employee, inventory costs, cycle times, etc. An investor focused organization will measure things like market share, stock price, return on assets, productivity growth rates, profits, etc.

A healthy organization has a balanced score card (measurement system). Within this balanced system, priorities are placed upon measurements that relate to the customer/consumer. For example, market share, customer satisfaction index, competitive price, reliability, etc. The second priority measurements focus on the organization's performance. For example, productivity, poor-quality costs, percent successful new products, value-added per employee, return on investment, etc.

The organization's measurement system is a meter of what the organization feels is important. It should reflect the organization's basic principles and identify how the organization is performing related to all of its stakeholders' values. Very few measurement systems consider anything other than the investor and management. As a result, the other stakeholders, the customer, the employees, the suppliers, and the community soon realize that the organization is using them. Humanity has evolved to the point today that all organizations need to develop an overall concern and sense of responsibility for all of their stakeholders. Developing a balanced measurement system is a major step in this direction.

There is no doubt about it. Case after case demonstrates that the lack of good measurements is a major obstacle to improvement. Every experienced manager knows that providing performance feedback to every employee is an essential part of any improvement process. They realize that if you cannot measure it, you cannot control it. And if you cannot control it, you cannot manage it. It's as simple as that. Customer-related measurements are important, but you need to probe deep into the process so that you can ensure the end output will be good, that all parts of the process are improving, and that the long-term performance has not degraded. In-process measurements provide windows through which the process can be observed and monitored. These windows must be dependable and provide a continuous view of all the critical processes. In addition, the measurement system needs to measure the important elements of all stakeholders' interests. Without dependable measurements, intelligent decisions cannot be made

This chapter looks at how measurement systems work, and how they are used. We will discuss the 11 Ws:

1. Why you should measure

2. Where you should measure

3. What you should measure

4. When you should measure

5. Who should be measured

6. Who should do the measuring
7. Who should provide feedback
8. Who should audit
9. Who should set business targets (standards)
10. Who should set challenge targets
11. What should be done to solve problems

Measurements are critical to:

- Understanding what is occurring
- Evaluating the need for change
- Evaluating the impact of change
- Ensuring that gains made are not lost
- Correcting out-of-control conditions
- Setting priorities
- Deciding when to increase responsibilities
- Determining when to provide additional training
- Planning to meet new customer expectations
- Providing realistic schedules
- Satisfying your stakeholders

■ Using Measurements

Measurements can be an effective tool for guiding an organization if used correctly, but if they are misused they can be destructive. It has been said that if you give a statistician enough data, they can prove anything. As the old story goes, three statisticians were hunting rabbit. A rabbit jumped out in front of them and the first statistician shot one foot in front of the rabbit. Then the second statistician shot one foot behind the rabbit. The third statistician did not shoot and put down his gun, saying, "Well, on an average, we got him."

Measurements are helpful tools to guide the organization. They can never replace good judgment, but they can greatly improve the decisions we make. Essentially, measurements can be divided into three categories.

1. Performance measurements
   — Return-on-assets
   — Profits
   — Percent product purchased
   — Costs
   — Inventory turns

— Cycle times
— After-sales service costs

2. Process Improvement
— Percent defective
— Poor-quality cost
— Process capability
— First-time yields
— Competitor comparisons

3. Forecasting
— Market trend analysis
— Product costs
— Customer expectations
— New product requirements
— Budgeting

Each of these three uses have unique requirements associated with them, each of which must be considered when designing a measurement system. The measurement system must be robust enough to provide guidance in all three categories and for any level in the organization. It should also include measurements related to the high priorities of each of the organization's six stakeholders (see the "Overview" chapter).

Measurements drive performance. A good measurement system drives an organization in a positive direction. A misused measurement system can be harmful. Measurements can be improved by:

- Improving the process that is being measured

- Focusing efforts on improving the measurement at the expense of other parts of the process that are not being measured

- Misuse of the data to give the false impression of improvement

We have seen all three approaches to improvement used by individuals in almost all organizations. The organization requires an effective, well-managed measurement system to ensure that the process is improved and that management is not given a false illusion of improvement.

■ Benefits of Measurement

Why is it that an employee who complains about how hard he or she has worked all day, will then go home and play three sets of tennis (where he or she will expend twice as much energy as during 8 hours at work), and love doing it? It is because he or she feels a sense of accomplishment when the measurement system gives direct feedback. Why don't people just run back and forth across their driveway, bouncing a ball with their $250 tennis racket? They would get the same amount of exercise, and the same experience of coordinating their movement to make contact with the ball. The thrill of bowling is not throwing the ball; it's knowing how many pins you knocked down.

The fact is that they don't get the same sense of accomplishment out of bouncing a ball in the driveway because there are no set rules, no one else is involved to see how they are performing, and there is no winning or losing. In short, there is no measurement of their performance. Without measurements, there is no sense of accomplishment and little or no reason to try to improve or even perform at the present level. Performance is directly related to how well the activity is measured and the interest other people exhibit in the measurement. Only about 1 percent of the people reach the self-actualization stage of personal development where other people's opinions are unimportant.

On a whole, management does a poor job of providing process measurements. A survey by the American Productivity and Quality Center found that only 38.7 percent of employees thought that there were good, fair performance measures where they worked. People want to be measured. They need to be measured. The only people who don't like to be measured are the poor performers. In fact, if management fails to establish appropriate systems, the good performers will develop ways to measure themselves to show you how well they are doing. Frequently, however, these measurements are not the ones that are really important to the business. Management must work with the employees to develop measurements that have meaning for both the employee and the organization.

In sports, the rules are well defined. No one can bowl more than 300 points in a game. The same is true of processes. The procedures (rules) limit how good we can be. Management's role is to develop a set of procedures and processes that allows employees to do their best. The employee's responsibility is to maximize his or her performance so that the output is as close to the process design limits as possible.

Why measure? Without it, you take away the individual's feeling of accomplishment, and you never know who to fire or who to promote. Measurements, and a good reward system, stimulate the individual and the team to make the additional effort that sets the organization apart from the ordinary.

As important as measurement is, by itself it is worthless. Unless an effective feedback system exists, measurement is a waste of time, effort, and money. Specific feedback enables an individual to react to the data, and correct any problems.

For example, when an IBM plant in Havant, England, began to provide each assembly operator with final unit test feedback on the units he or she produced, the defect level dropped 90 percent.

Such dramatic results are not limited to manufacturing. Another corporation that used weekly time-record logs to pay its employees, and to determine how much overtime and overtime meal allowances to pay, found that, initially, over 15 percent of these documents were either not turned in on time, or turned in with errors, causing a great deal of additional handling. A simple chart was provided to each manager documenting the percentage of defective time-record logs for each department within each function. As a result of this data feedback and the associated corrective action, the time recording errors dropped from 15 percent to 1 percent in less than six months.

■ Understanding Measurements

Why You Should Measure

Measurement is fundamental to our way of life. We measure everything. We measure our lives in seconds, minutes, hours, days, months, and years. We measure how far we travel in miles or kilometers, the food we buy in ounces or grams, the milk we drink in quarts or liters. Measurements are such an integral part of our lives that we couldn't survive without them.

As babies, the doctor measured our height and weight to be sure we were healthy. When we started school, our teachers measured us to understand our weaknesses and help us progress. Probably the single biggest thing that enabled Japan to progress so rapidly is the way it used measurements and competition between individuals to get children to excel in their education system. As we leave school and go to work, measurement continues. At work, our worth is measured by our salary, and how well we perform.

Where You Should Measure

The major problem with most processes is that performance is measured only at the end. In most cases, this provides little relative feedback about individual activities within the process or, when it does, it is too late.

The organization should establish measurement points close to each activity so that the people performing each separate activity receive direct, immediate, and relevant feedback. Consider, for example, how difficult it would be to manage your long distance telephone bill if all calls in the organization were charged to the same account number.

When You Should Measure

Measure as soon as the activity has been completed. Don't run your business like someone who does not record the amount of the checks he or she writes, but just waits until the bank statement arrives to determine what the balance is. Delaying measurement only allows additional errors to be made.

What You Should Measure

It is very hard to measure a manager's performance. How do you score people on their judgment? We have found that managers have to work within budget constraints, just like everyone else. And every organization, be it profit or non-profit, is accountable for financial performance. Even though most people are against objective measurement, they still want to be rewarded for good work. So bite the bullet. Every manager, every team, every person can be and should be objectively, equitably, and quantitatively measured.

While the information from the accounting system may not truly reflect a manager/team/individual's performance, it does provide some useful data. The best grouping of measures needs to be blended from a mix of data systems. You need to develop lists by function and process that will give a good mix of quantitative and qualitative measures. "Manage an organization as nature would; show neither malice nor pity."[1]

Make sure the system is fair, equitable and no favoritism shown, and you will produce outstanding improvement in effectiveness and efficiency.

It is important that we have not only financial measurements, but also measurements of the effectiveness, efficiency, and adaptability of our organization and its parts. For example, typical measurements for a sales process might be:

- Sales versus objective
- Percentage of bid errors
- Number of unpaid invoices over 20-days old
- Percentage of lost business
- Percentage of unanswered phone calls in four hours
- Percentage of letters retyped
- Dollars not paid versus dollars billed
- Time between order receipt and order entry
- Percentage of rush orders

For additional typical process measurements, see Dorsey J. Talley's book, *Total Quality Management—Cost and Performance Measures* or Ernst & Young's Technical Report TR 93.014.

The most important part of any plan or team operation is to define its objectives in quantitative, measurable terms. How does one begin to quantify and formulate objectives and measurements?

- Objectives should be formulated in terms of what is to be accomplished.
- Objectives should have specific deadlines so that measures can be time-referenced.
- Objectives should always be forceful; e.g., comply by, achieve, gain.
- Objectives and measurements should always be communicated to all personnel through various media. Post the objectives and results on an Improvement Board.
- Objectives should clearly express the key opportunities which all personnel can relate to. They should be expressed in understandable and meaningful terms.

Who Should Be Measured

Management is responsible for providing sound measurement systems, and appropriate feedback, to help everyone do their jobs better.

Management signals what is important by measuring its results. Many people figure that if it isn't measured, why do it? Consequently, every important job can and should be measured. Although, theoretically, each task should be measured and reported to the individual performing the task, this is not always practical. The best way to start the measurement system is to examine each of the organization's stakeholders' expectations and be sure that these key needs are being measured and reported. Typical stakeholder measurements are:

- Customer satisfaction—External customers
- Value added per employee—Management
- Job security—Employees
- Longer contracts—Suppliers
- Return on investment—Investors
- Increased tax base—Community

The second priority is to examine the critical process flowcharts, and identify those activities that significantly impact total process effectiveness, efficiency and adaptability. Then establish measurements for these critical activities, as well as for the total process.

Reviewing the internal customer satisfaction level will identify a third set of measurement priorities, focusing on activities that are not meeting internal customer expectations.

Give fourth priority to activities that require significant resources. You can measure performance in terms of effectiveness and efficiency and express it in physical terms (e.g., time to perform a task, cycle time, etc.), or in dollars, allowing a number of resources to be combined (e.g., value-added costs, labor costs, etc.).

The fifth priority is to measure the performance of each individual and provide personal, confidential feedback to each employee. This data is essential to individual excellence.

When developing your performance measurements, get help. Ask your customers what is important to them, ask your employees what is meaningful to them, ask your suppliers what is fair to them, and get their agreement with and support of the family of measurements you decide on. Robert Kaplan and David Norton call this grouping of the most important measurements a "balanced scorecard."[3]

Who Should Do the Measuring

The best person to do the measuring is the person performing the activity. There is immediate feedback, and he or she should have the best understanding of the job. In cases where the self-inspection error rate is too high, let the people performing the activity check each other's work. They will learn quickly this way. It is also a good way to exchange ideas and to begin standardizing. As a last resort, have someone who is not part of the activity check the output. Never rely on the external customer to be your last inspector, but if the internal process fails, be sure to have a data system that collects this information and reports it back so that the internal process and measurement system can be corrected. Whoever does the measuring should be well trained, and use documented criteria to evaluate the output.

Who Should Provide Feedback

This is the supplier-customer partnership theory we have talked about. Each output receiver (customer) should provide positive and negative feedback, and constructive suggestions to the person or people providing the product or service.

Who Should Audit

Quality Assurance audits manufacturing personnel to ensure they do their jobs right. We use external auditors to audit an organization's financial records to ensure that we are not being fleeced. Doesn't it make sense, then, to audit all of our business processes using an independent party? In addition, the management in each area should conduct regular, formal, documented audits using written audit procedures? They should then report the results of these audit procedures to management, and to the employees, along with appropriate corrective action.

Who Should Set Business Targets (Standards)

Consider business targets (standards), and what they mean. A standard sets the minimum acceptable performance for an individual or unit performing an activity. It should be the result that the present process will produce with a person who has been trained to do the job, has the necessary tools, and has the ability to do the job. It is not the present performance level. In most cases, the standard will be lower than the present performance level, if you have an experienced person doing the job. When the standard is not met, there is something wrong, and you should take action.

Let's look at the two key measurement types—effectiveness and efficiency—and how to set standards for each of them. First, let's address effectiveness (quality). The person or people who receive the output (the internal and external customers) should set the effectiveness standard. You should meet with these customers and determine exactly what they need. Then design a process that a below-average person can use and still meet the standard. Remember, 50 percent of your employees are below average, so design your processes to operate using below-average employees. You can be sure you will have below-average employees working on all your processes sooner or later.

Efficiency (productivity), in contrast, is not customer-driven. It is controlled by the process. All processes have an inherent efficiency—the least amount of resources required to provide the output when everything goes right. But things don't always go right. Waste and inefficiency are built into the process.

Consequently, the scientific approach of task-time analysis usually concedes to a much less exacting process of having the employee and manager review past performance data to establish an efficiency standard. The manager then must decide if the value-added content is worth the price. If it is not a good value, the process either is redesigned or the manager decides not to perform the activity. Such standards often are based on the present employee's skill in using the process, and are unrealistic when a less-skilled employee performs the activity.

Who Should Set Challenge Targets

Don't confuse business targets (standards) and challenge targets. They are very different. Business targets are set by the customers, since they reflect their expectations, or by management, since they define minimum acceptable performance. A challenge target, on the other hand, is an objective set by the team or the individual performing the activity. A challenge target is always more stringent than a business target, and supports the concept of continuous improvement. It is the means for providing customers with surprisingly good quality. Failing to meet a challenge target should not impact the business plan. Your objective always should be to be better than your customer expectations, better than your business plan, and to accomplish more with fewer resources.

Once a business target is met, the team, or the individual, should set challenge targets to stretch them to meet new, higher levels of performance.

■ Measurements Are Key to Improving

Dr. Charles Coonradt, a management consultant, believes that we should make business more like a game. He points out that the same workers who require special clothing to work in the cold storage section of a plant for only 20 minutes, with a 10-minute break to get warmed up, will spend an entire day ice fishing on a frozen lake. When the air-conditioning breaks down, and the temperature reaches 85 degrees, the office is closed. However, on the way home, employees will stop and play a round of golf in the same heat—and pay to do it.

Why is this? Sports succeed in exciting people because they have:

- Rules

- Measurements

- Rewards

Let's apply these principles to business:

Rules.   Every sport has rules that govern the game. People know them, and are penalized when they do not live up to them (e.g., a 15-yard penalty for clipping another player in football). Business, too, has its rules. They are the procedures and job descriptions. Business also has referees—the quality system auditors. When you do not play by the rules, you should be penalized.

Measurements.   We need to know how well we are doing, and it needs to be personal. How popular do you think golf would be if the only feedback you received is that, of the 200 people who played golf last Sunday, the average score was 93, and you weren't told what your score was. Yes, measurements are critical to maintain interest in an activity, particularly if you want to improve.

Measure both individual and team performance. A baseball team wins or loses the game, not an individual. Apply team measurements to small groups (20 maximum). For example, a football team is 11, a baseball team is 9, and a basketball team is 5.

Rewards.   Sports enthusiasts get their rewards from trying to improve their games. The professionals see big dollars rolling in, and the amateurs win trophies, but all are tied into a measurement system. In business, we should all be pros, or we should not have our jobs. Sure, it is nice to get the trophies, plaques, and dinners when we excel, but we also should receive financial rewards. Our pay should be tied directly to our personal measurement system.

Measurement is important for improvement for several reasons:

- It focuses attention on factors contributing to achieving the organization's mission.
- It shows how effectively we use our resources.
- It assists in setting goals and monitoring trends.
- It provides the input for analyzing root causes and sources of errors.
- It identifies opportunities for ongoing improvement.
- It gives employees a sense of accomplishment.
- It provides a means of knowing if you are winning or losing.
- It helps monitor progress.

Support organizations tend to think, and maybe even believe, that they cannot be measured. Consequently, in the past we measured products only, and ignored the support processes. This happened for several reasons:

- In the 1950s and 1960s, product cost (material and labor) constituted a significant portion of total cost. Hence, management was preoccupied with measuring and managing it. Support process cost was a smaller element of cost, and therefore was ignored.
- Traditional measures of input and output could not be applied easily to support processes.
- White-collar workers believed that their work was varied and unique. As a result it could not be measured.

Measurement trends are changing:

- From product measurement to process and service measurement
- From managing profits—to—managing assets
- From meeting targets—to—continuous improvement
- From quantity measurements—to—quality measurements focusing on effectiveness, efficiency, and adaptability
- From measurements based on engineering and business specifications—to—measurements based on internal and external customer expectations
- From a focus on the individual—to—a focus on both the process and the individual
- From a top-down process dictator approach–to—a team approach to developing measurements and managing performance

Obviously, all processes can and should be measured and managed in the same way manufacturing processes are.

■ Types of Measurement Data

As you establish your measurement system, the organization will be working with two types of data: attributes data and variables data.

Attributes Data

This kind of data is counted, not measured. Generally, attributes data requires large sample sizes to be useful. It is collected when all you need to know is "Yes" or "No," "Go" or "No-Go," or "Accept" or "Reject." Examples of attributes data include:

- Did an employee arrive at work on time?
- Was the letter typed with no errors?
- Is a department below budget?
- Did the meeting start on schedule?
- Was the report turned in on schedule?
- Was the phone answered on the second ring?

Variables Data

Variables data is used to provide a much more accurate measurement than attributes data provides. This involves collecting numeric values that quantify a measurement and therefore requires smaller samples. Examples of variables data include:

- Number of times a phone rings before it is answered
- Cost of overnight mail
- Number of hours to process an engineering change request
- Dollar value of stock
- Number of days employees call in sick per year
- Number of days it takes to solve a problem

Occasionally, it may seem difficult to establish meaningful measurements (e.g., How good was a presentation? How well did a document reproduce? How clean is the office?). In many situations, human judgment enters the picture, and you can compare relative values. For example, you can judge print quality by comparing a number of pictures.

In some cases, there is no other way but to ask your customers for their opinion on some of the softer measurements. After all, their opinions are the real measure of your organization.

■ Clear Performance Standards

For the past two years, *Industry Week* Magazine has commissioned two surveys on how organizations are being managed. The most significant conclusion to draw from the surveys is the consistency and clarity of the messages: "… the erosion of clear goals the further down an organization a manager works, and similar, though less-pronounced, drop in clear performance standards,… in top management, 80 percent have clear goals, in middle management the percentage is 70 percent, and in first-level management it drops to 61 percent."[2] These findings also provide insight into performance appraisal.

■ Measurement Characteristics

Measurements are essential. If you cannot measure it, you cannot control it. If you cannot control it, you cannot manage it. If you cannot manage it, you cannot improve it. Without measurements, every result is a surprise. Measurements are the starting point for improvements, because they enable you to understand where you are and set goals that help you get where you want to go. Without them, needed changes and improvements to the process are severely hindered. You need to develop effectiveness (quality), efficiency (productivity), and adaptability (flexibility) measurements and targets for all your critical processes.

The organization should keep in mind the following characteristics of a balanced scorecard:

1. It reflects the customer's "agenda."
2. It reflects management's "agenda."
3. It reflects the input of the doers in the process.
4. It is attainable, yet requires the organization to stretch.
5. It is easily measurable.
6. It is clearly stated and understandable.
7. It is aligned with the organization's vision and goals.

In setting process measures, it is important to remember that the reason processes are measured should be to improve, not to punish. It is helpful to inventory the measures currently used in the organization to see if there are any which would be useful as a process indicator. A family of 3 to 9 measures works best to indicate improvement, rather than a single measure.

There are several considerations when developing process indicators. First, there should be a balance of viewpoints. The customer's (external) viewpoint must be balanced with that of the doers in the process (internal). Second, the 3 to 9 measures should be a mix among effectiveness, efficiency, and adaptability indicators:

- *Effectiveness*—The extent to which the outputs of a process or activity meet the needs and expectations of its customers. In this case, the term "customer" refers not only to the person who receives the output, but to all future people who come in contact with or are affected by the output. Effectiveness is having the right output at the right place, at the right time, at the right price. You can see how it is more than quality. Examples of effectiveness measurements are:
  — Reliability
  — Useability
  — Serviceability
  — Responsiveness
  — Appearance
  — Errors per output
  — Percentage of on-time deliveries
  — Customer satisfaction

Internal Effectiveness Indicators show how well the suppliers are meeting their requirements. Sometimes the external customer is not aware of these types of problems. Examples are:

- Words that start with "re" (example: rework, repair, reinspect, etc.)

- Errors caught

- Changes required

- Interruptions

External Effectiveness Indicators relate to how well the product or service satisfies the external customer of the process. Examples are:

- Late deliveries

- Customer complaints

- Warranty returns

- Wrong paperwork

- *Efficiency*—A measure of the resources used to produce an output. Efficiency improves when it takes less resources to produce a specific output. Typical examples of efficiency measurements are:
  — Cycle time per unit
  — Processing time per unit
  — Resources used per unit
  — Value-added cost per unit
  — Transactions per hour
  — Tests per hour

— Reports per professional
— Parts per work hour

- *Adaptability (flexibility)*—The ability of a process or activity to handle future, changing customer expectations and today's individual, special customer requests. Adaptability measurements are often ignored, but they are critical for gaining a competitive edge. Customers always remember how you handled, or didn't handle, their special needs. Examples of adaptability measurements are:
  — Percentage of special orders entered in 8 hours
  — Percentage of special orders processed
  — Percentage of special orders granted at the employee level

- *Financial*—There is a group of financial measurements that has been the main measurement for most organizations. For example:
  — Return on investment (ROI)
  — Return on assets (ROA)
  — Profit percentage of sales
  — Assets
  — Annual Sales
  — Operating Costs
  — Value added per employee (VAE)
  — Stockholders' Equity

- *Business*—There is also a group of business-related, nonfinancial measurements that are important. For example:
  — Accidents per 10,000 hours
  — Market share
  — Security violations per month
  — Inventory turns per year

■ The Measurement Process

The process for developing a family of measurements is best formulated by a team of people who are involved in the following nine-step process:

1. Review important goals. The management and team should have some fairly specific improvement goals in mind. This comes from the vision, strategy, business plan, and objectives developed by management, with input from their employees. The measures should track directly to these objectives.

2. Review measurement principles. Review the types of measures described previously and also refer to Talley's TQM book appendix for 400 plus possibilities.

3. Conduct a brainstorming session. The team will brainstorm to define potential measurements. After the clarification and combination phase, the measurements are ranked.

4. Discussion and debate. The team should debate the relative merits of each proposed measurement, weeding out those which are redundant.

5. Present ranked list of potential measurements to management. The management and team should agree on the final balanced score card.

6. The team develops a plan to collect, track, and review the measurements.

7. Develop a baseline. The team should take baseline values and calculate the performance index if appropriate.

8. Ongoing measurement—period to period. The team measures periodically to check their progress in improving the process against stretch targets.

9. Economics of effectiveness and efficiency format fit. Will the measurements fit into the organization's overall poor-quality cost format? Can the team take the costing out of the financial chart of accounts or does the financial system need to be changed?

■ Poor-Quality Cost

One of the primary objectives of Total Improvement Management is to reduce the losses that are caused by waste or poor quality. Poor quality costs your organization money. Good quality saves your organization money. It's as simple as that. James E. Olson, former president of AT&T, said, "A lot of people say quality costs you too much. It does not. It will cost you less." But many organizations today do not measure the cost of poor quality; and if you do not measure it, you cannot control it. Why is it, then, that corporate management does not insist on the same good financial control over poor-quality cost (PQC) as they exercise over the purchase of materials, when often PQC exceeds the total materials budget?

To put it simply, the poor-quality cost reporting system is only one of the many tools needed in a comprehensive, organization-wide TIM process, but it is an important tool, since it directs management attention and measures the success of the organization's efforts to improve. It also provides management with the necessary tools to ensure that suboptimization does not have a negative effect on the total process.

Poor-quality cost is defined as: all the cost incurred to help the employee do the job right every time, the cost to determine if the output is acceptable, plus all the cost incurred by the organization and the customer because the output did not meet specifications and/or customer expectations. Figure 13.1 lists the elements of poor-quality cost.

Why Use Poor-Quality Cost?

PQC provides a very useful tool to change the way management and employees think about errors. PQC helps by:

1. Getting management attention. Talking to management in dollars provides them with information that they relate to. It takes quality out of the abstract and makes it a reality that can effectively compete with cost and schedule.

The Elements of Poor-Quality Cost

I. *Direct poor-quality cost*
 	
    A. Controllable poor-quality cost
        1. Prevention cost
        2. Appraisal cost
 	
    B. Resultant poor-quality cost
        1. Internal error cost
        2. External error cost
 	
    C. Equipment poor-quality cost

II. *Indirect poor-quality cost*
 	
    A. Customer-incurred cost
 	
    B. Customer-dissatisfaction cost
 	
    C. Loss-of-reputation cost

Figure 14.1. The elements of poor-quality cost.

2. Changing the way the employee thinks about errors. There is less impact on an employee's future performance when, as a result of his or her actions, an estimate has to be redone, than when the employee knows that it costs $3000 to redo the estimate. In one case, what is thrown away is a report; in the other case, it's 30 $100 bills that are discarded. Employees need to understand the cost of errors they make.

3. Providing better return on the problem-solving efforts. PQC *dollarizes* problems so that corrective action can be directed at the solutions that will bring maximum return. James R. Houghton, chairman of Corning Glass, has reported, "At Corning, cost of quality is being used to identify opportunities, to help prioritize those opportunities, and to set targets and measure progress. It's a tremendous tool, but we are taking great care to ensure that it is not used as a club."

4. Providing a means to measure the true impact of corrective action and changes made to improve the process. By focusing on poor-quality cost of the total process, suboptimization can be eliminated.

5. Providing a simple, understandable method of measuring what effect poor quality has on the organization and providing an effective way to measure the impact of the improvement process.

6. Providing a single measurement that brings together effectiveness and efficiency measurements.

Direct Poor-Quality Cost

Of the two major poor-quality cost categories, direct and indirect, direct PQC is better understood and traditionally used by management to run the business because the results are less subjective. Direct PQC can be found in the organization's ledger and can be verified by the organization's accountants. These costs include all the costs an organization incurs because management is afraid that people will make errors, all the costs incurred because people do make errors, and the costs related to training people so they can do their jobs effectively. Direct PQC encompasses three major expenditures: controllable PQC, resultant PQC, and equipment PQC.

Indirect Poor-Quality Cost

The other major part of the poor-quality cost system is indirect PQC, defined as those costs not directly measurable in the organization's ledger, but part of the product life-cycle PQC. Indirect PQC consists of three major categories:

- Customer-incurred PQC

- Customer-dissatisfaction PQC

- Loss-of-reputation PQC

We will discuss this part of poor-quality cost in more detail because few organizations have advanced to this level of refinement, and it is the most important part of poor-quality cost from the customer's standpoint.

Customer-incurred PQC appears when an output fails to meet the customer's expectations. Typical customer-incurred PQCs are:

- Loss of productivity because a service was not performed on schedule.

- Travel costs and time spent to return an item that does not owrk, and the repair costs if it is not under warranty.

- Overtime expended because input schedules were not adhered to.

- Cost of redundant equipment that is purchased as back-up in case the primary equipment fails to function.

Customer dissatisfaction PQC is the cost the organization incurs when their customer stops buying their product as a result of poor product performance. Customer dissatisfaction is a binary thing. Customers are either satisfied or dissatisfied. Seldom will you find one who is in between.

The quality level of United States and European products hasn't suddenly dropped. In fact, it has improved. What has happened is that the customer expectation level has changed. Customers now require a much better product to satisfy their expectations and demands. The customer dissatisfaction level has moved, but in many organizations the quality level has remained constant, or has not kept pace with customer expectations. These organizations may very well have been making parts to specifications, but the specifications were not good enough to keep their old customers, let alone attract new ones. Many of our business leaders understand that customer expectations are changing. For example, F. James McDonald, past president of General Motors, said, "At General Motors, we're proud of the gains we've made in quality and in the productivity associated with the improvements, and our customers are verifying the results of our efforts. But even as our own performance improves, the expectations of our customers continue to rise."

Loss-of-reputation PQC is even more difficult to measure and predict than are customer-dissatisfaction and customer-incurred PQC. Costs incurred because of loss of reputation differ from customer-dissatisfaction costs in that they reflect the customer's attitude toward an organization rather than toward an individual product line. The loss of a good reputation transcends all product lines manufactured by an organization. It can best be measured in loss of market share.

While vice president of General Electric, Armand V. Feigenbaum developed the concept of quality cost in the early 1950s. In 1987 he wrote,

> Our original development of the concept and quantification of quality cost has had the objective of equipping men and women throughout a company with the necessary practical tools and detailed economic know-how for identifying and managing their own costs of quality. These costs have successfully provided the common denominator in business terms both for managing quality as well as for communication among all who are involved in the quality process. Therefore, we have continued to develop, implement, and refine the cost of quality in companies the world over.

For more information on poor-quality cost and how to implement a poor-quality cost system in the business process areas, see H. J. Harrington's book, *Poor-Quality Cost*, published by Marcel-Dekker.

■ Surveys as a Measurement Tool

Another very effective measurement tool is surveys. Surveys are particularly effective at measuring the softer things that are so important to the organization's stakeholders. Questions like, "How do you feel about something?" . . . "What do you think is needed?" . . . "Which is better?" . . . all lend themselves to surveys.

All organizations should conduct regular surveys to measure the change in the attitudes, beliefs and perceptions of the following stakeholders:

1. External customer
2. Suppliers
3. Employees
4. Investors
5. Management
6. Community/Mankind

Most organizations regularly conduct customer surveys, a few conduct employee surveys, but very few use this valuable tool to understand and measure all of their stakeholders' requirements and concerns. The survey should be based upon the priorities that each of the stakeholders have. The top five priorities for each stakeholder can be found in the "Overview" chapter. The surveys should be designed so that the results of groups of questions can be combined to calculate key stakeholder indexes. These surveys should be conducted during the first six months of the TIM process, the earlier the better. These early surveys set the baseline for future measurements and help to identify improvement opportunities.

These opinion surveys provide another means to help management develop sensitivity and awareness. Through awareness of the stakeholders' overall attitudes, management can anticipate problems before they occur and take action to prevent them from developing. These opinion surveys should be approached carefully, keeping in mind that they will be repeated a number of times to measure trends.

Let's look at one type of opinion survey to understand its content. We will use the employee opinion survey for this purpose. Typically the survey would consist of between 60 to 120 questions covering 11 areas. They are:

1. Overall satisfaction with the organization
2. The job itself
3. Salary
4. Advancement opportunities
5. Management
6. Counseling and evaluation
7. Career development
8. Quality and productivity
9. Work environment
10. Benefits program
11. Improvement efforts

Possible survey questions might be:

1. Everything considered, how would you rate your overall satisfaction with the organization?
   a. Completely dissatisfied
   b. Very dissatisfied
   c. Dissatisfied
   d. Neither satisfied nor dissatisfied
   e. Satisfied
   f. Very satisfied
   g. Completely satisfied
2. How would you respond to the statement: "My job makes good use of my skills and abilities"?
   a. Strongly disagree
   b. Disagree

   c. Neither agree nor disagree

   d. Magazine Agree

   e. Strongly agree

3. How much trust and confidence do you have in your immediate manager?

   a. Very little or none

   b. A little

   c. Some

   d. Quite a bit

   e. A great deal

In addition, a section for write-in comments enables the employee to provide more detailed information and address concerns not covered.

It is imperative that confidentiality and anonymity are maintained if survey results are to be meaningful. Care must be exercised when the survey form is being filled out, during the data-analysis cycle, and when results are reported to the management team. Special care should be taken in providing feedback to small units.

To help define problem areas, each manager should be provided with a report showing how his or her people responded. This report should compare the work unit to the total function and organization for each question. A statistical analysis should be made to determine if the work unit is significantly positive or negative compared to the function and the total organization.

Each manager should conduct a "feedback session" in which the results of the survey are presented to the employees. These sessions are important because:

- The employees will be curious about the results in general and how the department compares to the rest of the organization in particular.

- It provides management with an opportunity to discuss employee concerns.

- It provides an excellent way to obtain ideas and suggestions.

- It shows that management is serious about the results.

- It allows the members of the department to develop corrective action.

■ Using National Quality Award Criteria to Measure Improvement

The U.S. Government presents the Malcolm Baldrige National Quality Award to the organizations that have proven that they are performing at the highest level of excellence in seven categories and have met an aggressive minimum performance level. It is the "Super Bowl" of business. This award is presented by the President of the United States to the winning organizations in November of each year.

Although a maximum of six awards can be presented each year (two in each of the three categories), normally only two to four organizations are honored because the others do not meet the stringent requirements. To provide guidance to organi-

Examination categories/items	Maximum points
1.0 Leadership	100
2.0 Information and analysis	60
3.0 Strategic quality planning	90
4.0 Human resource utilization	150
5.0 Quality assurance of products and services	150
6.0 Quality results	150
7.0 Customer satisfaction	300
Total points	1000

Figure 13.2. Malcolm Baldrige National Quality Award criteria.

zations interested in applying for the award and to establish an agreeable definition of excellence that would be accepted throughout the United States, required many months of diligent work by the top experts in the field of management excellence. As a result of this monumental effort, seven major categories were defined with over 30 different evaluation items. One thousand (1000) points were then divided among each of the items, with more points given to the priority items.

In discussing the Malcolm Baldrige National Quality Award, George Bush, then President of the United States said, "All American firms benefit by having a standard of excellence to match and perhaps, one day, to surpass." Even now, this measurement process is being further refined. Figure 13.2 is a representative list of the categories and items, with the maximum points for each.

Many countries around the world now give out national quality awards. These National Quality Awards provide another excellent way to measure the progress of one of the elements (quality) of your TIM process, but by themselves are inadequate because they focus on quality, not the total organization's performance. For example, one Houston, Texas organization that won the award went bankrupt just months after winning the award, while another winning organization's stock dropped 75 percent, let hundreds of thousands of employees go, and cut dividends by 75 percent after winning the award.

Here again, it is best to conduct this evaluation during the first six months of the improvement process so that the baseline is well established. At this point in the cycle the evaluation is not conducted to apply for the award. It will be used only to define areas that need improvement and to provide a quantitative excellence level of performance.

There are a number of approaches that can be used to conduct a national quality award type evaluation. Probably, the best way is to hire someone who is certified and trained as an auditor to collect the data and do the analysis. This approach minimizes the variation from evaluation to evaluation. Normally, an experienced auditor/evaluator will use surveys to collect some of the information required and to obtain the opinion of all levels within the organization regarding key categories in the evaluation. The result of this evaluation will be an overall excellence rating for the organization that will range between 0 and 1000, and a list of identified weaknesses. Any rating over 500 points is an excellent start, but is still far away from becoming an award winner.

Another approach that has proven to be successful is to have a team from within the organization fill out the application form and apply for the award. Applying for the award will result in a good evaluation of the excellence of your organization by your own team and the National Quality Award committee. Even if you don't win the award, you get feedback from the National Quality Award committee that will help direct your improvement process. The problem with this approach is the long cycle time involved (as long as a year). To offset this disadvantage, you can hire a certified auditor to perform the evaluation of the application prepared by your team.

■ Management Information System Measures

Every business function, manager and team needs measurements to run the operation. Being a pilot with many ratings and more than 7000 hours' flying time, Jim Talley has found it very important to really look at all the complex data that is available in a summary or condensed format. The grouping of instruments in the cockpit of an aircraft display that vital information for quick scanning, thus providing the indicators of where to look for something that is out of the normal operating range.

The balanced Management Information System (MIS) comes in four clusters:

1. Customers—How do our customers perceive the organization?

2. Owners/stockholders—How do we make a reasonable ROI to the owners of the business?

3. Productivity and quality—How efficiently we deploy our total resources and the excellence of our outputs.

4. People—How do we use the creative, innovative and educational talents of our employees?[3]

5. Other stakeholders—Measurements are important to our suppliers and community.

These five groupings or instruments in the cockpit can give you the most important summary information in a timely manner. This MIS is backed up by the lower tier measurements from the functions, processes and teams. The MIS can be manually generated and reported, but it is better to have it on electronic mail or a local area network (LAN). Whichever, it needs to be "real time." This scoreboard MIS minimizes data overlooked by emphasizing those important measures really required. The amount of paperwork, objectives and measures in most organizations rivals the "spotted owl" in using trees in our country. This scoreboard ties together the many measures previously discussed in this chapter. It also prevents top management from zeroing in on one facet or fighting fires. Too many times some managers try to get attention by: 1) developing a "fire," or 2) playing politics and suboptimizing their function or project. So let's examine the first four groupings of instruments in our airplane.

Customers

Most organizations have a vision, mission or quality policy that addresses customers. Let's use Tandy Electronics Computer Division as an example: ... "the policy ... is to ensure complete customer satisfaction through on-time delivery of defect-free products and services." A scoreboard for the customer should address delivery, quality, cost, complaints, etc. (See Fig. 13.3.)

Customer board		
Strategy	Objectives	Measures
Market share	Improve	• Percentage share of market
Customer base	Delight them	• Survey scores • Complaints • Warranty dollars • Delivery
New products	Quantity developed	• Timing to market • Percentage that met projected sales
Preference	Key accounts	• Percentage share of target accounts • Discounts
Partners	Collaboration	• Integrated product dev. teams • Proposals won • QFD analysis

Figure 13.3. Customer-related measurements.

Owners/Stockholders

How do we compare to our industry, competition and what is our rate of return to our stockholders? (These usually include our employees too.) Our vision continues ..."allowing us to prosper as a business and to provide a reasonable return for our stockholders." Are all these improvement activities we are doing improving the bottom line—ROI, profits, and dividends? (See Fig. 13.4.)

Owner/stockholder financial board		
Strategy	Objectives	Measures
Industry standard	Improve	• Sales/employee
Survive	Cash flow Inventory	• Receivables over 30 days • Turn 12 times
Win	Increase	• Profits/employee

Figure 13.4. Owner/stockholder-related measurements.

Productivity and Quality

After we find out what the customer really wants—possibly through partnerships and Quality Function Deployment—how do we translate these into the internal measures for continuous improvement? As our vision dictates ..." provide a system of continuous process improvement...." The Malcolm Baldrige National Quality Award criteria and ISO 9001 Quality Standard criteria are excellent guidelines to use. Analyzing all your processes enables your teams to identify the value-added-only tasks and remove all areas of waste. Every employee needs to be working on cycle-time reduction, first-time yields and cost reduction as they relate to their jobs. (See Fig. 13.5.)

Productivity/quality board		
Strategy	Objectives	Measures
New products	Design	• Avoidable engineering changes
Customer base	Delight them	• Release schedules
Manufacturing excellence	Reduce costs	• Scrap and rework costs • Cycle time • Unit costs • Yields
Administrative excellence	Business Process Improvement	• Cycle time • Effectiveness • Report reduction • Efficiency • Adaptability
Procurement	Suppliers	• Reduce number of suppliers • Percentage of suppliers qualified
	JIT	• Inventory accuracy • Inventory turnovers

Figure 13.5. Productivity/quality-related measurements.

People

An organization's ability to improve, innovate, and learn lies in the strengths and capabilities of its employees. The continuous improvement in effectiveness and efficiency lies in training, retraining and motivating the people. Setting measurement targets and resetting them to higher standards is the way organizations stretch themselves to higher levels of improvement. Our vision continues ... "our people are our source of strength. They provide our corporate intelligence and determine our reputation and vitality; involvement and teamwork are our core human values." Figure 13.6 provides typical people-related measurements.

People board		
Strategy	Objectives	Measures
Work force	Training	• Percentage trained • Percentage retrained • Hours/employee/year
	Attendance	• Percentage of absenteeism
Technology	Techniques	• Integrated product development • Quality function deployment • Statistics
Environment	Safety Morale	• Lost time • Grievances/employee • Suggestions/employee
Community	Charity	• Dollars given/employee • Families adopted/function or dept. • Dollars bond drive/employee

Figure 13.6. People-related measurements.

With TCM, TPM, TQM, TRM, and TTM, managers are learning that no longer does the bias of just the financial organization dictate upper management measurements. It takes a totally integrated, balanced scoreboard for an organization to achieve TIM. The balanced approach really helps move an organization toward being a *winner*!

■ Planning the TIM Measurement System

The ability to measure the status and effects of installing TIM varies as a function of time. This measurement system can logically be divided into three phases.

Phase 1—Activity measurement. Phase 1 starts the day that the management team agrees to undertake an improvement process and will be an ongoing measurement. It is used to measure the TIM-related activities that are underway. Typical activity measurements are:

▪ Number of people trained to use the improvement tool

▪ Number of department improvement teams formed

▪ Hours of top management time devoted to TIM

Phase 2—Improvement results. Phase 2 starts about six months into the TIM process and continues through the first 60 months. Prior to six months into the process there are little or no improvement results available and management should not expect to have this type of data presented to them. It does not mean that data of this nature should not be collected during the first six months of the process. Typical improvement results measurements are:

- Poor-quality cost
- Response time to suggestions and percent accepted
- Dollar savings from team activities

Phase 3—Performance (business) results. When it comes right down to it, this is the type of measurement results that must be improved for any of the stakeholders to benefit from the improvement activities. Performance results measurements should start about 18 months into the TIM process and will become ongoing measurements. Typical performance results measurements are:

- Percent of market share
- Profits
- Customer retention and acquisition
- Earnings per share

■ Summary

Why do improvement efforts fail? One of the major reasons that they fail is a lack of hard, measurable results. This includes all functions and areas of an organization. There is a real need to see the economic impact (e.g., return-on-investment) for both the short term and the long term. The strategy, the objectives and the associated measures are the keys to being a winner. When all employees understand the relationship between their performance and the organization's success, they will make the ultimate efforts to be winners and sustain their jobs and life style.

The following shows how the three types of organizations approach measurements.

Attitude About Measurement

- *Losers:* Not important.
- *Survivors:* Treated as an afterthought.
- *Winners:* Set up the measurement system at the beginning of the process so that a baseline is defined and progress can be measured.

Targets

- *Losers:* Management sets targets.
- *Survivors:* Employees set targets for themselves.
- *Winners:* Management sets business targets. The employees set more stringent challenge targets for themselves. The target is less important than the trend.

Measurement Communications

- *Losers:* Data is collected so that management can keep things under control.

- *Survivors:* Job-related measurements are shared with the employees.

- *Winners:* All improvement measurements are posted for everyone to see and reviewed with the employees as a team at least four times a year.

How Measurements Are Used

- *Losers:* To identify individuals that need to improve.

- *Survivors:* To define problems and measure progress.

- *Winners:* To help the individual understand his or her impact on the organization and align the individual goals to those of the organization.

■ References

1. Sloma, Richard S., *How To Measure Managerial Performance*, MacMillan Publishing Co., Inc., NY 1980.

2. Sommer, Dale W. and Frohmam, Dr. Mark, "American Management Still Missing Some Basics," *Industry Week*, Vol. 241, No. 14, July 20, 1992 pp. 36-38.

3. Kaplan, Robert S. and A. L. Dickinson, "The Balanced Scorecard—Measures That Drive Performance," HBR, Jan-Feb 1992.

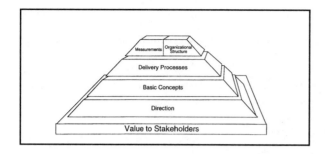

14

Organizational Structure: Restructuring the Organization for the Twenty-First Century

Paul C. Kikta
Principal, Ernst & Young

We trained very hard, but it seemed that every time we were beginning to form up into teams, we would be reorganized. I was to learn later in life that we tend to meet any new situation by reorganizing. A wonderful method it can be for creating the illusion of progress while producing confusion, inefficiency, and demoralization.

GAIUS PETRONIUS, ROMAN AUTHOR, 70 AD

■ Introduction

Reorganizing has been the silver bullet of management for years. The organization isn't performing the way it should? Reorganize! Centralize! Decentralize! Have the customer service unit, and every other critical function, report directly to the CEO! And the pressure to change is increasing.

Organizations have entered an era of unprecedented challenges and uncertainty. As a result, leaders are increasingly looking to structural alternatives because the traditionally reliable functional design models do not appear to be as effective in the ever-changing environment of today.

Specifically, there are three major trends in the environment forcing organizations to adapt structures that are more flexible, efficient and responsive:

Globalization.   The trend toward Globalization creates stiffer competition. Organizations are no longer safe in their own backyards from competitors whose headquarters are located half way around the world. Organizations no longer have "safe" markets to protect their more risky or more profitable ventures. As such, organizations need to be flexible and adopt structures that are appropriate in each competitive market. Globalization also forces organizations to be more flexible as new ways of doing business are developed and introduced around the world.

Increasing Pace of Technological Change.   The technical life cycles of products and services are rapidly decreasing and with them go the life cycles of various business processes. Computer manufacturers whose products used to have 2 to 3 year life cycles now are fortunate if they can enjoy a competitive advantage for 12 to 18 months before someone else is out there with a product that is better, quicker and cheaper. Each new generation of technology creates new opportunities to redesign or reengineer business processes. The acceleration of technological change requires flexible structures to adapt to rapid technological and process changes.

Shifting Demographics.   As increasing numbers of non-traditional workers (contract workers, part-timers, supplier/vendor partnerships) become the staple of the work force, organizations will have to become increasingly more flexible in order to work with these various groups. The concept of virtual organizations is a reality. In some organizations today, there are employees who never physically work with each other except through the medium of technology. In others, members of a project team may not even be employed by the same organization or work in the same geographic location.

These trends create a fast-paced, unpredictable and highly competitive economic environment. In order to compete, organizations must find a way to adapt quickly and efficiently.

Traditional organizational structures, based on a military command and control paradigm and a mechanistic view of business processes, have become inherently inflexible. Recently, new and more flexible models of structure are being added to the structure design repertoire.

■ The Evolution of Organizational Structure

Organizational structures are evolving. Organizations tend to adopt structures in response to new and different business pressures. To date there have been four major phases in the evolution of organizational structures.

1. Vertical

2. Bureaucratic

3. Decentralized

4. Networked

As corporations have evolved, they generally structured themselves through each of these phases and usually in the order presented. However, each organization may move through the phases differently. Some may even have different structures currently within their organization. As such, any or all of these structures may exist in an organization at any point in time.

■ The Vertical Organization

The vertical organization, prevalent in the 1950s and sixties, existed to manage specific functional competencies. It focused on functional performance and used a fairly small and centralized corporate staff to make decisions for the labor and production work force which made up 70 to 80 percent of the total work force. Planning was organizationally separated from execution. Steps in the work process were distinct and aligned with specific jobs.

The vertical organization is based on a military style of monitoring and controlling performance which originated centuries ago. Way back when Moses led the Jews out of Israel, he organized them in groups of ten. The leader of each group reported in to another group of ten, and so on. It rests on a very rigid and narrow definition of an individual's duties and a very strict translation of the proper span of control. In the traditional military model, each soldier is assigned a specific task. One soldier is responsible for loading the gun on the right side of the ship, another is responsible for aiming it. A third is responsible for firing the gun.

This process works smoothly, despite the fact that each soldier performs only a small part of the overall process. The tasks are sequential and discrete and there are clear and specific procedures and communications protocols to coordinate individuals and groups. So long as the process or technology remained stable and constant, this model worked extremely well (see Fig. 14.1).

In addition, commanding officers serve not only as vehicles for maintaining control and performance; they also are responsible for transferring knowledge. Commanding officers have worked their way up through the ranks and usually have technical and functional expertise which they transfer to their subordinates. Like the military model, hierarchy is the foundation of knowledge transfer in the vertical organization.

Since the required tasks in an industrial organization were seen as analogous to the division of labor in the military, the military structure seemed to be a perfect model for organizations to follow. In fact, industrial operations performed well for many years with this structure. However, with success came the need to expand and a new structure was needed to manage the more complex operations.

Advantages	Disadvantages
• Clearly defined scope of tasks	• Limited flexibility or exposure to other responsibility areas
• Limited range of knowledge or skills required to perform effectively	• Limited career development opportunities
• Knowledge transfer along the chain-of-command is enhanced	• Knowledge transfer across the organization is difficult
• Competency development within a vertical unit is enhanced	• Narrow skills base within a vertical unit
• Efficient in stable, predictable environments	• Ineffective in dynamic or unpredictable environments

Figure 14.1. Advantages and disadvantages of a vertical organization structure.

■ The Bureaucratic Organization

Over time, corporate organizations grew to accommodate their successes. For instance, automobile manufacturers increased production volume and added models and features. Additional machines were required to produce the increased volume and diversity of products. More machines required more people to run them, which, in turn, required more people to manage the people who were running them.

Next came the policies and procedures required to coordinate and control the additional size of the work force and the complexity of the tasks. And in a never-ending upward spiral, more people were needed to develop and manage the new policies and procedures. As such, success drove vertical corporate organizations into bureaucratic organizations.

The business needs which created the bureaucratic organization was just an extension of the vertical organization model. For the vertical organization, jobs were created to take responsibility for specific aspects of the work process. In the bureaucratic organization, jobs and additional layers of management were created to take responsibility for coordination of specific business processes and policies (see Fig. 14.2).

Bureaucratic organizations, characterized by multiple layers of management, and broad-reaching policies and procedures, were usually unable to effectively respond to rapid changes in the marketplace. Therefore, as the need to be more responsive to the market became evident, organizations took to restructuring again.

■ The Decentralized Organization

This time, corporations came to the realization that as the size and complexity of the organization increased, so did the costs of maintaining a centralized bureaucracy to support this organization. The next logical move was simply to break big organizations into smaller ones.

Advantages	Disadvantages
• Clear policies and procedures	• Policies and procedures create inflexibility
	• Potential for long cycle times when process crosses many responsibility areas
• Stable organization systems and processes	• Glacial responsiveness to change
• Consistent service and quality levels	• No individual judgment or empowerment
• Clear performance expectations	• Performance expectations tend to be internally focused
• Clear roles and responsibilities	• Cooperation across role and responsibility areas is difficult
• Enterprise-wide focus	• Internal focus
• Effective strategy deployment	• Difficult to dramatically change strategy
• Optimization at a micro level	• Potential for sub optimal performance at an enterprise level

Figure 14.2. Advantages and disadvantages of a bureaucratic organization structure.

As a result, organizations began to break themselves up into smaller decentralized units with each unit a profit center reporting directly to an operations manager. Generally, each unit had complete authority, within "corporate guidelines," to create whatever policies and procedures were needed to maintain profitability and generate adequate returns to shareholders. In the automotive industry, for example, they divided their organization into units, each responsible for a different make of car.

Decentralization provides several advantages (see Fig. 14.3). Smaller business units tend to be more flexible and are therefore more responsive to market demands. Additionally, smaller business units usually require less managerial overhead. However, the coordination is less effective between these independent units as compared to divisions within a more centralized organization. Responsiveness is gained at the costs of coordination.

As such, decentralized organizations often face a problem of coordinating with customers. A common scenario is multiple salespeople attempting to service the same account, with none of them able to provide a full-line of solutions in a seamless manner. Two divisions of a leading automotive parts supplier were not only competing with each other for the same customers, but were unknowingly being played off against each other to lower their prices.

Please note, however, that the move to a decentralized business unit structure creates little change in the basic operational structure of most organizations. It simply breaks big organizations into small ones. The guiding principles of organ-

Advantages	Disadvantages
• Strong customer focus	• Reduced enterprise focus
• Business units responsive to changes in customer needs and market demands	• Enterprise-wide ability to act in concert is difficult
• Business units are focused on the needs of their segments	• Business units are hard to coordinate when a customer is in multiple segments
• Self-sufficiency at the business unit level	• Duplication of resources and inefficiency at an enterprise level
• Accumulation of customer-related knowledge is enhanced	• Knowledge transfer across business units is difficult
• Business units empowered to focus competency development efforts in areas which support their own success	• Difficult to maintain consistent functional competency levels across the enterprise
• Business units empowered to develop own standards within corporate guidelines	• High potential for inconsistent processes, technologies applications and competence levels
• Accountability and control at business unit level	• Internal tension and competition for resources based on measurement system

Figure 14.3. Advantages and disadvantages of a decentralized organization structure.

izational structure—span of control, task specialization, functional silos, and knowledge transfer through the supervisor, do not normally change.

Despite its limitation, decentralization served organizations well until the advent of global competition and customer satisfaction became critical drivers of success. It was time for a change again.

■ The Network Organization

The latest stage in organizational structure development, the network organization, represents the first real innovation in the design of organization structure. In the network organization structure, the focus is on the customer and not on internal business functions. By focusing externally, rather than internally, networked organizations are better positioned to be more responsive to the total needs of their customers and changes in the market.

Network organizations are based on teams of people handling a process or serving a client, rather than on individuals performing functional tasks under several layers of management.

The emergence of network organizations is a direct consequence of two major changes in management thinking. The first change is an understanding of the importance of a multi-skilled work force. This change in management thinking rejects Adam Smith's notion that ultimate benefits come from of the division of labor.

The second change is a realization that business success is not based purely on technical or functional expertise, but rather on applying these to the processes and resources required to meet customer needs.

The network organization has its roots in the matrix structures of the past. In the matrix structure, the functional portions of the organization structure were loosely linked together along customer or product lines. More commonly found in project based organizations, the matrix organizational structure has had mixed results. Although the matrix organization allows for enhanced customer focus while maintaining functional integrity, the structure has typically been associated with increased organization tension and confusion surrounding dual reporting relationships.

Whether viewed as a success or as a failure, the matrix organization structure was hampered by the management thinking and technologies of the past. However, recent changes in each of these have allowed the network organization to emerge as a successful structural alternative.

Why Now?

The factors contributing to the need for flexible and efficient organizations are not new. Why then, has it taken so long to develop this structural response? There are three major trends which have set the scene for the evolution of the network structure:

1. Advances in information technology

2. The transition to a service economy

3. The introduction of total quality initiatives

Information Technology

Information technology is one of the most important enablers of network organizations because it:

1. Breaks the organization's dependence on "expert" managers.

2. Permits people to work together as a team regardless of geographic boundaries.

3. Empowers people to participate in and coordinate the work process at different points in time.

Traditionally, expertise has belonged to the employee, and knowledge transfer has been provided on a need-to-know basis by functional or technical experts. Employees became more valuable based on what knowledge or experiences they had. With the advent of expert systems and knowledge-based software packages, expertise is no longer monopolized by a few people. It is stored in computer databases and available to anyone with minimal computer knowledge and access to a PC. Expertise is shifted from being a personal asset to a corporate asset.

For example, Human Resource departments have typically required specialists in making compensation and benefits decisions. Now, a Human Resource generalist can take advantage of an expert system or a decision support tool to deal with a compensation or benefit issue. Similarly, workers on the factory floor who used to rely on their supervisor for work flow or inventory information now have that information available to them through terminals on the factory floor.

Technology has also allowed organizations to take spans of control to new extremes and dimensions. Managers can now oversee employees and business operations located in different parts of the world. Aided by advances in portable communications and computing technology, a sales manager can communicate and interface with sales personnel scattered across a broad geographic region.

Global communications, groupware and other technological innovations allow individuals and groups of individuals not only to work in a coordinated fashion from different points on the globe but at different points in time. Work which had previously required sequential processes can now be conducted in parallel through enhanced communications and coordination. Messages can be posted to an electronic mail system and be received and replied to at different points in time. A team of people can work on a project continuously from sites on opposite sides of the globe.

Through the sharing of expertise and information, managers no longer need to be involved with task control and resolution of short-term issues. Managers can then devote more time in planning and organizing. The end result is an organization which is flatter and more flexible.

As a result of the changes in technology, employees can make better business decisions and stay informed on the progress of the whole process. No longer are they limited to their "little corner of the world." As a result, their jobs can be broader. They can truly work "smarter."

Service Economy

The traditional models of organizational structure utilized in the industrial segment were soon discovered to be less than effective in the emerging service sector. This was partially due to the fact that tasks in service organizations are not normally as distinct or as repetitive as those in the industrial sector. This fluctuation in tasks is often a direct consequence of the interaction service employees have with their customers.

For example, bank tellers handle a variety of transactions, each of which can be complicated by the customer. One customer may not have a deposit slip. Another wants to split the deposit between two accounts. A third wants to transfer money from one account into another. Each transaction is technically a deposit, however, the actual tasks vary and require careful attention to the individual needs of each customer.

The shift from a manufacturing to a service economy has emphasized the need to have skilled employees who can rapidly and efficiently deal with an array of customer needs. Customers are no longer satisfied to move from point to point because of organizational bureaucracy, policies or procedures. Customers now demand that businesses meet their particular wants and needs on demand on the spot.

As such, rather than moving from station to station within a bank, customers expect the teller to be able to handle all their needs right then and there. The role of the service worker is now to serve as the interface between customers with random needs and the organization's processes, policies and procedures. In the context of the service economy, the narrowly defined jobs and roles of the past no longer make sense.

This shift to a service economy has also increased the need for organizations to better understand and satisfy the needs of their customers. Direct competition for customers has increased the importance of maintaining a strong sense of customer focus. The "one size fits all" model of business no longer makes competitive sense. The bureaucracies, processes and structures which were designed for efficient, mass production are no longer able to effectively satisfy customers on an individual basis.

Total Quality Initiatives

As organizations begin to reach out and try to meet the needs of each individual customer, they have found they need to change the way they do business. This desire to reach out and "satisfy and delight the customers" is known by many names.

In recent times, it seems that every organization has implemented a improvement initiative of some form or another. Although these initiatives vary widely, they usually address several consistent factors. The most consistent is the necessity to integrate people, processes and technology in order to "delight the customer."

This driver generally fosters network organizations in several ways. First, they promote focusing on the customer which is something that traditional, functionally separated organizations have had difficulty doing. Second, they promote the creation of cross-functional teams and the cross-functional development of individual workers. Third, they promote empowerment of workers.

■ Two Models of Network Structures

There are currently two popular models of network structure being implemented in organizations, *case management* and *horizontal process management*.

Case Management Network Structure

Case management is an increasingly popular organizational structure design to overcome many of the disadvantages of the more traditional structures (see Fig. 14.4). Case management was pioneered in the health care industry in response to the importance of patient service and relatively high levels of service and organizational complexity. In several health care organizations, individuals or teams have

Advantages	Disadvantages
• Strong customer focus	• Reduced enterprise focus
• Flexible	• Potential for variability and inefficiencies in the delivery of products and/or services
• Total accountability for customer satisfaction	• Stressful work environment due to high levels of interdependence
• Job diversity	• Requires highly skilled employees
• Eases coordination between functions	• Unclear roles and responsibilities
• Highly decentralized and cross-functional	• Dissipates resources
• Responsive to customer needs and market requirements	• Difficult to maintain strategic focus

Figure 14.4. Advantages and disadvantages of a case management network organization structure.

been assigned to track a patient through his or her stay in the hospital in order to manage the entire process of delivering patient care.

Case management does not eliminate functional divisions within an organization, rather it facilitates the coordination of the functions where one person takes the responsibility and ownership of an individual case from beginning to end.

In their August 1993 technical report entitled, *Case Managers: The End of Division of Labor*, Thomas Davenport and Nitin Nohria suggest four key components of successful case management:

1. A "closed-loop" work process that involves completion or management of an entire customer product or service.

2. A location in the organizational structure at the intersection of the customer and the various other functions or units which create or deliver the customer product or service.

3. Empowerment and role expansion of employees to make decisions and address customer issues.

4. Easy electronic access to information located throughout the organization, and the use of information technology to aid in decision-making.

Horizontal Process Management Network Structure

A step beyond case management towards cross-functional integration is horizontal process management with its own set of advantages (see Fig. 14.5).

Here an organization examines its core processes and designs a structure that encompasses all of the requirements necessary to meet the customer needs of that service or process. Some typical core processes are:

Advantages	Disadvantages
• Strategically aligned	• Duplication of resources
• Customer focused	• Competing goals across process teams and across levels of hierarchy
• Total accountability for process performance	• Stressful work environment due to high levels of interdependence
• Work force aligned along process lines	• Career pathing is complex and there is a potential for glass ceiling to be created
• All resources to do the job are available within the network	• Reduced critical mass or economies of scale
• Lower process cycle time	• Dissipation of knowledge
• Efficient process execution	

Figure 14.5. Advantages and disadvantages of a process-based network organization structure.

- New Product Design
- Order Fulfillment
- Order Entry
- New Product Introduction
- After Sale Service
- Supplier Partnerships
- Etc.

In some ways it is similar to a matrix organization, however, the solid line relationship is to the process and the dotted line relationship is to the function, rather than vice versa.

The goal is *not* to make a complete horizontal process organization. Rather, each organization should seek its own unique balance between the vertical and horizontal features needed to deliver performance.

Here is the list of 10 principles at the heart of horizontal process organizations:

1. Organize around process, not task.
2. Flatten hierarchy by minimizing the subdivision of work flows and nonvalue-added activities.
3. Assign ownership of process and process performance.
4. Link performance objectives and evaluation to customer satisfaction.
5. Make teams, not individuals, the principal building blocks of organization performance design.
6. Combine managerial and nonmanagerial activities as often as possible.

7. Treat multiple competencies as the rule, not the exception.

8. Inform and train people on a "just-in-time-to-perform" basis, not on a "need-to-know" basis.

9. Maximize supplier and customer contact.

10. Reward individual skill development and team performance, not just individual performance.

■ Implementation of Network Structure

Implementing a network structure has many ramifications for organizations. Besides just changing the structure, it affects many other aspects of the organizations as well, including:

1. Management style

2. Performance management system

3. Education and training programs

4. Communications processes

5. Compensation and rewards program

6. Management development

Once you decide to reorganize, the structural change is just the tip of the iceberg.

Management Style

Implementing any model of network structure in an organization requires a fundamental shift in management style for most people. For some, this shift often results in managers having to learn a different way of doing business in their organization. For others, particularly those adept at managing projects with cross functional teams, their management style will require little or no modification. But, for most, the answer rests in the middle.

This shift is not an elimination of a manager's responsibilities. Managers still need to have a high level of involvement, but with a different focus. Rather than peering over the shoulder of each individual, managers need to monitor the progress of the team, with an eye towards improving the teams decision-making ability. As Tom Davenport, partner at Ernst & Young, notes, "The case manager should not operate without any controls, but excessive control over case management defeats a key purpose of the initiative."

As a result, managers become leaders of teams by providing information and guidance to help them make decisions within the process. Process managers lead by supporting the team.

Performance Management

A network organization also requires a shift in the performance management system. In functional organizations employees are rewarded for moving up, or deeper into a particular specialty. In network organizations this need not be the case. Rather, the ultimate measure of success is "How satisfied is the customer?" Rather than being assessed only on individual competence, the measure may include how well individual competence was applied toward the total satisfaction of the customer as well as the team's contribution to their organization's strategic goals.

Education and Training

Organizations must provide additional training to support the network structure. If employees are required to perform broader tasks and are responsible for making on-the-spot decisions, additional training may be necessary not only for each new skill required, but also for improvement of communication skills and teamwork. This poses many potential problems for organizations, particularly when selecting or training entry-level employees. Who will be trained? What will they be trained in? How much is the optimal level of training? Do you buy or train the talent you need?

Communication System

For the teams within a network structure to properly function, an open communications environment must exist. Process information, business data and organizational strategy are just a few of the subjects that need to be at everyone's finger tips. When teams in a network organization are expected to make good business decisions, the quality of their decision making will be directly proportional to the quality of information they have. They need direct communications links with customers and suppliers as well.

Compensation and Benefits

Multi-skilled individuals with team and customer focuses are behaviors that need to be recognized, reinforced and rewarded in network organizations. As such, the new compensation and rewards system must be able to accommodate this. This has caused the growth of various pay for performance, pay for skill, and productivity bonus systems in the modern organization.

Additionally, in a network organization, whether using a case management or process management approach, employees and the team needs to be rewarded based on customer satisfaction, individual contribution to group performance and overall organizational performance.

Management Development

With flatter organizations becoming a reality, the chances for advancement "up the corporate ladder" are drastically reduced. So how does an organization provide career opportunities for its employees? First of all, by making lateral movement in an organization a "successful" career endeavor. Second, by realigning the goal setting, performance review, and career development process to select managers based on their ability to form and mobilize networks to serve customers rather than on building their own functional skills.

■ Organizational Structure Design

Dr. H. James Harrington states, "As much as I would like to tell you that everyone should evolve to a Network Organization, I cannot. My experience indicates that all four organizational structures and even the combination of them must be considered depending on the organization environment."

Once the decision to change your organization's structure has been made, the next question usually is to decide "What is the best organizational fit for my strategy and competitive environment and makes best use of my distinct core competencies?" The answer to this comes not from a single diagnostic tool, rather from a technique of "informed dialogue" which is a combination of analysis and dialogue conducted in an iterative way?

"What is the best way to decide on the 'right' structure and fit for my organization?"

The first step in this process is to look at the organization from three perspectives: The *strategic perspective* which looks at the organization from the top-down and determines the overall shape of the organization. It's a process of moving the big boxes around to determine the right fit.

The second is the *operational perspective* which deals with the strategic business units. In this case you look at the organization from two directions. You review the strategic fit with a look from the top-down. You ensure the appropriate mix of operational, managerial and support processes through a bottom-up review.

Finally, the *tactical perspective* is completed with a bottom-up approach and determines the work team and job designs.

All of this (strategic, operational, and tactical) comprise what is called organizational structure. It is the combination of strategic, operational and tactical decisions that will be the basis for determining the "right" organization structure.

Let's now look at what is the approach and questions that can be asked at each of the levels.

Strategic Perspective

The strategic perspective design process leads you through a series of questions to determine how to best allocate and focus resources to carry out the strategy of the organization.

To start this process, you should develop an understanding of the current state of the organization and its environment. Some questions you should explore are:

- How does the competitive environment shape the way we conduct business? Is a specific structure forced on us by competitors?

- How well do we meet our customers' demands? How does structure affect our ability to meet customer demands?

- What are the interrelationships among the different functions and units and how do they impact each other? How can structure support the coordination among them?

- What are our core competencies? How well do they support our strategy? How can structure facilitate the strategy?

- What impact does our history and culture have on how we have structured our organization? What barrier could it impose on a new structure?

The answers to these questions will have a dramatic impact on what you can and should do with your structure. For example, if reacting to a volatile market where the customer needs are constantly changing is the number one strategic issue, then a decentralized product design unit attached to horizontally based organization makes more sense than a centralized design center at corporate headquarters.

One way to focus the effort of answering these questions is to approach the process by looking at data input, influencing principles and output options. Figure 14.6 is an illustration of how one organization used that approach. This process is normally conducted over a period of weeks where information is collected on the data inputs. That is then presented and filtered through the principles with the resultant outputs used to help set the direction for the next step which is to address the operational perspective.

As a result of following the process below, one organization developed the following:

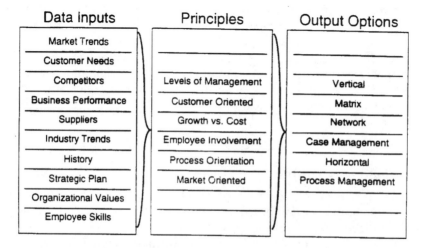

Figure 14.6. Organizational structure visioning matrix.

The goals of the new organizational structure are to achieve better coordination between functions, decentralize authority, increase employee involvement at lower levels of the organization, restructure the decision-making process, more clearly define responsibilities, and encourage more innovative, forward thinking.

The predominant need regarding organizational structure is for a cross-functional focus—reducing the isolation of functions between divisions. The final organizational structure will require features of a horizontal organization where structure is built around processes and teams. This will include cross-functional teams focused on the four basic processes, and a flatter organization to drive decision-making authority to a lower level.

Not all structured visions need to be so complete. One organization did theirs just by developing some bullet points:

- Manage by process rather than function.

- Achieve a common goal of satisfying the customer.

- Have departments interact with each other before making policy or procedures changes.

- Lessen finger pointing between departments and divisions.

- Give employees a chance to have an understanding of their own functions (job responsibilities) in respect to the entire system.

- Provide consistency in work between departments (i.e., every individual performing the same task in a similar manner).

- Help employees learn from each other.

Operational Perspective

The next step is to review the design impact of the strategic business unit level and how an organization can cluster its work to support the strategic intent and direction of the business.

One approach is to use three grouping options. These are *activity*, *output*, and *segment* (see Fig. 14.7).

Each grouping has relative advantages and disadvantages in terms of competitive response, market response and internal functioning and strategy implementation.

Dividing the organization by *activity* is similar to the traditional vertical organization where activity is defined as a function or knowledge group. Such an organization would have predominantly functional components at the highest level such as finance, operations, sales and marketing, etc.

Divisions based on activity usually promote high functional expertise and utilize staff efficiently. This is particularly effective where functional expertise and knowledge transfer are key to a strategy. However, since the work process tends to run across divisions, interdivisional tensions are likely to be observed.

Grouping option	Structural implications	Example
ACTIVITY —Function —Knowledge/Skill	All personnel who contribute to or accomplish similar activities or who perform similar functions are grouped together.	Auto manufacturers have historically used activity as the primary method of grouping i.e., marketing, manufacturing, and service were all separate divisions.
OUTPUT —Product —Service —Project	All specialists needed to produce a given product, service or project work together.	Auto manufacturers have moved to a more horizontal structure which includes cross-functional teams by output. For example, There are different divisions for minivan, luxury cars and compact cars.
SEGMENT —Market/Industrial —Users/Clients —Geography	All specialists needed to serve an industrial/market segment, or to meet user/client needs, or to serve distinct territories work together.	Banking typically is divided by region, with some divisions serving the east, central and western portion of the country.

Figure 14.7. Analysis of the three grouping options.

Dividing the organization along *output* lines allows each product group to focus on the efficient production of a specific product/service. Such an organization would have predominantly product/service components at the highest level such as consumer electronics, industrial products, warranty operators and components, etc.

Divisions based on product/service usually promote increased product innovation and productivity advantages. They tend to provide a rapid response to existing markets. This is highly effective in a highly competitive market where production efficiencies are key. On the other hand, coordination of marketing activities across different product groups is generally less effective. Also, any leverage that may be achieved with supplier and distribution channels through coordinated purchasing and logistics is generally less than that of activity designs.

Dividing the organization along *segment* lines allows each group to focus on the responsive delivery of products and services. Segments may be divided by geographic industrial/market segment or user client needs. This method results in specific structures like: Americas group, European operations, high net worth clients, etc.

Divisions based on segment typically promote faster time to market or enhanced customer sensitivity and focus. They tend to have well-integrated customer support systems and rapid response to customer needs. It is highly conducive to a market where customized products or services are the norm.

Most organizations use two or all of these groupings. For example, often within traditional, activity-grouped organizations where the largest divisions are operations, sales, and service, subdivisions are usually based on output or segment.

The key is in being able to identify the right combination. Most of the answers to doing this will come from the strategic perspective process and the development of the structural vision. Remember that the operational perspective is the bridge between the strategy of the organization and the way in which the work is performed.

It's important to keep in perspective that the groupings are not mutually exclusive of each other. In fact, the most effective organization uses all types of groupings. They may be grouped by activities at the senior management level, by market segment at the division level, by product at the plant management level, and by user at the work flow or process level.

Tactical Perspective

Whether we are dealing with a team of people or an individual, we still need to determine how to structure work at a tactical level. As with the previous two perspectives, there are some questions to ask, which depending on their answers will give us some direction in this process.

1. To what degree does the job have a clearly identifiable beginning and produce a meaningful product or service? What is the interval of time between completing a task and the completion of the work process?

2. What categories of work logically are grouped together for an individual or team? What is the relationship each task has with the task preceding and following it?

3. How should work be managed and coordinated? Who makes the decisions? What level of decision making is made? How should decisions be made?

4. How do information and knowledge flow? Is feedback complete, immediate, direct and individualized on tasks and operational completeness?

5. How routine is the work? Is it governed by standard methods and procedures. Does it change from day-to-day, customer-to-customer?

It is the cumulative knowledge of the three structural perspectives that finally provides you with the data necessary to put together a design plan for the organizational structure. But, just in case you feel that's the end of the process, there is still one more organizational frontier to address.

■ What Are the Barriers to Implementation?

There are numerous inherent barriers to the successful implementation of structural changes: senior management preferences, organizational culture, regulatory issues, current customers demands. These barriers can usually be divided into two types—organizational and transitional. For instance, if you were going to change from a traditional vertical organization to a team-based horizontal organization, you might encounter the following:

1. Management/Leadership Development Needs.   Reduction of levels of management will result in increased levels of accountability for managers. Successful implementation will require careful assessment of readiness of managers to take on additional responsibility. Senior management, which has carried out much of the day-to-day business decision making, needs to be ready to relinquish much of this authority to the business team level, and needs to have patience in the development of the team members to take on such responsibility.

2. Functional Walls.   For horizontal teams to work, the members need to shift some of their identity to the team and away from their functional area. The history of functional focus in the culture may present some tension in starting up teams. When functional organizations look to improve, they normally focus on improvements within their own functions.

3. Team Leadership.   Horizontal process teams need to have an owner who understands the complete process across all functions and has credibility to sponsor and sustain implementation.

Transition Barriers

1. Friction.   This type of barrier is passive and is a direct result of the change itself. This disruption of work, communications, decision-making and organizational power are direct consequences of the transition efforts. Although this kind of change should not directly halt the progress of the transition to team-based management, it can, if left unchecked, create an environment where active resistance will occur.

2. Resistance.   This type of barrier represents active resistance to the transition process. An example of this is the backlash that occurs as the result of increased organizational friction. Should the change process become too disruptive, what originally would be resistance to the process of change can eventually end up as the active resistance to the change itself.

Understanding the barriers to the transition to a new structure helps to determine how to get to your desired future state. Remember, structural transitions are evolutionary and similar to a learning curve process wherein each advancement represents refinement of the previous stage. In early stages of development, it is helpful for experts to have involvement in chartering teams, establishing objectives, defining processes, and monitoring performance. As the organization and its employees gain experience and new skills, the transitional activities can be successfully transferred to the operating levels of the organization.

Once the strategy is set, the design established and barriers identified, you are ready to begin implementation.

Some changes may be of such magnitude that it will take a Herculean effort on the part of management. For example, a successful movement from hierarchical, functional behavior to a team-based organization requires patience and long-term commitment, tolerance for error, and behavioral role models. The transition process can be measured in years, not months, and unless the organization is committed to this investment, the transition may fail and management is likely to lose credibility with its employees.

Dr. H. James Harrington points out, "A major restructuring change in any organization should only be undertaken when it will produce a very significant performance improvement and then it must be accompanied with an effective Organizational Change Management plan."

■ Summary

Motorola is a good example of an organization that is moving toward a network organization, but even Motorola has not adopted the full network concept and they probably will not. To provide the best total service to all of the organization's stakeholders, a combination of organizational structures need to be used and modified to meet the unique needs of the situation.

In order to achieve the exponential results from reorganization needed to survive in our rapidly changing environment, new structures must be considered. Tomorrow's successful organizations will be those making standard decisions based on processes, customers, and teams rather than the old models based on the specialization of labor and command hierarchies. Used properly, these alternatives offer a tremendous potential for success.

However, even the network organization model of structure may not reign for long. New paradigms of structure will be necessary as unimagined technology becomes commonplace and work force demographics shift dramatically. The organizational structures of tomorrow will be very different from those currently in use. Perhaps belonging to an organization may simply mean having an access code to its computer network? Now what are the structural implications of that?

The following shows how the three types of organizations approach organizational restructuring.

Losers: They frequently reorganize to take advantage of the capabilities and interests of their key managers. The organization is often reorganized as the result of a power play or to replace a poor-performing manager. Losers look at structure to solve all their business problems. If it "doesn't work," reorganize is their motto. They will usually select the latest fad in restructuring in hopes that it will be the "silver bullet" for them. Any changes they do make are usually with an internal focus with the desire to improve functionality rather than customer service.

Survivors: These organizations often reorganize to focus on specific products and to keep their business units small. Their corporate headquarters execute little control over the business units. Usually they use a decentralized structure.

Winners: Reorganizations are driven by customer opportunities and new product developments. The organizational structure is directed at obtaining the best, most effective use of their employees in support of their core capabilities and competencies. Many of these organizations have shifted their primary organizational structure approach away from small business units to focus more on their critical processes and are now experimenting with limited network structures. This reflects their willingness to change if new concepts prove to provide a competitive advantage. Winners do not use structure as the first step in solving their business problems. They make sure they have first defined the business problem's root cause and solved that first, preferably from a process perspective. They then select the right combination of organizational structures available at the strategic, operational and tactical levels and don't look for the "one size fits all" structure. They recognize that structure is designed to facilitate good decision making and adapt the style that most aptly provides for the most efficient decision-making processes.

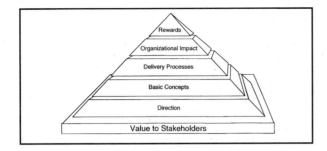

15

Rewards and Recognition: Rewarding Desired Behavior

All the good maxims have been written. It only remains to put them into practice.   BLAISE PASCAL

■ Introduction

The single most basic behavior is to perform in ways that we are rewarded for. The first lesson we learn is directly related to being rewarded for crying. A baby cries and it is fed or its diapers are changed. Later on in life, we learn that if we cry we get picked up and held. We grow a little older and we are told to clean our plates and we'll get dessert, or to be good, and Santa Claus will bring us a new toy. Later on we are told that if we clean our room, we can go to the movies.

Yes, all through our lives, we have been rewarded for acting out a desired behavior, and punished when our behaviors are undesirable as defined by someone else or even our own selves. These rewards make eating the asparagus and mowing the lawn and being good a little more worthwhile. It is important to note that all these rewards occur relatively close to the time that the desired behavior occurs. If we didn't get the dessert until the following Sunday or go to the movies until next month, or if mother tells us on February 2 that if we are not good, Santa Claus will not bring us a toy, the asparagus would stay on the plate, the clutter on the floor, and we probably would continue misbehaving.

Three factors affect the degree to which the desired behavior is reinforced. They are:

- Type of reward
- Elapsed time between when the desired behavior occurred and the time the reward is given
- The extent to which the behavior meets or exceeds the performance standard

Up to this point, we have discussed direct, tangible, positive stimuli that reward people for acting in a desired behavior mode. There are two other ways to encourage people to behave in a desired manner. They are:

- Negative stimulation
- Humanistic stimulation (recognition)

Negative stimulation takes the form of physical or mental pain within the individual(s) who do not perform in the desired manner. For example, telling a child who wants to go outside to play that they will have to stay at the table until their plate is clean, or spanking a child because their room is not picked up, or taking away television rights because they do not have their homework done, are all examples of negative stimulation. A manager applies negative mental stimulation to an individual when the manager explains why the person is not performing at an acceptable level. Often, employees will subject themselves to mental pain. We have all walked out of a meeting thinking, "Why did I say that? How dumb can I be?" Really good, conscientious employees will take themselves to task when they do something wrong far more than their managers will.

Humanistic stimulation, often called *recognition*, occurs when a person receives satisfaction because they recognize they have mastered a desired behavior, or when others are made aware of the individual's desirable behavior. Although recognition is an intangible reward, its positive impact on behavior is usually very effective and should never be overlooked. For example, we put a good report card up on the refrigerator door, recognizing that the child has done a good job, or a teacher displays a particularly good drawing in the classroom.

Whenever possible, direct and humanistic stimulation should be combined. For example, when someone is promoted, the promotion is indirect stimulation, and the increased salary is direct stimulation. Often, management thinks about rewards and recognition as two separate activities. In truth, recognition is just one element of a total reward structure that is needed to reinforce everyone's desired behavioral patterns. As Don Roux, a Minneapolis-based sales and marketing consultant, stated, "They (incentive programs) both motivate people to perform some task or achieve some goal by offering rewards. The desirable performance is rewarded, and rewarded behavior tends to be repeated."

In 1989 Xerox won the Malcolm Baldrige Award. The Xerox Business Products and Systems National Quality Award release stated:

Recognition and Rewards: Ensures that Xerox people are encouraged and motivated to practice the new behaviors and use the tools. Both individuals and groups are recognized for their quality improvements—whether that takes the form of a simple thank-you or a cash bonus.

That highlights a very important point. Up to now, we have been talking about rewarding individuals, but that is not enough. In today's environment, the organization needs to encourage teams of people to work together to provide the most efficient, effective and adaptable organization. If we reward only individuals, we develop an organization of prima donnas who are only interested in doing things that make them look good. It is for this reason that your reward process must include both individual and group rewards.

The complexity of today's environment and the sophistication of today's employees make it necessary to carefully design a reward process that provides the management team with many ways to say thank-you to each employee, because the things that are valued by one individual may have no impact upon another. In addition, the reward process needs to be closely aligned with the organization's personality. The reward process that functioned well in the 1970s is probably inadequate today because the personalities of most organizations have undergone major changes. The influx of women and various minority groups has had a major impact on the way the reward process needs to be structured.

In today's environment, men's attitudes have changed. The male population is aging, and men are often not the sole breadwinner for the family, causing them to be less financially driven. Because of this, it is easy to see, for example, why time-and-a-half pay is no longer a satisfactory reward for giving up a Saturday for many employees. It is for these reasons that we need to take a fresh look at our reward processes to upgrade them so that they meet the needs of today's organizations and their aggressive goals.

Vince Lombardi said, "Winning isn't everything. It's the only thing." This is true for many people, but for others it is enough to help someone else win. At the Olympics, only one man stood on the top platform to receive the Gold Medal for cross-country skiing, but without the many people standing along the route to give him water, he would not have won. To these little people (and there are a lot more of us little people than there are Gold Medal winners), often recognition is simply having someone else acknowledge your worth. Recognition is something everyone wants, needs, and strives to obtain. Studies have shown that people classify recognition as one of the things they value most.

■ Ingredients of an Organization's Reward Process

A good reward process has eight major objectives:

1. To provide recognition to employees who make unusual contributions to the organization and stimulate additional effort for further improvement.

2. To show the organization's appreciation for superior performance.

3. To ensure maximum benefits from the reward process by an effective communication system that highlights the individuals who were recognized.

4. To provide many ways to recognize employees for their efforts and stimulate management creativity in the reward process.

5. To ensure that management understands that variation enhances the impact of the reward process.

6. To improve morale through the proper use of rewards.

7. To reinforce behavioral patterns that management would like to see continued.

8. To ensure that the employees recognized are perceived as earning the recognition by their fellow employees.

Why does recognition matter? George Blomgren, President of Organizational Psychologists, put it this way, "Recognition lets people see themselves in a winning identity role. There's a universal need for recognition and most people are starved for it."

■ Reward Process Hierarchy

In this chapter, the word *reward* is defined as something given or offered for a special service; or to compensate for effort expended.

Rewards can be subdivided into the following categories:

- *Compensation*—To be financially reimbursed for service(s) provided
- *Award*—To bestow a gift for unusual performance or quality
- *Recognition*—To show appreciation for behaving in a desired way

■ Why Reward People?

After the organization has provided the employee with a paycheck and health coverage, what more can or should the organization do for the employee? Management is obligated to do more than just eliminate the employees' financial worries. Employees excel when they are happy, satisfied, and feel that someone else appreciates the efforts they are putting forth. A tangible and intangible reward process can go a long way to fulfilling these needs when properly used.

Research has proven that when management rewards employees for adopting desired behaviors, they work harder and provide better customer service. The benchmark service organizations are more likely to have well-defined and well-used approaches for telling their employees that they are important individuals. Individuals in the world-class organizations that go beyond expectations are held up as customer heroes and role models for the rest of the organization. This provides a continually more aggressive customer performance standard for the total organization.

■ Key Reward Rules

The reward process needs to be designed, taking into consideration the following points:

- Organization's culture
- Desired behavioral patterns
- Employee priorities
- Behavior/reward timing relationships
- Easy to use

The reward process must be designed to be compatible with the culture and personality of the organization. Things that may be very desirable in one organization can be quite inappropriate in another.

The reward process should be designed to reinforce existing desired behaviors and/or new desired behaviors. This means that the desired behaviors need to be defined first. Changes in behavioral patterns are much more difficult to implement than those behavioral patterns that are already part of the organization's culture and only need to be reinforced. Because of this fact, additional focus needs to be given to the reward process to encourage the employees to commit to the new behavioral patterns.

The employee must perceive the reward as being desirable if it is going to have the desired results. If the employee is a ski buff and is rewarded with tickets to Heavenly Valley Ski Resort, that's definitely a positive reinforcement of a behavioral pattern. If the employee is nearing retirement age, however, the tickets may have no meaning, but increasing his or her retirement benefit can be very motivating. I lecture extensively each year. Being given a plaque as a reward for presenting a paper is more of a bother to me than a reward, but giving me a Cross pen and pencil set fits a functional need that is very important to me and becomes an appreciated reward.

It is important to involve your employees in designing the reward process so that the rewards are meaningful to the employees. In designing your reward process, select a group of representative employees to help design this process. Tell them what your reward budget is, and they will tell you the best way to spend it.

There needs to be a very close relationship in time between when the desired behavior occurred and when the reward is given. To reinforce desired behavior, reward the employee(s) immediately if you can. It is best if you can reward them when they are performing the act. (For example: "I really appreciate you staying late to wait for that customer to come in and get his job. Why don't you take your spouse out for dinner this weekend, and the organization will pay for it.") Too often, managers hold off recognizing desired behavior, hoping that a more meaningful reward can be given later. We advise in these cases, giving a small reward right away and the bigger reward later on. With this approach, the employee gets immediate, positive reinforcement and an important reminder.

Cut the bureaucracy out of your reward process as much as possible. Give management general guidelines, and eliminate the checks and balances in all but the most significant rewards. For example, an individual should not be given more than three minor awards each year, and no more than 10 percent of the employees should receive major contribution awards each year. Be sure that employees are not given special awards for just doing their jobs.

Give the manager the power to give the reward and process the paperwork later. Major rewards should be processed through a Rewards Board to be sure that required standards are met, but there should be a long list of rewards that the manager can give to the employee on-the-spot. For example: dinners for two, movie tickets, theater tickets, $50 merchandise certificates, etc.

■ Types of Rewards

Everyone hears "thank-you" in different ways. The reward process must take these different needs into account. Some people want money, some want a pat on the back, others want to get exposure to upper management, while still others want to look good in front of their peers.

Paul Revere Insurance Company developed a program they called PEET (Program to Ensure that Everyone is Thanked). As part of this program, each of the 20 top executives are given a list of two employees that they will go out and visit at their work station to thank them for doing a good job each week. The same 20 top executives each month are given 5 recognition coins that are good for a free lunch in the organization's cafeteria. They are asked to identify 5 different employees who are exceeding requirements in their area each month and go to their work area and give them the recognition coin while thanking them for their personal contributions. This required the executives to gain a better understanding of their work force and got them out into the work area for a very positive experience. Once a month they reported back on who received the recognition coins. At first, only a small percentage of the coins were passed out, but when it was pointed out to the executives that it was too bad that they did not have 5 people who exceed requirements, things changed. Receiving a coin very soon was recognized as a real honor, and as a result, a high percentage of the coins were not redeemed, but kept as trophies.

American Express has one awards program that they call, "Great Performer Award Luncheon." Typical activities that won employees invitations to these luncheons were:

- One American Express employee bailed a French tourist out of jail in Columbus, Georgia.
- Another took food and blankets to travelers stranded at Kennedy Airport.
- Another got a non-English-speaking Japanese mother to Philadelphia when her plane was diverted to Boston. She drove the Japanese woman across town to the train station, bought her a ticket, and wrote instructions on how to get to her destination.

Unusual performance for employees to take on their own? Yes, but that's what we need if we want to have empowered employees and a truly world-class organization.

It's easy to see that the reward process is only limited by the creativity of your people and the individuals who design the process. The National Science Foundation study made this point: "The key to having workers who are both satisfied and productive is motivation; that is, arousing and maintaining the will to work effectively—having workers who are productive not because they are coerced, but because they are committed." The NSF study continued, "Of all the factors which help to create highly motivated/highly satisfied workers, the principal one appears to be that effective performance be recognized and rewarded—in whatever terms are meaningful to the individual, be it financial or psychological or both."

To help structure a reward process, let's divide the rewards into the following categories:

1. Financial Compensation
   a. Salary
   b. Commissions
   c. Piecework
   d. Organizational bonuses
   e. Team bonuses
   f. Gainsharing
   g. Goalsharing
   h. Stock options
   i. Stock purchase plans
   j. Benefit programs

2. Monetary Awards
   a. Suggestion awards
   b. Patent awards
   c. Contribution awards
   d. Best-in-category awards (example: best salesperson, employee of the year, etc.)
   e. Special awards (example: president's award)

3. Group/Team Rewards

4. Public Personal Recognition

5. Private Personal Recognition

6. Peer Rewards

7. Customer Rewards

8. Organizational Awards

Basically, the following media are used individually or in combination to reinforce desired behavior.

1. Money

2. Merchandise

3. Plaques/Trophies

4. Published Communications

5. Verbal Communication

6. Special Privileges

A well-designed reward process will use all six, because each has its own advantages and disadvantages. One of the biggest mistakes management makes is to use the same motivating factors for all employees. People are moved by different things because we all want different things. The reward process needs to be designed to meet the following basic classifications of needs:

- Money

- Status (Ego)

- Security

- Respect

Let's look at some of the advantages and disadvantages of the six reward media.

Advantages of Financial (Money) Rewards

- Its value is understood.

- It's easy to handle.

- The rewardee can select the way it is used.

Disadvantages of Financial (Money) Rewards

- Once it is spent, it is gone.

- If used too often, it can be perceived as part of the employee's basic compensation.

- It is hard to present in a real showmanship manner.

- Often it is not shared with the family.

Advantages of Merchandise Rewards

- It can appeal to the total family.

- They have a trophy value.

- They can be used as a progressive reward. (Example: The employee is given points that can be redeemed or accumulated for a higher-level merchandise reward.)

- They cannot be confused with basic compensation.

Disadvantages of Merchandise Rewards

- More administrative time is required.

- A choice of merchandise rewards must be kept in stock.

- There may be other things the rewardee would prefer.

Advantages of Plaques/Trophy Rewards

- Directly tied in to the desired behavior.
- Lasts a long time and is never used up.
- Customized to the employee (name is often engraved on them).

Disadvantages of Plaques/Trophy Rewards

- May not be valued by the recipient.
- Usually not useful.
- Often not valued by the family.

Advantages of Published Communications

- Can be very specific.
- Receives wide distribution.
- Inexpensive way of rewarding an individual.
- Document is long-lasting.

Disadvantages of Published Communications

- No tangible value.
- Can reduce cooperation and cause envy.
- Can be blown out of proportion by the individual in comparison to total contributions over a longer period of time.

Advantages of Verbal Communication

- Very personal.
- Can be given when the desired behavior occurs.
- Least expensive.

Disadvantages of Verbal Communication

- Not tangible.
- Has no visual reinforcement.
- It can be misunderstood.

Advantages of Special Privileges
(Example: attend a conference, take a business trip, additional vacation days, new equipment, etc.)

- The recipient can be involved in the selection.
- Can meet unique employee needs.
- Highly valued by the employees.
- Can be used to support organizational objectives.

Disadvantages of Special Privileges

- Can cause envy.
- Can impact the organization's operation when the individual is not available.
- Can be expensive.

To be the most effective, the reward process should combine different reward media, thereby taking advantage of the positive impacts that each have and offsetting their disadvantages. For example, an individual incident could be reinforced by presenting the recipient with a check and a plaque at a department meeting.

■ Financial Compensation

Most improvement gurus and personnel consultants downplay the importance of financial compensation in the behavioral modification process. Professor Frederick Herzberg wrote in the publication, *Psychology*, "Money is not a great motivator unless, perhaps, workers are kept below the subsistence level so they must work harder to eat more. The promise of money can move a man to work, but it cannot motivate him."

We agree with experts like L. Porter and E. Lawler that money does more than provide food and shelter. It provides the individual with a sense of self-worth because the individual views his or her financial compensation as a direct measure of how much the organization values his or her efforts. Financial compensation can be divided into two parts; i.e., that part that is required to meet the family's needs for food, clothing, shelter, and medical services (what Herzberg would call "subsistence requirements"); and the remainder that rewards the family by providing them with status, pleasure, pride and power. Certainly the importance of money to provide the basics is undeniable, but the importance of money to reward the individual(s) for their efforts can be equally as important.

The difference in the reward value of money is demonstrated in two extreme cases: 1) The individual who is content receiving a welfare check each month and devotes none of his or her time in the pursuit of money; and 2) the corporate vice president who has dedicated his or her life to the pursuit of power and money. The first example provided little-value-added effort to humanity. The second worked 60 to 70 hours per week, although his or her basic needs were met in the first 2 or 3 hours of work.

It is important to remember that money can buy the individual more than food, shelter, and possessions. It can earn the individual power, respect and self-confidence. This does not mean that money can make up for personal inadequacies, but the absence of it can, and often does, keep the individual from reaching his or her full potential.

There are still those individuals in the Human Resource functions that feel that financial compensation is not an important motivator. I ask them to explain, "Why is it that 438 out of the Fortune 500 CEOs receive bonuses based on how well their organizations perform?" Certainly these CEOs don't need the money more than the individual employee working on the back dock who is trying to pay for his or her refrigerator needs it. If money motivates the CEO, it motivates the rest of the employees. As Bryan King, President of Delta Business Systems, when discussing financial compensation, put it, "If we can see how fast someone's canoe is moving in the water, we provide an incentive for him to improve."

■ Monetary Awards

The word "award" indicates that it is a unique reward for an individual or small group who have made unusual contributions to the organization's goals. Monetary awards are one-time bonuses paid to the recipients immediately following an unusual or far-exceeds expectations type of contribution. They may also be given to individuals for long-term, high-level performance or unique leadership. The award should be specific and the person or persons who receive the award should be perceived by management and fellow employees as special. The amount of the monetary award should vary based on the magnitude of the contribution. Monetary awards can be subdivided into four classifications. They are:

- Suggestion Awards
- Patent Awards
- Contribution Awards
- Best-In-Category and Special Awards

■ Group/Team Rewards

One key part of any improvement process is the use of teams/groups to solve problems and implement change. In this environment, it would be a major mistake to omit the team structure from the reward process. Very often, team rewards center around the three "P's"; i.e., Presentations to upper management, Plaques to hang on the wall, and Pats on the back. But don't limit your team reward process to just the three P's. Some organizations establish criteria that is used to recognize and identify teams that have made unusual contributions. Teams that wish to apply for this reward submit a write-up explaining the process they went through and the results they achieved. These write-ups are then submitted to a Reward Assessment Group that evaluates the individual contributions to select winners, who make presentations at Corporate Recognition Events and to executives. Frequently, these presentations are filmed and replayed over the organization's communication systems.

On a national level, Japan has regional conferences to select the best quality circle, who in turn come to a national conference to make presentations to determine the outstanding quality circle of the year. Heavy emphasis is placed on how the circle used improvement methodologies to solve the problem, with a lower emphasis on the actual savings that result from solving the problem. This approach allows every team to compete, regardless of the opportunities they have for making a major contribution to the organization. In the United States, the Association for Quality and Participation conducts similar annual run-offs and presents an annual award.

Whenever possible, team members should be allowed to choose their own rewards, as long as the reward is within budgetary limits. As Linda Goldzimmer, a Melville, New York consultant and author of the book, *I'm First: Your Customer's Message To You*, says, "A sophisticated reward system allows for those differences. If the team accomplishes X, and X is worth $1000 to each member, the form of reward should be decided by the recipient. The program must be feasible from an administration standpoint, but team members will work harder for rewards they themselves have selected."

It is for this reason that catalog-type merchandise rewards, travel, time off, luncheons, plaques, trophies, or common stock are usually the rewards that group participants receive. Catalog programs are very popular and are used by many organizations. Most consultants recommend that cash awards not be used for teams/groups, but we disagree. Cash awards meet the requirements for the individuals to select their own gifts, while providing an opportunity for public recognition of the individuals involved. It is important that you separate the cash awards from the normal incentive and compensation programs. Cash awards are only one type of group/team reward that should be considered.

Group/team rewards provide an objective way of building esprit de corps within the team or group. It is an important incentive that has a major impact on creating and encouraging cooperation among team members and at the same time, builds a sense of interdependence within the members of the group.

■ Public Personal Recognition

Many experts contend that competition awards are destructive to team building. In practice this is not true. Competition among groups or organizations develops a drive to improve and win that in the long run benefits all parties involved. Just look at how the Olympics drives the individual to strive for supreme excellence without having a negative impact on the other competitors. We contend that the four-minute mile would never have been broken if records weren't kept to challenge the individual to put forth the additional effort. There's an old saying: The difference between playing to win and playing not to lose is the difference between success and mediocrity.

Napoleon was a master at public personal recognition. He used small pieces of metal on colored ribbon to motivate individuals to sacrifice their lives. Military organizations around the world recognize the importance of public personal recognition to motivate and reward people for exceptional performance.

In Tom Peters' and Robert Waterman, Jr.'s book, *In Search of Excellence*, it stated, "We still have Boy Scout merit badges, trophies gathering dust, and a medal or two from some insignificant ski race held decades ago. Nothing is more powerful than positive reinforcement. Everybody uses it." Public personal recognition, in concert with a trophy, ribbon, money or merchandise, all have a very positive effect on reinforcing desired behavior, not only of the individual who receives the recognition, but all the people that view the recognition.

At Stacoswitch Corporation in Costa Mesa, California, all supervisors were provided with badges that read, "We do it right or we won't do it at all." Supervisors with the lowest reject rates got gold stars on their badges. National Car Rental uses plaques with the individual employee's name on them hung in their business offices around the United States to recognize outstanding performance. Major hotel chains have done the same. Football players get to put stars on their helmets for each outstanding play they make. High school athletes get letter sweaters for each time they make the team.

A key factor when you consider public personal recognition is the impact it has on the individual. Public personal recognition has a tendency to commit the individual to a performance level that they must live up to on an ongoing basis. An individual who has the best parking spot in the lot reserved for her with her name on a plaque in front of it reading, "Reserved for Ruth Gow—Employee of the Month," has a hard time not continuing to try to live up to the expectation that has been instilled in all the other employees' minds.

LensCrafters makes good use of individual public recognition. They hold an annual Laboratory Olympics where technicians put their know-how to the test and compete on a regional basis to win a trip to Hawaii, where a corporate-wide competition is held. This contest includes cash prizes, gold, silver, and bronze medals, and lots of wining and dining. The winners receive diamond-studded gold rings, presented to them by the president of the organization.

■ Private Personal Recognition

Just saying, "You did a great job on this report" instills a very special sense of pride in most employees. Their reaction may on the surface seem embarrassed or even negative, but deep down inside it makes them try even harder the next time. Praise is habit forming; the more you get, the harder you work to get more. As Charles Schwab, founder of the investment brokerage firm, Charles Schwab & Co., Inc., put it, "I have yet to find a man, however exalted his station, who did not do better work and put forth greater effort under a spirit of approval than under a spirit of criticism."

We train managers how to handle inadequate performers, but not how to say "thank-you." We train managers how to look for errors and to be critical when anything is wrong. Praise is much more effective in changing undesirable behavioral patterns than criticism. When employees react to criticism, they do it because they have to. When you praise someone, they try harder because they want to. If you have been having problems with a behavioral pattern, look for a time when the individual behavior is in line or near to the desired behavior, and praise the individual with words like, "Now that's really good." People want to feel they are doing a good job and are wanted. Vol. D/No.2A of *Bits and Pieces* puts it this way: "Anything scarce is valuable: praise, for example!" Equipment works a lot better if it is oiled and serviced regularly. So do people.

Private personal recognition is the reward activity that should be used most. In the other reward recognition methods, we have recommended that rewards be given only to outstanding performers. This is not the case with individual private recognition. In this case, everyone should be rewarded, and rewarded frequently. Managers should look for ways to say thank-you to each employee each and every day.

Everyone has jobs that are undesirable. Unfortunately, some people have more undesirable jobs than others, but without someone doing the undesirable jobs, the total process would come to a halt. One thing that makes these undesirable jobs palatable is a manager who recognizes these tasks and thanks you for doing them. Management can never express their sincere appreciation for a job well-done, too often. These pats on the back are what keep all of us moving ahead. Without these small, positive encouragements, we have a tendency to lose momentum and flounder in a pool of indecision. These frequent positive feedbacks help employees establish their priorities and assure them that they are going in the right direction.

Our experience reveals that managers who provide an ongoing deluge of positive feedback to their workers are the managers who are just lucky enough to have the best employees. These are the departments that always get their work out on schedule, exerting seemingly little effort, and when the organization has a problem, it is these departments that are first to step forward and volunteer to help. Their absenteeism is down and their productivity and morale are up.

Most important of all, private personal recognition must come sincerely from the manager's heart. The employee is quick to realize insincere comments and concerns. To quote Leo Buscaglia, "Too often we underestimate the power of a touch, a smile, a kind word, a listening ear, an honest compliment, or the smallest act of caring, all of which have the potential to turn a life around."

■ Peer Rewards

One of the most significant rewards is the sincere appreciation of your peers for a job well-done. Frequently, these types of rewards are associated with professional societies, where the practicing professionals select the brightest and best in the profession to honor at an awards luncheon. Today, the same concept has become a very positive motivating force within many organizations. These award recipients are selected by the employees, not by management.

Often, management establishes the basic ground rules and financial constraints related to peer recognition rewards. Representatives of the employees should then define what types of behavior should be recognized and establish how these behaviors will be rewarded. Empowering the employees to select the reward frequently provides management with very pleasant surprises related to the creative way employees apply the limited reward budget. Our experience indicates that employees usually plan something that is fun and the winners are usually very touched. Typically when management prepares a reward ceremony, it turns out to be a ceremony. When employees plan the same event, it turns out to be a party, and the gifts that are presented turn out to be from the heart rather than the pocketbook.

Customer Rewards

One of the major stakeholders often left out of an organization's reward process is the external customer. After all, without them we wouldn't have a job. Shouldn't we reward them for their loyalty? Remember, it costs ten times as much to find a new customer as it does to keep a present customer. Certainly the organization should be willing to share this savings with the loyal customer.

Early programs to reward customers started in the 1930s with movies and gas stations giving away dishes to reward customers for their continued loyalty. This rolled forward, and super markets began to reward loyal customers in the 1960s and seventies by giving them stamps that could be redeemed for merchandise. Today, credit cards, airlines, car rentals and hotels all reward frequent users of their services by giving them free travel and/or merchandise. General Motors reduces the price of their cars based on how much the customer uses their credit card. This type of approach needs to be considered to reward all customers of all organizations for their loyalty.

Organizational Awards

The importance of building organizational trust and pride in our employees' minds is a major objective of most improvement processes. Unfortunately, organizational pride has slipped in most Western organizations over the past 40 years.

We need to start to rebuild the pride our employees had in our organizations, for when individuals start to take pride in the organizations they work for, they take more pride in the things they do, because they do not want to tarnish the organization's reputation. One of the chief advantages that Japan, Inc. has over Western countries is the pride and dedication their employees have toward their organizations.

An effective way of building your employees' pride in your organization is to have the organization recognized by its peers as being outstanding. This is what the recent series of organizational improvement awards are accomplishing, as well as setting benchmarks for other organizations. Today, there are many award programs implemented throughout the world to recognize excellence in individual organizations. Some of them are:

- Deming Prize—Japan
- Japan Quality Control Prize—Japan
- Malcolm Baldrige National Quality Award—United States
- Shingo Prize—United States
- NASA Award—United States
- President's (Federal Government) Award—United States
- European Quality Award—Europe
- Australian Quality Award—Australia
- Best Hardware Laboratory—IBM

■ Implementation of the Reward Process

To develop an effective reward process, many factors have to be taken into consideration. The following steps will help you avoid some of the pitfalls that face most reward processes.

1. *Reward Fund*—The organization should set aside a specific amount of money that the reward process will use. This amount will set the boundaries that the reward process will operate within.

2. *Reward Task Team (RTT)*—This team will be used to design and/or update the reward process.

3. *Present Reward Process*—The RTT should pull together a list of all the formal and informal rewards that are used within the organization today.

4. *Desired Behaviors*—The RTT should prepare a list of the desired behaviors.

5. *Present Reward Process Analysis*—The present reward process should be reviewed to identify the rewards that are not in keeping with the organization's present and projected future culture and visions.

6. *Desired Behavior Analysis*—Each desired behavior is now compared to the reward categories to see which category or categories should be used to reinforce the desired behavior. Each behavior should have at least two ways of rewarding people that practice the behavior.

7. *Reward Usage Guide*—When the reward process is defined, a reward usage guide should be prepared. This guide should define the purpose of each of the reward categories and the procedures that are used to formally process the reward. This guide will be used to help management and employees understand the reward process, and to help standardize the way rewards are used throughout the organization.

8. *Management Training*—One of the most neglected parts of most management training processes is how to use the reward process. As a result, most managers are far too conservative with their approach to rewards, while others misuse them.

In creating a reward process, consider the following:

- Always have it reinforce desired behaviors.
- Reward for exceptional customer service and performance.
- Publish why rewards are given.
- Create a point system that can be used to recognize teams and individuals for small and large contributions. The employee should be able to accumulate points over time to receive a higher level reward.
- Structure the reward process so that 50 percent of the employees will receive at least a first-level reward each year.

- Structure the reward process so that the managers can exercise their creativity and personal knowledge of the recipient in selecting the reward.
- Provide ways that anyone can recognize a person for their contributions.
- Provide an instant reward mechanism.

In a paper by Shelley Sweet (a Palo Alto, California, quality consultant), entitled "Reinforcing Quality," she warns us to avoid the following seven pitfalls:

1. Cumbersome procedures are costly to administer.
2. Executives or middle managers are not consistently supporting the program.
3. Awards are applied inconsistently.
4. Unexpected behaviors result.
5. Employees perceive that the same employees are rewarded repeatedly.
6. Enthusiasm wanes.
7. Company cost-cutting curtails the program.

■ Summary

We have positioned rewards as the last building block in the Total Improvement Management process, not because it is the least important, but because it reinforces all the other building blocks. It is the glue that holds the improvement process together. It is the special jewel that crowns the top of the improvement pyramid. Without a good reward process, employees may think that they are doing the right things, but with it, they get positive reinforcement that the changes they are making meet the organization's expectations.

Now let's compare the way different types of organizations approach the reward process.

Planning the Reward Process

Losers: It grows out of the individual situations and is defined by management.

Survivors: It is well documented and the process is controlled.

Winners: It is built to support the desired behaviors that the organization wants to reinforce and is designed to help implement the organization's business plan.

Management Training for the Reward Process

Losers: No training required because management believes that all managers know how to use the reward process.

Survivors: Handled as all other procedure training is handled.

Winners: Treated as a basic management training class that uses role playing and is required for all managers.

Ways Rewards Are Used

Losers: Rewards are given only when there is no doubt that they are earned.

Survivors: Everyone gets rewarded equally.

Winners: Management actively seeks out ways and individuals to reward. They consider the reward process as an important tool in the management process that cannot be over-used when it is merited. These organizations' employees accomplish more because they are rewarded more. Employees believe that management cares about what they do and how they do it.

Our reward to you for reading this far is to make the closing brief. *Thank you for letting us share our ideas.*

(For more detailed information on rewards, see McGraw-Hill's book, *The Improvement Process*, or Ernst & Young's Technical Report TR 93.016.)

Index

Affinity diagram, 260
American Productivity Center, Inc.
   · (APC), 26
American Society for Quality Control (ASQC), 408-409
Appraisals, 278-279
Area Activity Analysis (AAA), 258-259
Arrow diagram, 261
"As is" organizational structure, assessment of, 108-109
Assessments, 80-82
   employee opinion survey, 81-82
   improvement-needs assessment, 80
   international quality study assessment, 81
   quality award assessment, 80-81
Attributes data, 427

Banking/financial industry, 386-392
   customers, 390
   foreign competitors, 386-387
   improvement activities, status of, 387
   improvements, 389
      tools for, 390-391
      typical improvement process, 391-392
   major problems facing, 386
   quality/productivity measurements and trends, 387
   quality strategies for, 387-389
Benchmarking, 43-48
   of business process, 349-350
   failure of improvement efforts, 44-48
      root causes of, 45-48
   and the health care industry, 400
   and supply management, 317-318
BPI, *See* Business Process Improvement (BPI) Brainstorming, 256
Bureaucratic organization, 447

Business objectives, 94, 95-96
Business planning, 87-103
   defining actions, 96-98
      budgets, 97-98
      performance plans, 98
      strategies, 97
      tactics, 97
   direction setting, 91-94
      critical success factors/obstacles to success, 93-94
      mission, 91-92
      strategic focus, 93
      values, 92
      vision, 91
   effectiveness of, 98-101
      and communications, 100-101
      and consensus on implementation, 101
      and customers/competitors/capabilities, 99-100
      and implementors, 100
   expectations (measurements), 94-96
      business objectives, 94, 95-96
      performance goals, 94, 96
   good plans, elements of, 89
   market focus, 90
   purpose of, 88-89, 90
   use of, 101-102
Business plans, environmental change plans vs., 105-106
Business Process Improvement (BPI), 258, 339-355
   phases of, 340-353
      continuous improvement (phase V), 353
      implementation/measurements/controls (phase IV), 350-353
      organizing (phase I), 340-343
      streamlining (phase III), 345-350
      understanding (phase II), 343-345
   success of, 353-354

Business targets, 424
Buying cycle, health care industry, 398

Career building, 281-282
Career growth training, 276
Case management network structure, 452-453
Cause-and-Effect Diagrams, 258
Certification, 366, 367-368
   ISO 9000, 187
Challenge targets, 425
Change agents, preparing, 128-129
Change management plan, developing, 342-343
Commitment levels, recognizing, 135-137
Commodity Team, 326-327
   areas of responsibility, 327
Concurrent Engineering (CE), 362-363
Creativity, 296-305
   basis for change, 298-299
   continuous exercise, 300, 302-304
   curiosity, 299, 301
   paradigm shifting, 299-300, 302
   perseverance, 300, 304
   risk taking, 299, 301-302
Critical business processes, defining, 340-341
Crosby, Philip B., 20
Cross-discipline training, 285-286
Customer data, 154-156
   baseline studies, 155
   long-term studies, 156
   monitoring studies, 155-156
Customer focus, 142-143
Customer partnerships:
   customer's executive team, presenting proposal to, 165-166
   developing, 161-166
   joint cooperation, defining opportunities for, 165

Customer partnerships (*Continued*)
  partner's direction, understanding, 165
  potential partners, meeting with, 164-165
  target organizations, selecting, 163-164
Customer perception, 145-146
Customer-related measurements, 143
Customer rewards, 479
Customers:
  banking/financial industry, 390
  health care industry, 395
  utilities industry, 402-403
Customer satisfaction, 148-149, 154
  designing for, 161
  satisfaction measurements, 157-158
Cycle time analysis, performing, 344-345

Decentralized organization, 447-449
Delivery processes, 33, 37
Deming, W. Edwards, 20-21, 22, 59
Department Improvement Team (DIT), 250-251, 352
Design for Manufacturability and Assembly (DFM/A), 361-362
Direction setting, 91-94
  critical success factors/obstacles to success, 93-94
  mission, 91-92
  strategic focus, 93
  values, 92
  vision, 91
Direct poor-quality cost, 432-433
Downsizing, in utilities industry, 406-407

Education, management, 224-226
Eisner, Michael, 409-410
EIT, *See* Executive Improvement Team (EIT) Employees:
  job descriptions, 227
  opinion surveys, 81-82
  and management, 234
Empowerment, 294-295
  in health care industry, 400-401
Environmental change plans, 104-140
  advocates, preparing in required skills, 129-130

"as is" organizational structure, assessment of, 108-109
business plans vs., 105-106
change agents, preparing, 128-129
commitment levels, recognizing, 135-137
desired behavior/habit patterns, 114-116
environmental vision statements
  establishing, 111-114
  preparing, 113-114
  stockholder' involvement with, 112-113
impact of change, 126-127
key roles, identifying/orchestrating, 125-132
need for, 106-107
organizational change management (OCM), 120-125
organizational culture:
  creation of, 107-108
  strategic importance of, 137-138
organized labor involvement in, 110-111
performance improvement goals, setting, 114
resistance, 132-135
rolling ninety-day improvement action plan, 119-120
sponsorship of change objectives, building, 127-128
synergistic relationships:
  enabling, 130-132
  generation of, 131-132
  implementing, 132
three-year improvement plans, 116-118
  combining, 118-119
Environmental vision statements:
  establishing, 111-114
  preparing, 113-114
  stakeholder involvement, 112-113
Excellence:
  individual, 269-312
  product process, 356-379
  service process, 380-415
Executive Improvement Team (EIT), 67, 76-79, 111-112, 240-245, 350
  basic team effectiveness training, 241
  board of directors, 78

and establishment of environmental vision statements, 111-114
improvement champion (czar), 77
improvement leaders, 77-78
improvement steering council, 77
mission, establishing, 242-243
organized labor involvement, 78-79
project approval, 244-245
resources, providing to, 243-244
responsibilities of, 76-77
support to teams, 245
team process, establishing, 241-242
team structure, setting/approving, 242
Expectations (measurements), 94-96
  business objectives, 94, 95-96
  performance goals, 94, 96
Experiment design, 375
External customers, 141-167
  customer complaint handling, 158-159
  customer data, 154-156
    baseline studies, 155
    long-term studies, 156
    monitoring studies, 155-156
  and customer focus, 142-143
  customer perception, 145-146
  customer-related measurements, 143
  customer satisfaction, 148-149, 154
    designing for, 161
    satisfaction measurements, 157-158
  data systems, 156-157
  getting, 160-161
  and managers, 209-210
  marketing impact on, 149-151
    inadequate follow-through, 150-151
    lack of accountability, 151
    poor forecasting, 149-150
  needs vs. expectations vs. desires, 148-147
  sales and delivery staff impact on, 151-153
  singling out, 141-142
  staying close to, 160-161
  strategic customer partnerships, developing, 161-166

External customers (*Continued*)
  words, choosing carefully, 143-
  145

Feedback, 204-205
  providers of, 424
Feedback systems, 352-353
Feigenbaum, Armand V., 21, 59
Finalized implementation plan, 352
Five-way (star) communication,
  215-216
Flowcharting, 259
  of business process, 343-344
Force Field Analysis, 257
Ford Motor Company, quality im-
  provement program, 4-6
Formal classroom training, in
  teams, 253

General Motors, downward trend
  at, 3-4
Government, need to improve, 13-
  16
Graphs, 256-257
Group/team rewards, 475-476

Health care industry, 393-401
  benchmarking, 400
  buying cycle, 398
  customers, 395
  empowerment, 400-401
  leadership, 398-399
    skills development, 396-397
  measuring quality in, 396
  quality solutions, 393-395
  teams, 399-400
  technology applications, 397
  training, 399
Horizontal process management
    network structure, 453-455
House of Quality, 359-360

IBM, downward trend at, 6-7
Idea sharing, 291
Idea submittal training, 290-291
Improvement, 9-57
  and customers, 19
  impact on stakeholders, 38-41
  long-term analysis, 12
  management:
    confusion of, 19-23
    improvement dilemma, 23-24

rating the organization, 12-13
  monopolies/government, 13
short-term analysis, 11-12
Total Cost Management (TCM),
  24-25
  methodology phases, 24
winners:
  characteristics of, 16-19
  classification of, 10-11
Improvement champion (czar), 77
Improvement Effectiveness Pro-
    gram (IEP), 287
Improvement expenditures, ROI
  for, 52-54
Improvement leaders, 77-78
Improvement measures, 79-83
  assessments, 80-82
  poor-quality cost (PQC), 82-83
Improvement methodologies:
  blending of, 32-33
  impact on other methodologies,
    31-32
  relationships of, 30
  success of, 42-43
Improvement-needs assessment, 80
Improvement Process, 22-23
Improvement pyramid, 32-38
  basic concepts (tier 2), 33, 36-37
  delivery processes (tier 3), 33, 37
  direction (tier 1), 32, 34-36
  organizational impact (tier 4), 33,
    37-38
  rewards and recognition (tier 5),
    33, 38
Improvement-related training, 275
Improvement steering council, 77
Indirect poor-quality cost, 433-434
Individual excellence, 269-312
  appraisals, 278-279
  building bond with manager, 282
  career building, 281-282
  creativity, 296-305
  employee complaints, turning
    into profits, 286-291
  empowerment, 294-295
  idea sharing, 291
  new employees, 279-281
  reinforcement of desired behav-
    ior, 283-285
    salary as reinforcer, 284-285
  Request for Corrective Action
    (RCA), 292
  safety, 293-294
  Self-Managed Employees, 305-
    308

Speak-Up Program, 292-293
  start of, 295
  training, 272-277
    cross-discipline, 285-286
Individual performance plans, de-
    veloping, 277-278
Information technology, 375-377
  information systems, 375-376
  measurements, 376
In-process measurements, 352
International quality study assess-
    ment, 81
Interrelationship diagram, 260
Ishikawa, Kaoru, 22, 59
ISO 9000:
  certification, 187
  change mechanism, 185-187
  implementation, 179-182
  phased implementation, 184-185
  requirements framework, 177-179
  series standards, 174-177
  system structures, 183-184

Japan:
  quality improvement in, 49-51
  suggestion programs, 289-290
Job descriptions, 226-227
  employees, 227
  management, 226-227
Job-related training, 275
Juran, Joseph M., 21-22, 59

Leadership, health care industry,
  398-399
Loser classification, 10-11

Macro improvement tools, 258-259
Management:
  accountability of, 196-198
  basic principles of, 205
  change process, 222-223
  desire to change, developing,
    223-224
  education, 224-226
  and employee opinion surveys,
    234
  and external customers, 209-210
  fear of improvement process of,
    199
  job descriptions, 226-227
  negative impact of improvement
    on, 232-234

Management (*Continued*)
new role of, 199-201
and participation/employee involvement, 210-216
problems caused by, 198
self-assessments, 234-235
styles of, 207-209
tomorrow's managers, 205-209
and trust building, 201-205
what to call them, 192-193
See also Top management; Trust building
Management Improvement Teams (MITs), 230-232
Management Information Systems (MIS):
customers, 439
measures, 438-441
owners/stockholders, 439-440
people, 440-441
productivity/quality, 439
Management participation, 190-237
and error-free output, 227-228
five-way (star) communication, 215-216
getting employees to work, 212-215
importance of, 194-195
Management Improvement Teams (MITs), 230-232
measurement systems/performance plans, 228-229
new middle managers, 221-222
organized labor's involvement in, 216-218
push-pull of management, 211-212
town meetings, 220-221
upward appraisals, 229-230
Management tools, 259-261
Market focus, 90
Matrix diagram, 260
Measurement process, 38, 416-443
attributes data, 427
audits, 424
benefits of, 419-420
business targets, 424
categories of, 418-419
challenge targets, 425
characteristics of, 428-430
delaying, 421
feedback, providers of, 424
and improvement, 425-427
measurement points, establishing, 421

National Quality Award criteria, 436-438
performance standards, 428
Poor-Quality Cost (PQC), 431-434
process, 430-431
responsibility for taking, 421-422
surveys, 434-436
types of data, 427-428
understanding, 421-425
variables data, 427-428
who measures, 423
who should be measured, 422-423
Measurements/controls, establishing, 342
Mission, 91-92
establishing, 242-243
Monetary rewards, 475
Motorola, 376-377

National Quality Award, as criteria to measure improvement, 436-438
Network organization/structure, 449-452
case management network structure, 452-453
communication system, 456
compensation/benefits, 456
education/training, 456
evolution of, 450
horizontal process management network structure, 453-455
implementation of, 455-457
and information technology, 450-451
management development, 456-457
management style, 455
performance management, 456
service economy, 451-452
total quality initiatives, 452
New employees, 279-281
New process, design, 347-349
implementation, 352
Nominal Group Technique, 257

On-the-job training, of teams, 253-254
Organizational change management (OCM), 120-125
best practices, 122-125
working definition of, 121-122

Organizational culture:
creation of, 107-108
strategic importance of, 137-138
Organizational policies/procedures, 68-76
corporate instructions/directives, 74
employment security, 74-76
improvement beliefs/concepts, 69-71
improvement policy, 71-72
new performance standards, 72-74
Organizational rewards, 479
Organizational structure, 38, 444-464
"as is," assessment of, 108-109
bureaucratic organization, 447
decentralized organization, 447-449
design, 457-461
operational perspective, 459-461
strategic perspective, 457-459
tactical perspective, 461
evolution of, 445-446
implementation, barriers to, 461-463
network organization, 449-452
vertical organization, 446-447
visioning matrix, 458

Peer rewards, 478
Performance evaluations, 278-279
Performance improvement goals, 94, 96
setting, 114
Poor-Quality Cost (PQC), 82-83, 259, 353, 431
direct, 432-433
indirect, 433-434
Preferred process selection, 350
Preliminary boundaries, defining, 341
Preliminary implementation plan, 350
Prioritization matrix, 260
Private personal recognition, 477-478
Problem-solving tools, teams, 256-259
Process alignment, 345
Process control, 373-374
Process control cycle, 372-373

Process cost analysis, performing, 344-345
Process decision program chart (PDPC), 260
Process design concepts:
    manufacturing, 369-372
    continuous flow, 370
    error-proofing, 371
    one-minute change-over, 370-371
    production control systems, 371-372
Process Improvement Team (PIT), 24, 251
    forming/training, 341-342
Process owners, selecting, 341
Process redesign, 345-346
Process walk-through, conducting, 344
Product development phase, 358-369
    Concurrent Engineering (CE), 362-363
    Core Competence or Capability, 358
    Design for Manufacturability and Assembly (DFM/A), 361-362
    innovation:
        and process design, 364-365
        and time to market, 363-364
    Process Capability, 365
    Process Qualification, 365-366
    Product Planning Matrix (House of Quality), 359-360
    qualification plan, 366-369
    Quality Function Deployment (QFD), 358-360
    reliability, 360-361
Production phase, 372-375
    experiment design, 375
    process control, 373-374
    process control cycle, 372-373
Product Planning Matrix (House of Quality), 359-360
Product/process design and innovation, 369-372
    process design concepts, manufacturing, 369-372
Product process excellence, 356-379
    information technology, 375-377
    product design phase, steps in, 357
    product development phase, 358-369

production phase, 372-375
    product/process design and innovation, 369-372
Project approval, 244-245
Project plan, developing, 342-343
Public personal recognition, 476-477

Qualification plan, 366-369
    certification, 366, 367-368
    independent process audit, 369
    qualification, definition of, 366
    qualification lots, 368-369
Quality award assessment, 80-81
Quality Control Circles (QCC), 251-252
Quality Function Deployment (QFD), 260, 358-360
Quality Management Systems (QMS), 168-189
    benefits of, 170-111
    definition of, 169-167
    development of, 171-174
    implementation, 179-182
    ISO 9000:
        certification, 187
        change mechanism, 185-187
        implementation, 179-182
        phased implementation, 184-185
        requirements framework, 177-179
        series standards, 174-177
        system structures, 183-184
    procedures, 182-183
    Quality Manual, 182-183
Quick fixes, implementing, 345

Ramp up strategy, 323
Relative performance analysis table, 11
Request for Corrective Action (RCA), 292
Resistance, 132-135
Rewards and recognition process, 33, 38, 465-482
    customer rewards, 479
    financial compensation, 474-475
    group/team rewards, 475-476
    hierarchy, 468
    implementation of, 480-481
    key reward rules, 469-470
    monetary rewards, 475
    objectives of, 467-468

organizational rewards, 479
peer rewards, 478
private personal recognition, 477-478
public personal recognition, 476-477
purpose of rewards, 468
types of rewards, 470-474
Rolling ninety-day improvement action plan, 119-120

Safety, 293-294
Scientific management, 206
Self-assessments, 234-235
Self-Managed Employees, 305-308
Self-Managed Work Teams, 252
Senior management, See Top management
Series standards, ISO 9000, 174-177
Service industries:
    banking/financial industry, 386-392
    best service organizations, 412
    definition of, 381-382
    enlightened vs. in-the-dark organizations, 410-411
    health care industry, 393-401
    importance to U.S. economy, 382
    improvement in, 408-412
    major classifications of, 385
    problems in, 382-384
    service characteristics, 384-385
    utilities industry, 401-408
Service process excellence, 380-415
Simulation model, preparing, 344
Smokestack industries, quality improvements in, 2-3
Speak-Up Program, 292-293
Star communication, 215-216
Step progression strategy, 322
Strategic focus, 93
Suggestion programs, 288-291
    idea submittal training, 290-291
    Japan, 289-290
    U.S., 290
Supply management, 313-338
    benchmarking, 317-318
    classification schemes, 319-321
    Commodity Team, 326-327
        yearly activities/responsibilities of, 332-333
    current state assessment, 314-315
    definition/scope of, 316
    evolutionary strategy, 322

Supply management (*Continued*)
  generic model, 325-326
  history of, 317
  implementation:
    guidelines/models for, 333-334
    pitfalls to avoid during, 335-336
  initial supplier qualification, 329-331
  material goals/strategies, 315-316
  minimum certification criteria, 322
  ramp up strategy, 323
  step progression strategy, 322
  supplier certification, 331-332
  supplier surveys, 327-329
  supply management process (SMP), 324-325
    application of, 327-332
  tools/techniques, 323-324
Surveys, 434-436
Survivor classification, 10-11
Synergistic relationships, 130-132

Task forces, 253
Task teams, 252
Team facilitator, 246-247
Team leader, 245-246
Team meetings, 254-256
  evaluating, 255-256
Team members, 246
Teams:
  building, 238-268
  decision making by, 261-262
  elements of, 240-247
  Executive Improvement Team (EIT), 76-79, 111-112, 240-245
  future of, 265-266
  macro improvement tools, 258-259
  management tools, 259-261
  meetings, 254-256
  problem-solving process, 247-250
    opportunity cycle, 248-250
  problem-solving tools, 256-258
  problem teams, dealing with, 264-265
  rewards, 475-476
  success, measuring, 263-264
  team facilitator, 246-247
  team leader, 245-246
  team members, 246

  team process, implementing, 262-263
  training, 253-254
  types of, 250-253
Ten Benchmarks for Quality Success (Feigenbaum), 21
360 degree appraisals, 230
Three-year improvement plans, 116-118
  combining, 118-119
Top management, 59-86
  commitment to improvement, demonstrations of, 64-65
  Executive Improvement Team (EIT), 76-79
  involvement in improvement process, 61-62
  and measurements of improvement, 79-83
  organizational impact, 76-79
  personal performance indicators, 66-67
  personal support of improvement process, 67-68
  and quality/productivity improvements, 60-61
  required resources, supplying, 68
  supporting policies/procedures, releasing, 68-76
  time, dedicating to process, 65
  winning over, 62-63
  See also Management
Total Business Management (TBM), 31
Total Cost Management (TCM), 24
Total Improvement Management (TIM), 15
  measurement systems, planning, 441-442
  organization, effect on, 41
  pyramid, 32-38
  See also Improvement pyramid
Total Management Methodologies, *See* Improvement methodologies
Total Productivity Management (TPM), 25-26
  Productivity Improvement Program, phases of, 25
Total Quality Management (TQM), 23, 26-28, 42-43
  basic elements of, 27
  sample improvement results, 28

Total Resource Management (TRM), 28
Total Technology Management (TTM), 29
Training, 272-277
  career growth, 276
  cross-discipline, 285-286
  idea submittal, 290-291
  improvement-related, 275
  job-related, 275
  and network structure, 456
  teams, 253-254
Tree diagram, 260
Trust building, 201-205
  doing it with a smile, 202-203
  and feedback process, 204-205
  listening, 203
  telling employees "why," 202
  urgency/persistence, conveying, 203-204

United States:
  moving production back to, 52
  quality improvement in, 48-49
  suggestion programs, 290
Upward appraisals, 229-230
Urgency, sense of, conveying, 203-204
Utilities industry, 401-408
  continuous improvement, 404-405
  culture change in, 403-406
  customers, 402-403
  downsizing, 406-407
  improvement, 403-404
    centralized efforts, 406
    industry characteristics impeding on, 401-402

Variables data, 427-428
Vertical organization, 446-447
Vision, 91

Wallace, Inc., 7
Winners:
  characteristics of, 16-19
  classification of, 10-11
Win-Win Square, 8
Words, choosing carefully, 143-145

About the Authors

H. James Harrington has had a long and distinguished career in the quality field, including 40 years at IBM, and is widely considered a leading authority on the subject. He is currently a principal at Ernst & Young and serves as its international quality advisor. He is Chairman of the prestigious International Academy for Quality, Director-at-Large for the Association for Quality and Participation, and honorary advisor to the China Quality Control Association. He is past President and Chairman of the Board of the American Society for Quality Control, and lifetime honorary President of the Asia Pacific Quality Control Organization. A much sought-after international speaker, he is the author of seven books, including the best-sellers *The Improvement Process* and *Business Process Improvement*.

James S. Harrington is presently a journalist with the *Gilroy Dispatch*. He holds a BA degree in education from San Francisco State University and an MBA from Santa Clara University. He has over 10 years' in-depth training in improvement methodologies and has traveled throughout the world discussing quality issues with leading quality practitioners.